TITAN OF CISLUNAR SPACE

- by -

Peter J. Schubert

AF808119

Prologue – Bitten by the Kyoto Dragon

When the rest of the Larsen C ice shelf plumed into the Weddell Sea, there was no immediate danger. Even when the Ronne ice shelf calved huge bergs at the end of a warm antarctic summer, sea levels did not rise, since the ice was already floating. Cool melt water to offset the warming seas was generally considered a good thing. Nobody lived near enough to be threatened, and ships were easily able to circumnavigate the large sections drifting north and east.

Warmed from above by UV rays slipping through the ozone hole, huge icebergs holding the contour of the Earth's curvature sloughed off megatons of ice-cold freshwater. The cold water sank in vast ribbons that stirred the ocean depths. Sea creatures were displaced or killed, some thrived. Still, the impact on humans was minimal. Scientists were curious. Most people considered this evidence of Earth's self-correcting mechanisms for climate change. College students threw parties in honor of the event.

Neverthless, Parks Vambolent's father, a climatologist at the University of South Florida, tried to raise an alarm. He had been famous once, briefly, when he published equations proving that increasing levels of greenhouse gasses would one day set up a "hurricane mother" that would spawn tropical storms year-round. He even predicted a latest possible date at which this would occur – but it was still many years distant. When the media buzz died, Dr. Vambolent's ideas did too. Now, when he tried to raise a warning again, nobody listened.

Parks skipped school to run the Hilly Hundred with his elder cousin, Bryce Vambolent. The two had competed on bikes since they were children in neighborhood Memorial Day parades – Parks had always come in second. He left Florida early to train on the Indiana hills, but in truth he just wanted to get away from his ranting father.

That which the elder Vambolent predicted came true. Vertical currents splashed against the ocean floor, creating turbulence on a vast scale. Sworls of cold water surged back up to the surface, creating complex tapestries of warm and cool. Superheated winds pushed out from the blistering Sahara Desert sucked moisture from the coastal forests of Africa and her tidal waters. Drifting over the open ocean, the wet air was pushed and pulled by high and low pressure regions over the cold and warm surfaces of the sea.

Dry air is more dense than moist air. So the air pressure over warm water is low. Low pressure draws air in. Where cool waters surround warm spots which are roughly circular, the indrawn air begins to swirl. Just like fluids down a drain, the Earth's Coriolis force twists the flow. If the hot and cold regions are far apart in temperature, and if the air above them is laden with moisture, a tropical storm can spiral into a hurricane.

Dozens of storms began building above the turbulent Atlantic that late summer. Two mighty sworls formed hurricanes. Between them, torn against nature by the spin of bracketing storms, a counter-rotating tropical storm formed – the first in history. Geosynchronous weather satellites photographed the massive storm complex. From above, it looked like two huge eyes. The space between was torn and violent, curved like intertwining claws. Dr. Vambolent was finally contacted by a reporter who wanted a quick sound byte. He gave it; a name for the storm: The Kyoto Dragon. It killed him three days later, leaving his son an orphan.

PART I
12 years after the Kyoto Dragon

Chapter 1

The Chalbi Desert

Parks Vambolent rode shotgun for his best friend and business partner, Cloy "Horse" Morgan. They left the seventh corner of the rectenna they were surveying, and drove toward the eighth, and last, almost a kilometer distant. Behind them, Jim Ibaduni and three laborers dug the pit and mixed cement for the anchor.

"I'll be glad to wash away this thirst with a chilly beer when we get back to Mombasa," Horse said, gesturing large with his hands, and disregarding the drifting turn of the wheel. "It's hotter than Cairo here. How close did we get to this god-forsaken place in that Tour D'Afrique?"

"The route came within a hundred miles or so," Parks said. "But I didn't make it that far, remember?"

"Oh, yeah. How could I forget! You fell, didn't you?"

"My spokes got spiked by one of Hababek's goons. It was in Luxor, at the end of that first day. I still have the scars." Parks plucked at his trouser legs which covered pink and white stripes of scar tissue from mid-thigh to heel.

Horse laughed. "You quit after that, didn't you?" he asked.

"I went back to finish college."

"Yeah, and work for your Uncle Scrooge. Bo-ring! We would have taken you back the next season, Parks. You were the best *domestique* I ever had!"

"Yeah, but then we'd *both* still be hawking soap and butter."

"True, indeed! But now look at us Parks! We're masters of the universe – or at least – well on our way. Yes, we are." Horse yanked the wheel back over towards the red flag roughly marking the final corner for the octagonal receiving antenna.

"Where would you be, Horse?"

The elder man looked sidelong at his partner, and measured his response. "Oh, working in my pappy's sweatshop, no doubt. But look at us, man, look at us! We're gainfully employed, we're building a stepping stone to the stars, and we've got your patents to carry us there. If we don't end up the most powerful men on the planet, I'll eat my hat." He pulled off his cap and bit the bill. He spat dust.

"Then what?" Parks asked.

Horse was slowing, and circling part way around the red flag. He stomped the brakes of the Range Rover to a sudden halt and looked directly at Parks.

"Revolution, my boy. Revolution!"

"Huh?"

"Look at the world, Parks, look at it! After making such a mess in Iraq, now we're battling the Iranians. If there's a Muslim in this part of the world that doesn't hate Americans, I've yet to meet him."

"So you plan to overthrow the government?"

"Of course! Only I won't do it myself. Almost everybody hates our stupid government. After the right eye of the Dragon wiped out DC and they moved the whole kit and kaboodle to the Springs, we've been on the brink of martial law. We've got more soldiers taking away people's God-given liberties at home than we've got playing whack-a-mole with all the kooks that pop up with nukes and WMDs, threatening to finish the job the hurricanes started."

Parks opened his mouth to speak, and they heard three sharp cracks. Gun shots. They opened their doors together, got out, and looked back towards the pit crew. Jim and his people were piling into the other Rover. Bearing down on them from the east was a jeep with a mini-gun mounted on top, plus a squat truck with a dark canvas roof. From their vantage, they could see red and orange fire jumping from the business end of the mini-gun. Sounds came after a few more seconds, much louder than the three warning shots. The Rover, looking small in the distance, caught fire. One dark figure scrambled out and ran. The man went suddenly limp and crashed into the dust.

"Let's get out of Dodge!" Horse shouted.

They jumped inside and Horse floored the accelerator. He headed directly away from the melee, breathing heavily. Parks looked at him. Pure white showed around the oval of yellow bloodshot in his eyes. Parks had watched his cycle team leader many times. Usually his eyes were heavy, nearly shut, as he gasped for air after a *hors categorie* climb. Seeing the white of panic, Parks himself felt a sudden calm. Something passed between them that only Parks noticed.

They reached maximum speed

"Look behind us," Horse commanded.

Parks twisted around in his seat. The dirty back window showed only swirling shadows of dust and salt from the floor of the dry lake bed. He turned the other way and looked out the window. A bullet whizzed by.

"They're pretty close," he yelled over to Horse.

Cloy Morgan pushed himself up in the seat as he pressed the floorboards with the still-forceful thighs of a former professional cyclist. Parks Vambolent leaned out the window again, stealing a quick look back. Their plume of powdered dust, sand and salt cut off any view behind. He sat back down and stared ahead. Horse was holding the wheel steady as they bee-lined across the desert floor with a jeep in pursuit.

"I'm turning things off," Parks announced, and turned off all the knobs and buttons of the dashboard with strong but shaky fingers. As a passable mechanic in a pinch, during a race he could fix simple things quickly, and had the presence of mind to hurry without tripping himself up. But this was far beyond what he had grown accustomed to, and it showed.

"Easy, Parks," Horse said. His confidence and full voice had returned. He was in the lead again. He put a hand on Parks' shoulder and gave him a look and a nod that said they were in this together to the end, whatever may come. Parks nodded in return, assuming he knew what his friend meant.

"Just keep straight; keep your head down," Parks said.

Horse grunted theatrically and doubled over into bikers pose, his legs still stressing the accelerator pedal into the steel floorboard of the Range Rover. Parks ducked down too.

A bullet pierced rear and front windows.

They both flinched from the sound, delayed a quarter second. The adrenalin hit within the second, and their hearts raced. Another bullet pinged into metal at the rear.

"All they have to do is hit our gas tank, and we're gonners," Parks said.

"No, they only need to hit a tire. And there's four of them," Horse said.

"Can we go any faster?"

"Can we shoot back? Can we drop something to flatten *their* tires?"

Parks cast about, taking a quick mental inventory and filtering it for sharp, angular objects. He pulled a camera, a plumb bob and a pick within seconds and palmed them. With an arm out the window, he tossed the handful up and over half the roof's width, scattering them. He came in and looked for more. Another bullet hit. Rather than a sharp sound, this one screeched as if it slowed gradually.

"They're getting better aim," Horse said.

Parks pulled the extra keys from the keychain and pulled a penknife from his trouser pocket, opened it, and threw them upward, out the window. He looked about, but didn't see anything else. He could climb into the back and get the surveying gear, but that would eliminate one layer between him and the gun. Not that seats were much protection from high-caliber bullets.

"Roll up your window," Parks said, "Wind." He cranked his own handle. Horse did too.

The howl of wind died. Dust froze and started to settle. For a space of several breaths there were no more shots.

Horse sighed and relaxed into his seat. He merely pressed the pedal down completely; did not try to push it through the floor. That's when the bullet struck him.

Nairobi

Elise Vambolent stood inside her leased manse and squinted in concentration. Perspiration tickled the back of her neck. The box with their financial records eluded her search. A forwarded letter from the Internal Revenue Service had followed them to Kenya. She would have thought their income too small to audit.

It was the heat of the day and she heard knocking. All morning long, workers had been hammering and stapling mosquito netting around the veranda. She tacitly assumed the workers had returned from their rest in the shade, and had started up again.

Passing the glass doors that led to the rear she did a double take. A man stood there, arms at his side. He raised a hand and waved. Her eyes cut with suspicion.

The man was well-dressed in a light suit jacket over khaki trousers. All the seams were crisp, with no wrinkles between. His off-brown hair could have been touched by gray. It set off the boyish softness around his angular features. He was tall with broad thin shoulders, from which his jacket hung straight to a tapered waist. Elise guessed him a trial lawyer, or maybe a fast-track businessman. Too fancy to be a salesman, but you never know. She wondered why he was in her back yard, sizing him up, not sure whether to call the police, or to run for a kitchen knife.

He smiled genially, then pointed back the way he must have come. He mouthed something to her. She couldn't make out words, despite his exaggerated diction. She'd been in Africa only a month, but she was pretty sure she had spotted an American. He left the patio on long strides.

Elise stepped to the front door of her adopted home. Both kids were at a neighbor's swimming pool. She tracked his movement around the house, and saw his shadow pass by veiled curtains on the windows. She beat him to the main entrance and slid the chain into place. He knocked.

"Yes?" she said, opening the door an inch.

"Mrs. Vambolent, I presume," he said cordially. "Please forgive my trespass around back. I knocked for several minutes, but there was no answer."

Elise's shoulders un-tensed just a bit.

"I thought that was the workmen, banging their hammers," she said.

The man laughed easily, a light banter with a basso rumble. The thrum it left in her chest felt good, and she wanted him to laugh again. She relaxed a little more.

His face went serious, and sincere. "Truly, ma'am, I'm sorry. It's just that I was very eager to meet Mr. Vambolent. I've come a long way to see him."

"And you are?" she asked him.

"Jenks, ma'am," and he spelled out the name for her. "Winnimer Jenks, actually. But nobody uses my first name except my parents." He laughed again.

Elise made her face more suspicious to hide the easy smile that wanted to form there.

"And your business with my husband?" she prompted.

His face instantly became all business. The smile vanished and he leaned forward. Elise felt a tinge of regret for being so rough on him.

"Actually, it has to do with his patents," Jenks told her. "Is he here, that I may speak with him?"

Elise hesitated just a fraction of a second before answering.

"He's taking care of a few things, and should be back any time now."

"Excellent. I am most eager to speak with him on commercialization of his ideas. Do you know when I might be able to meet him?"

"Does he know you?" Elise said, still speaking through the crack in the door.

"We were at a conference together several years ago, a business plan development competition at Stanford University."

"Oh, I remember," she said. "That was with Frank, Ned and Horse – I mean Cloy. Just a minute." She closed the door, slid off the chain, then opened the door wide. "Won't you come in please, out of the sun."

"Thank you," he said, stepping past her. His cologne was both floral and musky, quite pleasing, and she found herself taking an extra deep breath through her nose to get another sniff. "What a nice home," he said, looking around.

"Thank you," she said, brushing a wisp of hair from her face. "We're still unpacking, and I'm a little bit disorganized. Now, if you'll tell me a little bit more about whatever it is you want to speak to Parks about, and a phone number, I can have him call you."

Jenks produced a business card, which Elise studied. Winnimer Jenks, it said, then 'New Business Development Manager', and a company called NexGenSys. The address was in Timonium, Maryland. She had never heard of the company name or the city, but he was already talking casually as she studied the card.

"I was a judge at that conference," he said. "Although I pushed hard for your husband's plan, others on the selection committee felt it was too remote, too dependent upon research and development, and therefore the risk of cost overruns were too high. They dismissed it in the first round. I convinced them to allow a re-submission at a future competition, perhaps when more work had been completed. But he never came back to us."

"I see," Elise said, now studying him closely. His eyes were either grey or green, she decided, and they were set pleasingly deep under his brow. He possessed a strong nose over generous lips, framed by deep creases which moved enchantingly as he spoke.

"When my news clipping service caught an article from the *Daily Nation* with your husband's name, I became intrigued at once. It seems he has attracted more attention overseas than he was able to at home. I must admit, I was a bit alarmed, being a strong patriot myself. Still, I flew here directly to meet with him."

“What for?” Elise asked, frowning at the implication that her husband might be betraying his country.

“Why, to help him accomplish the same – and more – back home. You see,” he said with a sardonic lilt to his voice, “our generation of Americans are risk-averse. Sometimes we need someone else to recognize the potential in an idea before we can see it for ourselves. This is precisely what is happening in the case of your remarkable husband. What I’ve come for, Mrs. Vambolent, is to offer to buy the rights to your husband’s patents.”

“But that’s not what he wants,” she replied at once. “He wants to be the one to make it happen. That’s why we’re here in Kenya.”

“A very ambitious undertaking,” Jenks observed. “Especially for one man.”

Elise paused to look at him; to judge if he was being sincere or sarcastic.

“I hope you’re not implying that he is not capable,” she finally said.

“Oh, no, not at all, Mrs. Vambolent. Not at all. But developing an idea for commercialization can be a lengthy process. Even with a sizeable team, space technology can take many years, maybe even decades. Tell me,” he said, lowering his voice slightly, “how much time is left before your husband’s patents expire?”

“I don’t know the exact dates,” she hedged, then admitted, “but it’s not that far off, in space terms.”

“Precisely,” he said, opening his long-fingered hands. “What your husband needs is a go-fast team, to make all this happen before the patents run out.”

“And you can help do that?” Elise said.

Jenks nodded, smiling gently. “I have access to venture capital, and through a professional placement firm, I can assemble a team of exactly the experts your husband needs to turn his dream into a reality.”

“Really?” Elise said, as a scene formed in her mind. She saw Parks accepting a check for a million dollars, then going to work at a research lab to supervise the construction of his isotope separator. All the hard stuff, like getting approvals, launches, licensing, and the like would be handled by the company, this NexGenSys.

“He would still want to be involved,” she said firmly.

"Absolutely," Jenks said emphatically. "We would need him as the managing project director – Chief Scientist, or a title like that. After all, he is the mastermind behind the whole concept. That is what we do, Mrs. Vambolent. We find game-changing ideas like your husband's, and help turn them into a thriving business."

"What do you get out of the deal?" she asked.

"Most inventors are highly technical and deeply skilled. It is a rare inventor who is also qualified to take command of a large business operation. What we do is help the inventor focus on the technology, where their strengths are the greatest. Then, we provide professional business managers to run the operations. We hire marketing and sales agents to bring in contracts, and enlist accountants to file the tax returns. Together, we make a going concern, which benefits everybody."

"What if things don't work the way you plan?" she said. "What if you, or he, changes their mind and decides they don't like the deal?"

Jenks nodded his understanding. "You would be surprised how often that question comes up," he said. "Our standard approach is to offer the inventor a modest lump sum up-front, and then an offer of on-going employment at a comfortable salary."

"A lump sum, you say. Like in the case of my husband, how much would you be talking?" Elise had gently suggested to Parks about a thousand times that he simply sell his inventions outright.

"Well, I really should be talking with your husband directly about details like that," Jenks said reasonably.

"I understand," said Elise, but she felt let down, and wanted to probe further. She held her tongue out of deference to her husband. Then, she couldn't contain her curiosity, and blurted out:

"But approximately, Mr. Jenks. Can you give me a ballpark figure… to entice Parks to give you a call back as quickly as possible."

Jenks smiled, and for a heartbeat she wondered if he had somehow maneuvered her into asking this very question. Then, the moment passed.

"Remember," he said, "that our ultimate interest is the development of a new company, so we are more inclined to make the salary portion as generous as possible. Any lump-sum payment detracts from our investment capital, so we want to keep that within a reasonable range."

“Oh,” Elise said, deflating.

“I doubt,” Jenks continued smoothly, “that we could offer much more than something in the six digit range.”

Elise did the math quickly, and then said something that made her feel both bold and ashamed at the same time.

“Well, we were always thinking about a number with just one more digit than that.”

Jenks looked at her carefully, as if sizing her up for the first time. In a measured voice, he countered.

“I see. Well, the upper range of possible values might approach such a figure. I would of course, need to consult with my team before negotiating further with your husband.”

“Of course,” she said, backing off the subject.

“How soon might I expect a call from him?” Jenks asked, glancing down at his business card in Elise’s hand.

“He should be calling in tomorrow,” she said. “I’ll be sure to give him your contact information. I think he will really look forward to talking with you.”

“And I with him. In the meanwhile, Mrs. Vambolent, please don’t hesitate to call me if I can be of any assistance whatsoever.”

“Thank you, Mr. Jenks,” she said.

“Just Jenks, please. It’s so much less formal,” he said, extending his arm.

Elise shook hands, and the man’s cologne crept into her nostrils again. She breathed deeply and smiled.

“You can call me Elise,” she told him.

He bowed over her hand, repeated her name, and looked deeply into her eyes. She blinked and broke contact, moving him to the door.

Elise let him out, then held the door open as she watched him walk down the pathway. She looked at the business card in her hand as she slowly closed the door, and her heart swelled with excitement. Who cared if the IRS wanted to audit them. They were about to become millionaires!

Chalbi Desert

Horse slumped against the steering wheel, and listed. His foot came off the gas, and they began to slow.

“Horse!” Parks yelled. “Horse!”

Parks moved over. He grabbed his friend's shoulders and pushed him upright, straightening the wheel. Wiggling his foot over, he kicked away Horse's limp leg and stomped on the pedal. As soon as their speed picked up, a hail of bullets sounded all around them. The windscreen went white with spiderwebs.

"Parks." Horse whispered.

Parks was already nearby, but leaned his head closer. With one hand he steadied the wheel. With the other, he guided his friend gently to a reclining position. He couldn't see a bullet wound and asked.

Horse ignored the question and began whispering.

"I haven't shared everything, Parks."

"You can tell me later, at the hospital."

Cloy "Horse" Morgan gave one sidelong glance at his junior partner. Parks closed his mouth.

"There is a regulus, somewhere in the inner solar system. Find it, Parks. It could be at the bottom of the ocean. Maybe on the backside of the moon. Probably, it's on Mars. You've got to get to Mars, Parks. All your inventions. They…"

Another fusillade of gunfire drowned his next words. Parks strained to hear, all the while keeping the wheel straight and the pedal down.

"Yes?" he asked. Horse had gone quiet, and Parks thought him dead.

His friend finished taking a deep breath, which evidently caused him considerable pain. He spoke again as he let the air out.

"Business plan, Parks. After Stanford, I kept working on it. Without the regulus, you'll never make it. With it…"

Horse drew in another breath. His great face winced hard with the effort.

"With it you can build an empire."

"What's the regulus?" Parks asked quickly.

"The plan," Horse said, almost inaudibly. "At my flat. Get it before anyone else does."

"I will, Cloy."

"And start the revolution." A wry smile flashed across his pale lips, and he slumped lower in the seat.

Parks could smell the blood now. He looked around Horse's body but didn't see any. He looked over the back of the seat. In the footwell, a pool of glistening maroon liquid sloshed and rippled.

There was a lot of it. Parks looked closely at his friend's face, which had turned pasty and cool.

"What's the regulus, Horse?" Then, he cursed himself for being greedy. His friend was dying, maybe already dead. His own life was in grave danger. What should he be doing?

"Phaeton was the fifth planet," Horse whispered at the top of his next breath. "It broke apart. At the center, heavy elements gathered. Gold, Parks. Platinum. When the planet was destroyed the regulus was freed. Find it!"

More air came out of his lungs, but no more words. The chest didn't rise. Parks considered mouth-to-mouth resuscitation. First, he felt for a pulse. Nothing, but the Rover was bouncing and noisy. He leaned down and put his head against the other man's chest. No sound was there. No heart beat. No inspiration.

"Horse!" he shouted. Then, softly. "Horse."

For many minutes, the Rover drove straight on. The gunfire had ceased, but Parks didn't notice. More minutes passed. They hit a sudden bump, and he was startled out of his shock. Blinking, he worked to recover his presence of mind and act. Back on Team Agrifort, Horse had drilled that into their team: "Assess. Decide. Go." Immediately after any crash or delay, Horse wanted his team first up and back on the road.

"Threat assessment," Parks said to himself, and craned his neck to look out the rear window. He could see nothing but dust. Ahead of the vehicle was more desert. Far away, mountains. Were they still being pursued? Which direction should he go? "Decide," he told himself.

Stopping the vehicle was not an option. Yet sooner or later it would stop on its own – when the fuel ran out, when the engine overheated, when a tire blew out. He would be walking then. Best to gather what was needed, in case a quick escape was needed.

He first put his instrument-laden rucksack onto the accelerator pedal to hold it down. With a bungee cord from the center console, he rigged the steering wheel in position. From the second row of seats he pulled a hip pack and opened all the pockets. A mostly-empty water bottle was inside, a knife, an over-ripe mango, and a small first-aid kit. He would need more water.

With the front secured, Parks climbed backwards, keeping a wary eye out the back and ready to leap forward at the next sound of gunshots. Searching about for water bottles, Parks noticed a tall metal sign flash past. A sign in the middle of the desert?

He froze, and looked forward. Another sign, yellow, loomed ahead. They approached it at full speed and ran it down.

It was in English, and he only saw it for an instant. His eye read the essential words before his brain registered the information. "Mine field," it had read, then crumpled.

"Go!" he said to himself, and started for the front to stop the car. The blast happened faster than Parks could think. He didn't even register the event, feeling only bewilderment as he was thrown around the inside of the Rover as it rolled over and over through the mine field.

Blinking, stunned, Parks lay in a heap among the survey equipment in the back of the Rover. It was quiet, but for a crackling sound, like the clear plastic wrapper on cheap candy. Where was Horse? The front of the vehicle was open and twisted. The crackling sound came from there. He saw smoke rise inside the cabin.

With one hand on the hip pack, he dragged himself out the back, cutting himself on the safety glass. As the foam, carpet and upholstery caught fire, Parks moved further away. An arm protruded from under the burning vehicle. Was it attached? Parks gagged and almost retched.

Plastics inside the car joined the blaze and Parks stepped back from the heat. He stared, then broke his gaze away. Were the raiders waiting at a safe distance? Scanning the horizon, he saw nothing but dim purple mountains in the very far-off distance.

Taking a survey in all directions, Parks decided to exit the minefield the way they had come in. "Go." He said to himself, but his voice was feeble. His legs quaked.

What about Horse? Parks hesitated and looked back. The fire caught the petrol tank which howled and spurted flame high into the air. The lone arm lay burning, and the hand curled as he watched it. It made a fist, turned black, and caught fire.

Parks turned his back and started walking south across the hardpan. Nairobi was hundreds of miles off. His office in Mombasa similarly remote. Remembering the hours of high-speed driving to arrive at this remote rectenna site, his determination began to soften. Horse's voice of old arose in his mind: "Only survivors win." He tightened up and marched on.

And on.

As dusk faded to night, a waning gibbous moon rose like a flattening pumpkin. His legs were sore, and his cuts stung. It was dark and he let himself collapse. He curled into a fetal ball on the still-warm ground and drifted towards an exhausted sleep. Passing into the hypnogogic prelude to REM sleep he formed mental intentions of his rescue.

All night long, he dreamed of the countless ways it wouldn't happen.

Nairobi

"When will Daddy be home?" asked Eric Vambolent, as his mother tucked him into bed.

"Let's see," she said. "Today is Tuesday, so he and Mr. Morgan were supposed to do their survey work today. Your father should be driving back to Mombasa by now."

"Is he okay?"

"Well, of course he's okay! What kind of question is that?" she said with a light laugh. She smoothed his hair, but he shook off her hand. "Tomorrow, they file their papers, and then I guess the construction can begin. On Thursday, he takes the train back here, so he should be home the night after tomorrow." She pinned his arms down with the sheets and smothered his face with kisses, while he struggled to avoid them. "Now you get some sleep, young man."

She straightened his covers, making an elaborate job of it. Now and then, she poked a rib with her finger. He giggled, and kicked his feet, making her fix the covers all over again.

He finally settled down. Clouds moved away from the nearly-full moon, and the beam of light caught Eric's face through the skylight. She saw his wide eyes glisten. He was staring straight into it, just like his father does.

"Will we really be billionaires?" he asked.

Elise smiled with her mouth only. She didn't like the way Parks involved the kids in his fantasies. Abruptly, she remembered their visitor that day. The smile spread up her face. "Well, Eric," she said carefully, "maybe we'll be millionaires. And that's almost as good."

Eric lay very still, blinking a few times.

"I don't think that's good enough for Dad."

Elise sighed, suddenly feeling tired herself. "Well, you can ask him yourself on Friday. He promised you and Mindy a trip to the Olduvai Gorge for an archeological dig."

"Cool," Eric said.

Elise kissed him once more, on the forehead, and bade him good night. She pulled the door nearly closed, then went to Mindy's room.

Her daughter lay prone on the bed, reading a book by the indirect light of a table lamp across the room. Elise flipped on the overhead bulb.

"Your eyes will go bad if you read in poor light. Just like Abraham Lincoln," she scolded.

"Yeah, but he went on to be President," Mindy quipped back.

"Okay, sweetheart. Time for lights out."

"Okay, Mom. Just 'till the end of this chapter."

Elise sat on the edge of the bed, and reached over Mindy's arm, pulling back a page. Then another. No chapter end in sight. Her feeling of fatigue deepened. "Okay, but if I come back and find you still reading…"

"I know, I know, you'll twist out all my light bulbs."

"But you'd still find a way to read, wouldn't you?"

Mindy took her eyes off the page, and gave a quick glance to her mother. "I'm getting tired, Mom. Don't worry."

Elise kissed the top of her head, and said "Love you".

Mindy said "Love you, too" into her book.

Elise shook her head silently, closed the door all but a crack, and straightened up.

Leaving the children's wing, she walked into the main part of the house. The open living area was well-lit, but felt empty, lonely. Their housekeeper had left for the day, and only the muted sound of crickets and frogs competed with the sqeaky ceiling fan. She poured a glass of wine from the wet bar, and flopped down onto the couch. With one hand, she absently fingered through magazines that were already months old.

Her thoughts came back to Jenks, and she looked towards the front door where he had entered. *A million dollars*, she thought. In a way, it was a watershed. Over the years of their marriage, her respect for Parks had eroded.

At first, he seemed like a great prospect. He had a great job, big dreams, and an uncle wealthy enough to have a building named after him on a college campus. But despite his outward generosity, her uncle-in-law did not practice charity in the home. Parks got tuition paid only by promising to work off the debt in the service of his uncle's company. For room and board, he was on his own. After 3 years, he left school to be a cyclist. His patents for pyramid soap and cone butter earned him a tidy sum from his sponsor. When he suffered a terrible crash, he'd gone back to school, and agreed to work for his uncle in China.

Elise met Parks in Indianapolis after he returned from Egypt. She had been doting on Parks' cousin, Bryce Vambolent, and was waiting on his return when news came of his heroic death. She was still around the house, consoling with Mrs. Vambolent when Parks came back from the Tour d'Afrique, bronzed and muscular, albeit covered with oozing gauze bandages. Eventually they began dating while he finished college, and she followed him to Beijing. There, they lived for almost three years as the Vambolents helped the Chinese expand their nuclear generating capacity.

Parks came up with his space inventions while in China, and promptly quit working for the his uncle about the same time Mindy was conceived. For the next nine months, back in the states, he tried getting work in the aerospace industry. But aerospace was in another downturn. He had the wrong degree, he had no job, so no reasonable offers came.

Absent any financial help from either side of the family, Parks finally took a job as a technical writer in New Jersey. They dragged their newborn daughter across four states, and lived in a run-down house while Parks pursued his space dreams on evenings and weekends.

Again and again, he was turned down. Their meager savings were spent on his conferences. The money set aside from his Agrifort patents was earmarked to pay maintenance fees on his space inventions. She started looking at her husband in a new light. These were not the characteristics of a winner. And she did not want to be shackled to a loser.

Eric's conception came as a surprise to both of them. Reading the fine print on her birth control, they learned the effectiveness is only 99.3%. Month upon month, their relationship began to fray and weaken. Those were bad days for the New Jersey Vambolents. No matter how much she begged him to return home and take his old job back, he would have none of it.

Shortly after Eric was born, there was a sudden change for the better. She and Parks had flown to Hawaii for their seventh wedding anniversary. He left her alone on the beach for four days. Not that she minded terribly, but the thought of it rankled.

Then, he had returned with great news. Suddenly, people were listening; and it was in Japan, of all places. She was no international expert, but she didn't think the Japanese could be easily fooled. In fact, she had read somewhere that Japanese companies often developed products out of American research that never reached a market. And now, they were funding her husband's ideas.

Yes, he would want to go all the way. But she was more practical than her husband. A million dollars in hand is worth a billion in the bush.

That's what she would tell him when he returned.

Chalbi Desert

In the morning, Parks awoke stiff and cold. His mouth was parched, and lined with salty dust. He had to urinate urgently, but was reluctant to give up the water. He remembered reading that Hindu yogis would drink the middle third of their own urine for its healthful healing properties. After only a short hesitation, he decided to at least *save* this liquid, using the empty water bottle. Feeling sheepish about the operation, he allowed the first part of his stream to fall on the ground. It was dark amber, and the moisture vanished almost instantly in the dust. Then, he started to fill the bottle, stopping at what felt like one third of the remainder. He decided to hold in the rest, since an empty bladder would probably draw more water from his body tissues. With the cap screwed on tightly, he held the elixir up to his eye. It sparkled in the early morning sun like a crisp Chardonnay. He liked that fantasy, and put the warm bottle into his pocket, deciding to think no more about it.

Taking his bearings, he started off south. Despair battled with his willpower. With an effort, he straightened his sore back and set his shoulders square to the ground. His head, he held lightly upright, walking as if suspended from above by an infinitely long marionnette wire. To keep his panicky ego from dwelling on the thirst, hunger and fatique – not to mention the grief that was waiting to well up inside him, Parks turned his thoughts to the event which had brought him to what Horse had called a "god-forsaken place".

With Elise as his cheerleader, he'd applied for jobs across a wide swath of the aerospace industry. None of the big firms were interested in his inventions, and his credentials did not match their needs. Working instead writing manuals and technical reports, he strove on the side to gain credibility in the academic community. He wrote papers on a wide range of subjects, sending abstracts out to conferences, and doing the preliminary research to prepare papers on his inventions and how they could be used in a cislunar architecture of space economy. The goal he set for himself was to make space solar power a reality, and to displace the nuclear technology his uncle had compelled him to promote for so long.

While he did find a few fellow enthusiasts at the conferences he could afford to attend, most of them were struggling solo on their own pet projects. He worried about being grouped into the legion of quasi-scientific people referred to within the space community as crackpots doing cartoon engineering.

Once, he collaborated with his old team leader, Cloy Morgan, to build a business case for a cislunar economy, with the help of some of Parks' school buddies Franklin Juneau and Ned Austin. As part of his MBA curriculum, Horse was required to enter a business plan competition. The four of them had put together what seemed like a great story, but they had been eliminated in the first round. Some too-young looking judge named Jenks had delivered the bad news to their team. To this day, Parks still did not believe the ersatz sincerity with which the verdict was delivered. Then, yesterday he had learned Horse had continued to work on it. Parks wondered what that was all about.

Then he remembered Horse's description of the "regulus".

Shaking his head to clear the dozens of questions that raised, he let out a shuddering sigh. His shoulders drooped as thoughts of the day before threatened to pull him down. His pace faltered, and he could little more than shuffle his feet across the ground.

Fed up with the lack of success in his own country, Vambolent switched tactics entirely. Tired of broadcasting his ideas like seeds, hoping for one to sprout, he decided to select a specific target. Who in all the world, he asked himself, had a hunger for energy, and the perspicacity to pursue a space-based solution?

The answer was clear: Japan.

Stretching half-way round the globe, linking Japan to the Persian Gulf, oil and gas tankers were strung like a chain of pearls.

On a clear day, the captain in one tanker's wheelhouse could see diesel smoke from the ship ahead and the ship astern. In Tokyo, planners in the Diet were keenly aware of the imminent peak in oil production. With new sources of oil becoming more expensive to extract, and as old sources were exhausted, energy prices would begin a steady climb upward.

Nuclear power was increasingly the power source of choice for Japan's electrical demands. Many of his uncle's customers were there. But after Fukushima Dai-ichi, public opinion turned hard against nukes. Environmental groups criticized the practice of sending Japan's radioactive waste to developing countries for burial.

JAXA, Japan's consolidated space agency, was working with several universities on space solar power, the end goal of Vambolent's inventions. Japan was also keen to expand their role in space, having fallen far behind China as the next space power. Furthermore, Japan's technological lead in clean fuels, hybrid vehicles and fuel cell cars suggested to Vambolent that he'd have better luck there.

His first letter to the Director-General at Japan's Ministry of International Trade and Industry (MITI) received no reply. Neither did the second. For his third attempt, Vambolent solicited the advice of a Japanese-American co-worker to place a call to Director-General Sugeto Mifune. An assistant took the call, listened to the explanation, and politely refused.

Finally, Vambolent arranged a visit. He and Elise flew to Honolulu for a combined seven-year wedding anniversary, and for Parks to participate in a Geological and Planetary Sciences conference at the university. Leaving his wife alone on Waikiki beach, Parks flew on to Nagoya Airport, and took an airport limo into Tokyo. His co-worker had coached him, and Parks followed the advice carefully.

The first day, he announced himself to the secretary, accepted her dismissal, then simply sat down in Mifune's outer office, quietly waiting. She scorched him with foul faces every time she looked up. But he simply sat rigid, hands on his thighs.

Each time the door would open, or Mifune-san would leave for lunch or the WC, he would see Parks Vambolent, dressed in a conservative chalk-stripe business suit. Vambolent would bow his head deeply as the man passed, not even looking to see if it was returned. On the third day of waiting, Mifune's curiosity finally rose.

He invited Vambolent to join him on a cigarette break in the building basement.

*   *   *

Mifune spoke about the wet weather, and the desperate state of affairs in remote corners of the world. He seemed to imply that while Japan was soggy from the many typhoons, it was still better to be here than anywhere else. Vambolent mostly nodded in response, and did not initiate any conversation.

Mifune exhaled a thick cloud of tobacco smoke, aiming it politely out the side of his mouth. When he spoke, his voice was still thick with the remaining fumes.

"So, you have some idea? About space?" His English was clear, and he had apparently learned of Vambolent's queries.

"Yes, Mifune-san."

"You are a patient man," Mifune smiled briefly.

"Yes, Mifune-san. Thank you, sir, for meeting with me." Vambolent smiled back.

"Well…?" Mifune prompted.

Vambolent drew a breath and began speaking. "You have brought many new technologies to Japan through the Science and Technology Agency. Ductile high-temperature superconductors, optical switching technology, holographic memory storage. The country owes you a great debt."

"Yes, yes. Please get on with it," said Mifune. But he had sat back in his chair, and appeared to enjoy his cigarette more. He twirled it idly between his fingers.

"Your institution is very important to Japan," Vambolent said earnestly. "It is your foresight and wise decisions that make Japan the pre-eminent technology leader in the world today."

"Yes, okay. You are flattering me. What is your name again?"

"Parks Vambolent, sir."

"So, Vambolent-san. What is *your* idea?"

"Actually, it is *your* idea that I came to see you about, sir."

"My idea? What idea of mine do you refer?" His face showed puzzlement, but he ground-out his cigarette butt, and leaned closer to the table.

"You were interviewed by the magazine *The Forge* last year. In that article you stressed the importance of diversification and sustainability, especially in energy technology."

"Yes, this is so. It is no secret that Japan has few natural resources. Energy is vital to our economic growth. I fear the rising price of oil is limiting our ability to overcome this long recession."

"Exactly, sir. And while nuclear energy provides much of Japan's needs, the citizens are opposed. Plus, the cost of shipment and storage of the waste products is growing higher every year."

"Yes, these facts are unfortunate, but all too true." Mifune pulled another cigarette from his breast pocket, and fired the end with a slim piezoelectric spark lighter.

Vambolent waited. His primary objective was to get Mifune to come up with the idea himself. He hoped he had not overestimated the man.

"A filthy habit, no?" he said, looking from his fresh cigarette to Vambolent. "They say it is not the tobacco, but the impurities which kill you." He paused to take a drag, and then held the cigarette in front of him, examining it closely as he exhaled. "It is like our society. There are a few bad ones who ruin it for everyone else."

Vambolent paused, not sure if a response was needed. Finally, he spoke. "I'm not sure if I know what you mean, sir."

"No, of course not. I was thinking of something else. Now, as for your idea. My office received a number of letters from you. You have some idea about space and energy, right? Some way to get solar power from space, like the SPS ideas from 1968?"

Vambolent did not respond directly. "MITI is working with automakers in the development of hybrid power vehicles," he said, "where batteries, fuel cells and flywheels complement a small gasoline engine to make a more efficient car." Mifune nodded in reply, saying a quick *'hai'*, impatient for Vambolent to continue. "Your organization is also working on hydrogen storage techniques, is that right?"

"Yes, that is correct. You are very well informed."

"Sir, I think you have prepared the Japanese economy for a major leap forward."

Mifune narrowed his eyes and looked piercingly at Vambolent. He was not sure how to take this. Did the *gaijin* know something that he, Mifune, did not? Surely there were many competing factions in the Japanese military-industrial complex, and there could easily be other groups looking to trump his own efforts. Was this man a spy? Could he be a plant from the mysterious Hiyate Kindoh, who seemed to anticipate so many of MITI's new developments, and then profit from them? Or, could this wide-eyed American have some new idea, and be trying to manipulate Mifune to accept it? He took a guess.

"So, Vambolent-san. You have an idea to make hydrogen using energy from space. Is that it?"

From the growing smile on Vambolent's face, Mifune knew he had hit the bull's-eye. Mifune's own smile grew as well. He had actually been the first to say the idea. From here and on into history, he could say that the basic idea was his, at least in Japan. And if this man before him knew of a clever way to make it happen, well, then, he was not one to turn away a gift.

"Fine then. I will hear *your* idea now."

Vambolent described the details of the step-wise implementation path he had developed, with a heavy slant towards a Japanese initiative. He explained that Japanese commercial development in space would give them an edge over both the US and China. Over more cigarettes, and cups of coffee, the two men talked until Mifune's secretary came down to find him.

"I have told you many important points which must be put into your presentation to make it acceptable," Mifune said. "I will assign one of my project managers, Yoshi Hanazawa, to help you with this. You will be primarily responsible for the development work with Kenya, if that is where you think this must be done. I would prefer Singapore because it is closer, but for many reasons, that would not be a wise decision at this time."

"Thank you, sir. I am honored to have worked with you." Vambolent bowed, and Mifune bowed in return. They both smiled, and then Mifune reached his hand forward for a Western handshake. As Vambolent gently grasped, Mifune instead gave a big Texas shake, squeezing Vambolent's hand and pumping it up and down. The secretary's eyebrows rose high on her forehead. She had never seen her boss act this way.

"This is the most fun I have had in many days, Vambolent-san. I look forward to talking with you again soon."

"Thank you, sir," replied Vambolent. He took a deep breath of the smoke-polluted air. It tasted wonderful.

*   *   *

Now, shuffling across the salt flat, those same feelings were rekindled. He enjoyed them for a brief moment before remembering all the things that had gone wrong since then. Fatique and dizziness pulled down on him suddenly, hard. His joints ached and his walking rhythm was reduced to a lurch. It was beginning to look as if Kenya was a huge mistake. Perhaps he should have approached Singapore or Malaysia; or maybe even Ecuador.

His mood turned dark, and his mind slowed to a crawl. Fragmented thoughts fired at random: plans to uproot his nascent infrastructure, march straight out of Africa, and start the whole shebang all over again somewhere else. So absorbed was he that he didn't notice the tufts of brown grasses that surrounded him now. When a man and a young boy crossed his path, he was completely taken by surprise at their greeting.

"Jambo," said the man. He looked suspiciously at Vambolent, whose dusty face was set and grim.

Vambolent stopped abruptly. Moving equilibrium carried him forward, and he nearly toppled, until he pulled himself by degrees back to vertical. He looked at the father and son, wondering if he was delirious, and this was a mirage or a hallucination. He decided to play along, for now.

"Jambo," he replied, and smiled. It looked like a grimace, but he was trying.

"You lost?" the man offered.

Vambolent nodded. They certainly looked real enough. And they had water. A gourd was hung around the boy's neck, held with a leather thong. Vambolent could not take his eyes off the gourd.

"You thirsty?" the man asked him.

Vambolent nodded again. He began to recover his manners, but it was an effort.

"I'm..." But he couldn't complete whatever sentence he thought he had started. It took all of his mental effort just to stand upright.

The man nodded to the boy, who looked rather spooked at this stranger's odd behavior. He unslung the gourd and passed it towards Vambolent, holding it out at arm's length.

Vambolent accepted the gourd, and held it. He looked gratefully in the man's eyes, and his face softened. Then he focused on his dire need for water, and pulled the twist of leather from the hole in the top of the gourd. Forcing himself to start small, he tipped up the gourd, and poured a small portion of water into his mouth. The effect it had on him was immediate. As he swirled the water in his mouth, his brain started to work normally. Little sips trickled down the back of his throat, and he began to feel his body again. He had a terrible headache, he realized. He took another sip, a little larger this time.

"Thank you," he said, and held the gourd back out towards the boy.

"Come to my home. We will care for you," said the man.

Vambolent nodded, and the two of them turned back from wherever they were going originally. He followed them through the sparse grass. The water had energized him, but he was more keenly aware of how desperate his situation really was. Walking was agony. His joints felt swollen, and he craved ibuprofen. As he walked behind the two locals, they both kept turning halfway around to look at him. He could tell from their faces that they were very concerned, and were probably expecting him to collapse at any moment. Their apparent care and compassion were a balm to his soul. At their hut, no doubt, they would know how to help him recover from dehydration and heat stroke. The light of hope this offered kept him going, despite the agony of each step.

It was miles away before they came to their home. Two skeletal-looking cows were tied to a small scrub tree, which had its leaves eaten off to the height of their bovine reach. Both cows had saggy udders hanging nearly to the ground, and were lapping water from a shallow trough. The cows looked up as the three approached. A strange wound showed on the haunches of one animal. The other made as if to low, but no sound came out. It was a surreal scene to Vambolent. The hut was made of poles and thatch, held together with what could have been mud, but must have been manure. A woman came out of the doorway, wrapped around the waist in a colorful khanga. She noticed the stranger, and trotted over to help him into her hut.

Vambolent leaned on her gratefully, and let her lead him to a single stool at a low wooden table. She fussed over him, pinching the skin on the back of his hand, and then looking hard into his eyes. She peeled back his eyelids, inspecting the lining. He just sat there, and let her minister to him. She bent over her cooking corner, and turned back to him with a bowl of pink liquid, setting it on the table. Afraid to trust his muscles, Vambolent bent his mouth down to the bowl, and tipped the liquid into his mouth. The taste was so unfamiliar, he could not recognize that it must have been milk. He looked up to her, with deep gratitude in his eyes. They would have sprouted tears if he had any.

"Thank you," he said simply.

She just smiled. It was a beam of pure white in the semi-darkness. Her teeth suddenly made everything else darker, and this was how he remembered her later – like the Cheshire Cat when it's about to disappear.

"What is this?" he asked, looking back to the bowl, and bending down for another sip.

"Blood and milk," she said, after a pause.

Vambolent almost spit out his mouthful. Somehow he held it in. He hoped she did not see his reaction, but he doubted this woman could have missed a thing. He held the liquid in his mouth, not swallowing, until he remembered the bottle of urine still in his pocket.

He swallowed.

It felt good going down. He could almost feel the nutrition spreading throughout his body. He suddenly felt overwhelmingly tired. His eyelids drooped, and he just could not hold them open. Then, he felt her hands on his shoulders, guiding him. He half rose from the chair, and staggered with her to the other side of the hut. There was a patchwork of khanga cloth squares stitched together and laid over a bed of straw grass. He laid down like a sack of yams, and was asleep by the time his head came to rest.

Chapter 2

Voi, Kenya

Under the direction of Chuck Schoenien, the Flying Doctor who brought Parks Vambolent in from the desert, two porters lifted his patient towards the train. On the steep stairs to the sleeper car, the rear porter, carrying almost the whole deadweight of their burden, slipped and dropped a handle.

Parks was exasperated with lying still as he had been ordered, and tried to get up. When Dr. Schoenien called "No, Parks, you're still too weak!" it spurred him to try even harder. Both Parks and the porter collapsed down the stairs in a tangle of limbs not unlike a crash on a cycle race.

Parks stood up immediately, intending to mount the stairs on his own. The doctor reached for his arm as Parks' field of vision narrowed to a tunnel and his head swam. His foot reached up and fell down once, twice, not coming even close to the step. Schoenien tried to guide him to the stretcher that the porters were readying again, but Parks was pushing people away at the same time he was trying to lean on them.

Another passenger waiting for the train stepped into the confusion and hooked a hand under Parks' armpit. As the man pushed Parks forward, the searching foot found purchase. The man heaved him up, past the lead porter, and Parks found the railing. At the top of the stairs, Parks leaned heavily into the wall. The good Samaritan moved into the car as the doctor came bounding up the stairs.

"You need to lay down now, Parks. A fall here could make you worse off than you are already."

"Awright, doc," Parks said, his voice thin and reedy.

With an arm around the doctor's shoulders, Parks made it to his cabin and collapsed into the bottom bunk.

Dr. Schoenien straightened his legs and smoothed out creases in the hospital gown Parks wore. Then, he worked to mount a hook on the upper bunk. Hanging a bag of saline solution there, he plugged the other end into a capped needle taped to the back of his patient's hand, and inserting into the raised veins there.

"Okay, Parks. You're all set. Nairobi's less than eight hours away. You're to stay in bed the whole time, understand?"

"And if I have to go to the WC?" Parks asked.

"Your tissues are so dry, I doubt you'll need to," the doctor said. "But if you have to, take the bag with you. Either wrap it around your neck, or better yet, get somebody to help you."

"Like the guy who helped me up."

"Yeah, well, maybe him, or you can ask the conductor. The cable right here will ring for help. Don't try any heroics, Parks. Promise me."

"I promise."

"Good. Now I've got to get back to my rounds. I've notified the local authorities here in Voi about the Shifta raiders who attacked your party, but you had better fill out a full report with the federal authorities when you get to the capital."

"Okay, doc."

The doctor waved goodbye, then stopped outside Parks' cabin to speak with the conductor. Parks watched as the doctor pulled a fat coin from his pocket and pressed it low into the palm of the uniformed man. Patting the conductor's shoulder, Schoenien left. The conductor turned and looked through the grimy glass. He smiled a wide, toothless smile.

Parks returned a smirk and flopped his head down on the pillow.

Not a minute later someone knocked on the door of his cabin.

"What!"

Parks looked up, and saw the old man who had helped him mount the stairs. "Oh, it's you." He sat up with an effort. "Please come in."

The man opened the door but stood with his toes at the threshold. Parks watched the man's rheumy eyes take in the saline bag, then narrow in understanding.

"May I come in?"

Parks waved him in. He had just cleared the door frame when the train jerked forward. The old man collapsed gracefully, coming to rest at the foot of the bottom bunk. Parks drew his legs up quickly to make room.

The man patted Parks' leg and left his hand there. "I am Mahmed Kwapili."

"Parks Vambolent." They tried to shake hands, but the IV got in the way. Instead, Kwapili just patted the leg again.

"How are you? I heard the doctor mention the Shifta." The man shuddered. When Parks did not reply at once, the man went on, talking out into the tiny room. "They do not get to this small village, but each year they come further west and south. Some of the local people, they join the raiders and attack their former neighbors for food and petrol. Their bases reach beyond Somalia now. The border is in dispute."

"They killed my best friend," Parks said. "And our guide, and three workers."

Kwapili patted the leg again. The touch was too familiar to Parks at first, but he was growing used to it. When Kwapili did not reply at once, Parks told his story. The man nodded, and Parks watched the ancient face as he talked: wrinkled brown sacs of skin shadowed each fading eye. Wiry brows tangled in every direction, and tilted expressively as he listened. The man's nose was wide, with a pronounced shallow in his upper lip. The lower lip was thick and poised above a matted beard of steel and ebony.

"You are lucky to be alive," the man said when Parks had finished. "Surely you are blessed by Allah."

Parks held in a sarcastic comment, as the image of Horse's burning hand came into his mind. They sat quietly for a long time, letting the rhythm of the train mark the passing of seconds.

"Well, I should be going," Kwapili said, but he did not leave. "If you need anything, you can bang on that wall, I am just on the other side."

"Thank you," said Parks. Still the man did not leave. Just every few seconds his hand would stir, patting absently, until Parks began to think him a madman.

Finally the man gathered himself to leave.

"Wait," he said.

The old man turned and looked directly at Parks. The bluish eyes seemed not to focus, like they were aimed slightly *into* him.

"There *is* something you could do."

"Anything to help," said the man.

"I've been through a lot…" Parks began. "I'm kind of unsure… I don't know what I'm trying to say, but I wonder if you might just… well, stay here and talk."

Kwapili slapped his thigh and laughed. "Of course. Let's talk." The man pulled himself up using the handle on the top bunk and moved his seat onto the armless chair at the sliver of a desk. His robes fell around it and he sat like he was posing for an alabaster statue.

"What would you like to talk about?"

"My business," Parks said.

"Ah, business. I am also a business man," Kwapili said, holding a thumb and middle finger together over his breast. "A landowner," he added, nodding solemly.

"What kind of land?"

The motile eyebrows of Mahmed Kwapili fell, darkening the eyes beneath. Then, an instant later, they bouyed up again. "It is of no consequence," he said dismissively. "We are here for your story. Tell me about your business, Parks Vambolent."

Parks folded his arms behind his head and stared at the bunk above him.

"I made a deal with President Mtonge. And now I cannot honor it."

"That is a very serious thing to say. Are you in trouble?"

"No. Not yet, anyway."

"Tell me about your deal with Jemmat Mtongwe," Kwapili said.

Parks told him how, with the help of a Japanese diplomat, he had timidly brought forth his proposal for a solar power receiving antenna. Surprisingly, Mtongwe had agreed to every request with only trivial adjustments. He had even donated national lands to the project, in the Chalbi desert. Not a prime location, but free of the many considerations involved with using a more populated area. Giddy with this success early in the negotiations, he agreed easily to the demands Mtongwe made in return: double foreign investment of hard currency, and reduce the country's illiteracy rate in half. As Parks found out later, such were the conditions the National Assembly had imposed for Mtongwe to stay in power. He had successfully pawned these responsibilities off onto Parks Vambolent and his project with MITI.

He told Kwapili how he felt sure his Chalbi Energy Project could eventually help achieve the capital influx goals. But the literacy issue completely bewildered him. He related the phone call to Elise which had given them a glimpse into a solution.

* * *

"Just build some schools," she suggested at once.

"I've been over all the simple answers, Elise. Mtongwe offers no help financially. The existing infrastructure is archaic and apathetic everywhere except in the bigger cities. Trained teachers are scarce in the bushlands and in the high desert, books are hard to come by, and the students lack the foundation for any sort of radically new program."

"Why don't you teach them with computers," Elise said, sensing his hesitation.

"What do you mean?"

"Give them each a computer," she said, "and then you only need a few teachers, no paper, and they could even learn at home if they wanted to."

A moment tottered between sarcasm and breakthrough. In a flash, the entire scheme formed itself in his imagination. He simply sat there on the bed, a quarter of a world away, as the complete outline of his new plan seemed to download into his conscious mind.

"Do you realize what you're saying?" he asked her.

"Mm-hmm." But she seemed oblivious to the revelation occurring within her husband. Or was she?

It had all seemed so clear then. Elise had followed him to Africa, uprooting their two small children, and agreed to spearhead the education effort. They had arrived just over a month before.

* * *

"And now, it's all ruined," he finished.

"Ruined?" Kwapili asked in surprise. "Your partner is dead, and that is a sad thing. But your business, it is not also dead."

"You don't understand, Mahmed. We can't build anything in the Chalbi where these Shifta raiders can come at will and take whatever they want. Besides, Horse – I mean Cloy – he was the brains behind the business. I'm just the inventor."

"Just the inventor, hah!" Kwapili said forcefully. "If you are smart enough to invent, you are smart enough to run a business!"

"Huh." Parks rolled onto his side, and propped up on his elbow. "But what about the land? The Chalbi is too dangerous. It's also too remote." Parks shook his head in dismay and settled back down on his side, punching the meager pillow with a balled fist to fluff it.

Mahmed Kwapili watched Parks with pursed lips. "Hmm," he said softly. Then again, "Hmmm."

"What?" Parks asked abruptly.

"I was thinking. Hmm." He rubbed his jaw through the beard and Parks heard the raspy retaliation of the thick, kinked hairs. "You need land. I have land. Maybe we can work out a deal of our own?"

"I need land that is nearly desert."

"Yes, I have that."

"It needs to be away from population centers."

"Well, that is true now, Voo and Rei are empty," he said, then added: "Besides a few squatters." He turned his head and spat, spitlessly.

"I need a large area, 10 kilometers across in all directions, at the very least."

"Hmm, I think that we meet all your needs, Parks. Would you like to come see the lands you will need?"

"But my best friend is gone. Who will run things?"

Kwapili shrugged his shoulders clear up to his large ears. "That is a problem, but a smart inventor like you, he will find an answer. Just ask Allah to put the answers into your heart, where you can find them. You cannot fail."

Parks sat up on the bunk, feeling dizzy, but ignoring it. "Just where is your land?"

Kwapili explained the land, giving the history and the various uses the land had enjoyed while his family had owned them. He talked about his son, who did not want the land, and now it was only sitting idle, wasted, useless.

"It will be good for my land to be productive again."

As they discussed more and more details, night came and still they talked. When the train arrived in Nairobi at 8:33 the next morning, Parks finally stood up, filled with new energy, and hydrated at last.

"Um, Mahmed? There's one last favor I need to ask you?"

"Certainly, anything. What do you need?"

"Wait. No, never mind." Parks yanked off the drip tube. The bag was empty and shriveled. Using a round bandage from his shirt pocket, Parks covered the hole in the back of his hand as he pulled out the needle.

"There. I'm ready to get started."

"Fine. We will meet in Mombasa when you return. I'm sure everything will work out fine."

Mahmed led Parks by the hand around the corner and down the steep stairs. Parks did not need the help, but welcomed it. They parted with a hug, and after a quick stop in the terminal, Parks caught a taxi home.

Horse was dead, and Parks would need to go back for him. After he saw his friend's body returned to the US, he would carry on the tradition as he knew Horse wanted him to do. Then, he remembered. Back in Mombasa, at Horse's apartment, was apparently a business plan that would make everything clear. He had to get there as soon as possible.

Nairobi

"Elise!" Parks said, walking up to the house. He had called ahead on the cab's cell phone to tell her he was coming back. He had told her about Horse.

She came out in a fluster, apron strings streaming behind her as she launched herself onto her husband, hugged him fiercely, then stood holding him at arm's length to inspect. Parks felt her gaze like an electric probe analyzing him like some lightning medical product of a deep shaman. His muscles twitched, and he felt sparks passing between them. They had been separated over a week and both missed their shared intimacy. They hugged again, closely, smelling each other, feeling each other's bodies, and feeling their own body hugged in return. Then, the moment passed.

"We're leaving, of course," she said.

"No, we have to stay. We promised the President himself."

"There's no way you're going back there!" she stated, stepping back from him, putting exactly one meter between their hearts. The electricity stopped and Parks felt oddly cold. After a breath the feeling passed, and he looked at his wife. Legs firmly planted, arms akimbo, lips set, she was determined. Parks muttered an oath under his breath.

"What was that?" she demanded.

"Look, Elise. I'm not going back to the Chalbi. Except to collect Cloy's body. Then, it's to Voo."

"Who?"

"This great old man named Mahmed Kwapili has offered me land much, much closer to the ocean. We can add desalination to the Energy Project."

She wasn't listening, because he wasn't telling her what she wanted to hear.

"We're leaving, Parks. You, me, Mindy, Eric, and that lizard of his if we can get it through customs. This is a dangerous place, and it's no place for you or your inventions. This is not space, but the most backwards backwater the Earth has to offer, and they don't even care about their education system, and those crazy people killed Horse, and without him the whole deal with the Japanese is a bust. Parks, we're going home!"

Parks reeled at the force of her speech. It echoed in the space between homes, and there was that kind of hollow amplification you notice when someone else tunes in to your discussion. They both stood their ground, she breathing heavily, he stunned and feeling suddenly sleepy. He had been up all night, and until a moment ago, flying high on adrenalin. Now his body was sending him shutdown signals. His knees felt suddenly adjusted to a heavier gravity, and he crouched a bit.

"Come on, Parks. The kids want to see you," Elise said. She put an arm across his stooped shoulders and guided him into the house, out of the gathering heat.

"Dad!" Mindy cried. "Daddy!" Eric said, and they both sailed into him, hugging his waist and knees. He reached down to their backs and rubbed them, pressing them into him. They bounded back and he went down on one knee. Mindy sat on the other, and Eric stood next to him, almost face-to-face. Parks kissed him on the cheek, and Eric didn't flinch. Then he caught a goofy smile and knocked his little head into his dad's. Hard forehead caught lacerated cheekbone.

"Ow!"

"Sorry," Eric said, and they hugged, just father and son. Then, Eric pulled away and ran off into the house.

Mindy recovered from being jostled, and said formally:

"Welcome home, Daddy."

He hugged her close, and she curled up against his chest, loving the arms around her.

"Okay, young lady," Elise told her. "Time to go play."

Mindy gave her mom a look, then turned to her dad again.

"Mom told us about Mr. Morgan. I'm so sorry, Daddy."

"Thanks, sweetheart. I miss him. I know you miss him too."

She nodded, her eyes large with tears, and then she sobbed. Parks caught the sob, and for the first time really released his grief. His reaction was strong and caught Mindy by surprise. She stood straighter, and hugged him as fiercely as she could. He bucked in her grasp as deep sobs racked him. Mindy held on tight and Elise watched as the emotions peaked and then subsided.

Elise instinctively took Parks from Mindy, and the two guided him into the master suite doorway, Mindy pressing herself extra close to fit through at the same time. She let go, and watched from the door jamb as her mother laid her father on the mattress. Before undressing him, she looked over her shoulder. Mindy took the hint, turned, and closed the door. She stayed on the other side, listening, for a few more minutes.

"Elise, it's a perfect location. We can still do this. Horse mentioned a new B-plan. Maybe I can do it on my own."

She looked at him closely, saw the determination in his face, and hers softened.

"How?"

"Well, um… I'm going to need a lot of help."

"Yes?"

"Like somebody to start a business for me. Somebody to do taxes. A lawyer. Those kind of folk."

"Parks, there are companies out there that do just that."

"Do what?" he asked.

"Set up your business. Let you focus on the technical parts. Take on some of the burden, and most of the risk."

"Hmm. I'd be very suspicious of any company offering to do all that."

"What if you knew someone, personally, in that line of work."

Parks peered closely at Elise. "Who do you mean?"

"Winnimer Jenks," she said, and handed him the business card she'd been given. "He stopped by on Tuesday. The company he works for helps inventors like you to make a profitable business. They pay you a lump sum, up front, in case anything goes wrong. Then, you get your pick of jobs in your company, and they pay you a fair salary."

Parks furrowed his brow.

"Jenks was here? In Nairobi?"

"Yes! He said he heard about your work here through a news service and flew right over to meet with you. I told him you'd call back. Oh, do, Parks. It's everything you ever hoped for."

"Well, not everything," he said, but she kept talking.

"Think about it, Parks. Working in an American company instead of working for the Japanese government. Working back home in Indiana, instead of in Kenya. Raising our kids in safety, instead of a country where people can just drive up and massacre you. Horse is dead, Parks. He was your partner for a reason. He had the business smarts, the B-plan as you call it. Give that plan to Jenks, and let him make you a millionaire."

Mindy tiptoed away to go get Eric, so he could listen too.

"But I have a promise to President Mtongwe. And I have a well-paying contract with MITI. And I'm doing what I love."

"Parks, what you love is inventing and making space work for people. Doing surveys in some god-forsaken desert is not for you. Working for the Japanese, when they control everything, is not for you. Parks, do you really think they are going to let you run an agency, or a business? No, they're only letting you work on their project so they don't have to pay royalties on your patents. As soon as they figure out a way around that, they are going to drop you like leftover sushi. Parks, you fill the childrens' heads with your fantasies, and then when opportunity knocks, you're not home. Call him, Parks. Call."

"I will, Elise. But something doesn't add up. What time did Jenks stop by on Tuesday?"

"About noon, no, a little later. It was after lunch. Maybe 1:30."

"Huh. That's when Horse and I were attacked."

"You're not suggesting Jenks had anything to do with it!" she spat. "He was right here."

"Mahmed, the guy on the train, said that even though Shifta raiders are common in northeast Kenya, he was surprised they had come as far as the Chalbi."

"Parks, you're being paranoid. Didn't Jenks help you at that competition thing?"

"Well, he seemed to help. I don't actually know if he did or not."

"Now you're going too far. Look, it's a golden opportunity, Parks. You want to get rich, you're not going to make it on your own. You need partners. I hate to say this with dear Horse still not in his grave, but with him gone, Jenks is your best bet. Don't screw this up, Parks."

Parks pursed his lips and concentrated.

"I feel like you're bullying me into calling him."

"Fine!" she yelled. "Do whatever you want, Parks. Take your golden opportunity and piss it all away."

She got up and stamped her way towards the door. Soft, rapid footsteps faded away from the other side. Elise banged open the door and yelled the children's names. Parks sat on the bed, business card in hand, and thought about what she had said.

He picked up the telephone and was about to call Jenks. Instead, he went to his office, careful to avoid Elise, and called Sugeto Mifune, who had returned to Japan. With a voice command, the videophone assigned to him as he worked on the MITI Energy Project placed the call.

"Hello, Parks-san. I heard the news. Are you okay?"

"You heard about Horse, too, then, right?"

"Yes, yes. It is very sad. Very sad. I am so sorry for you, Parks. If you need anything at all, I will help."

"We need to pick up his body," Parks said in a whisper.

"It is already being taken care of," said Mifune. Then, his face turned hard. "And we are tracking the people who did this. Rest assured they will be brought to justice and made to pay for their evil crimes."

Parks had not felt vengence until he heard of Mifune's swift action. It made him feel good. He felt more powerful in a way he did not understand.

"Thank you, Mifune-sama. You are kind and very generous."

"It is the right thing to do. No thanks are needed."

"I know this is an awkward time, but I wanted to talk to you about the future of the project now that Cloy is gone."

"Ah, I see. Are you finished then, with the project?"

Mifune had asked casually, with no real emotion behind such a statement. Parks wondered if the implied lack of concern was honest, or if it was a cultural thing that he did not understand.

“I’m still committed to the project,” Parks said. “but I want to make sure that your long-term vision for my role in it is what I want for my own career.”

Mifune sucked breath through his teeth, and looked off camera for a moment. Then, looking straight at Parks from the vidphone, tapped his finger for emphasis.

“This project is very important for Japan, and also for me. I need top people to make it a success. Your inventions and technical work are a vital part of that. I can see you progressing in position and authority as you develop and demonstrate your capabilities.”

“Well, that sounds pretty fair,” Parks said. “What do you see as my capabilities?”

“You do not know them yourself?”

“Of course, sir. But I wanted to know what you think, how you perceive me.”

“I see your work, and I value it,” Mifune said, still looking at Parks through the screen.

“But I mean, where do you see me, ultimately. For example, can you see me as a director within MITI, leading space energy projects? If we eventually turn this into a business, can you see me as a Chief Technical Officer, for example?”

Mifune sat back as he absorbed this. Parks could not read his face, which could have been wallpaper, except for an occasional blink.

“Chief Technical Officer,” he repeated at last. “That would be a position with a great deal of authority and responsibility. I could see this being a possibility. But such a position would have to be earned.”

“Of course,” said Parks, then pressed a bit further. “But how long might it take to reach such a position?”

“Oh, that depends on the man,” Mifune said. “We do not groom people for such positions. Those who make it to such a level rise there on their own merits. Your patents are very valuable to our work Parks, but they do not qualify you as a CTO. You will have to rise through the organization. Just like everyone else.”

“What about my nationality? Would it take me longer, because I am not Japanese?”

Mifune sucked his teeth again, and this time Parks interpreted it as a response to an insult. He almost apologized, but instead waited to find the answer to his most important question.

Mifune's voice had changed, now lower and more professional.

"The obstacles are greater for you, yes. But it is not your nationality at issue. You do not speak our language. You do not know our culture. Yes, you can learn these things, but they take time and considerable effort. That is as much as I can tell you."

"I see. Thank you, sir. I appreciate you talking to me about this. You see I've been made an offer to start a business back in the States. I wanted to see how I stand with you before telling them 'no thanks'."

"I see," Mifune said evenly. To Parks he looked upset, so Parks changed the subject.

"I'm assuming the Chalbi is no longer a suitable site for the rectenna?"

"Correct," Mifune said. "We will need to pick an alternate site. Ecuador, perhaps. I have already notified the staff here in Mombasa, and we are making preparations to leave Kenya."

"But wait!" Parks said, caught off-guard by the suddeness of his boss' decision. "There's another site we can use. It's actually much closer to the port and to the ocean."

"What is this?"

"Near the towns of Voo and Rei there is a large tract of land which which is available for lease. I haven't visited the site yet, but from what I've heard it meets all of our needs. It is even close enough to the ocean to pump water directly to desalination vats. We could have irrigation to grow crops right under the rectenna!"

"Who owns this land?"

"A man named Mahmed Kwapili. He offered to lease the lands to us for the Energy Project."

Mifune frowned deeply. "Private lands," he said. "The government will never allow it."

"Why not?"

"Mtongwe-san and his cabinet do not allow foreign governments, or their agencies, to own land in Kenya."

Parks puzzled over this, and an answer popped into his heart.

"*I* will lease the land," he blurted out.

"Pardon?"

"I'll form a company – a private company. You will be my silent partner. Then, I can lease the land from Mr. Kwapili and use it to build the rectenna."

Mifune's frown grew so severe Parks thought his skin would split. Then, he relaxed, and looked almost amused.

"Then, you can be CTO *and* president!"

The two men laughed. Mifune told Parks he would look into the idea and get back to him. They ended the call, and Parks bounced off to tell Elise the news.

Chapter 3

Mombasa

A battered and blue-smoking taxi wove through parked cars and vans, then deposited Vambolent at the law offices of Nbele & Associates. They were in the heart of the industrial section of the city. A partial solar eclipse edged the shadows with a violet tint. At one time, Parks would have known about events like this, and awaited them eagerly. Now, he barely gave it a passing glance.

The time was a few minutes before four o'clock, after the mid-day siesta. Now, the island city was already back in full swing, noisy and congested. The odor of decay rose up from the sewers and the thick fumes of a pesticide plant were even worse than the body odor inside the musty taxi cab. It was stiflingly hot, despite the partly-blocked sunlight. Whatever the sun lacked from above was readily supplied by the baking slabs of cracked concrete below. Vambolent was already soaking wet inside his cream-colored business suit. A uniformed guard with white gloves opened the door of the office building and smiled wide, showing a one-tooth grin beneath bloodshot eyes.

"Jambo," Parks said to him.

The man bowed his head and returned the salute.

He climbed the stairs and found the glass-paneled door painted with the law firm name. Yoshi Hanazawa had helped him arrange the permits and pay the necessary gratuities, but as the president of Vambolent Space Industries he came alone. He opened the door and stepped across the threshold. An attractive secretary sat slouched at a dilapidated desk, and looked up at Vambolent under dark-lidded eyes.

"I'm here to see Mr. Nbele," he offered.

"Just a minute," she drawled, and dragged an ancient plastic intercom box toward her across the desk. She pushed so hard on the wide brown button that her finger bent backwards nearly in half. She leaned down over the device, scowling at it.

"Your appointment is here," she shouted.

There was an audible click, and a voice replied over the speaker, crackled and hissing with static.

"Send him in, Natala."

She got up with exaggerated effort and walked the two steps to the inner office door. She twisted the handle, and pushed the door inward.

"He is expecting you," she said, watching Vambolent's progress across the room. As he stepped inside, she pulled the door closed, barely missing his heels.

Inside the office, Vambolent was greeted at once by Barakko Nbele, who rose from his desk chair and came around to shake hands.

"Welcome, Mr. Vambolent, welcome! It is so nice to see you today. How are you?"

"Fine, thank you."

Vambolent let go of Nbele's hand and turned to greet Mahmed Kwapili. The two smiled and embraced.

"Parks, I want you to meet my first born son, Afren. Afren, this is Mr. Parks Vambolent, the same one I have told you so much about. He is here to lease our grazing lands."

"Hello, Mr. Vambolent," said the handsome man standing beside his father.

"Please call me Parks."

"Parks, then. And you may call me Afren."

They shook hands, and everyone sat.

"Well, now that we are all here together, we may process this excellent transaction," the attorney began. "First, there are a few formalities. I hope you will excuse my indiscretion, but this is required by our esteemed government. Mr. Vambolent, I am obliged to check your personal documents and identification letters from the Ministry of Commerce."

"Of course," Vambolent said, setting his briefcase on his knees and snapping open the latches. He produced the papers, and handed them to Nbele.

"Fine, fine. Everything is in order. And now, your papers of incorporation, please?"

Vambolent extracted a blue folder with the words 'Vambolent Space Industries' emblazoned across a white label. There was a stylish logo of two concentric circles inscribed with square cross and a upward triangle with the letters *VSI* beneath the symbol.

"Here you are," Vambolent said, turning the folder around, and placing it on Nbele's desk. "VSI is incorporated in the state of Delaware, in the United States of America."

"Very good, very good, indeed."

With the abject failure of the Chalbi desert as a viable site, the project got at once easier and harder. The logistics, timing and investment would be greatly reduced, since the Kwapili lands were much closer to the ocean and the seaport. However, they would be running pipes and wires through populated districts, and building a giant microwave receiving antenna just a few tens of miles from some small towns. Officials at the Ministry of Finance seemed to know that MITI was behind the up-start VSI, and threw every conceivable roadblock at the proposed transaction. Of course, several generous donations were required to ensure this flimsy arrangement was approved, despite encouragement from President Mtongwe. The Ministry letters had come at considerable expense to the project. It greatly rankled Vambolent that the wheel of industrial progress required the grease of political bribery. But these were the costs of doing business in Kenya.

As Nbele reviewed the stack of papers, the younger Kwapili spoke to Vambolent. His voice was quiet, but his words clear and distinct. His tongue carried a British accent, making his speech melodic, and a pleasure to the ear.

"Kenya is a proud land, Parks Vambolent. Your project is a symbol of our progress."

"Thank you," Parks said.

"Yet, I notice you have chosen an unusual symbol for your firm. Could you please explain it?"

"Why, yes, certainly," Vambolent answered, caught somewhat off guard. He had drawn the symbol while doodling, as he spoke on the phone setting up his letters of incorporation. "The symbol of a circle with a square cross inside is the astronomical symbol for the planet Earth. In the VSI logo, the second circle outside represents an orbit."

"And the cross?" Afren Kwapili said. "Is it not a Christian symbol?"

Although the question had been asked casually, the little motions and noises that people constantly make went suddenly still throughout the room. The father was obviously Muslim, but Vambolent had no clue as to the younger man's spiritual orientation. Religious tensions in Africa were always high, and any answer he provided might lead to unexpected responses.

"Well, the cross, as a symbol, predates Christianity," Vambolent explained cautiously. "Within Africa itself, the ankh was a powerful symbol of the Ancient Egyptians. Many other ancient societies around the globe have used the cross as a symbol of Man, where the vertical line is the spine, and the lateral line is the shoulders. The cross inside the circle represents Man on Earth. The triangle I have added is for Woman."

"I see. Please do not think me antagonistic, Parks. But this symbol will probably appear on the buildings, pipelines, and trucks that you will use here in Kenya. The people will ask about its meaning," Afren explained. "You know that women are not liberated here, like in America. This may offend some people." He looked quickly at his father, perhaps as a representative of the Muslims who might take such offense. "On the other hand, the triangle could represent a pyramid, which could symbolize Egypt. This could also offend. The madman Hababek has made many people very angry with Egypt. In fact, one could look at this logo, and imagine a gun sight trained on the Great Pyramid."

"Oh. I never thought of that," Vambolent replied thoughtfully, staring at the logo with new eyes. He paused and looked around for help, but the older men seemed quite content to let this discussion continue. "I did have another idea," he said suddenly, and pulled a pen from his briefcase, and drew the folder to him. He sketched the double circle, but within put two crosses. "I called this the double Earth symbol. It was actually my first choice."

"Why did you not select this?"

"I thought the two crosses could be misconstrued as a 'double cross'."

The younger Kwapili laughed lightly, and nodded in agreement. "Oh, I see. Perhaps you could use two colors. Not one Man, but two Men, working together in partnership to help the Earth."

Vambolent looked at Afren Kwapili for a long moment, wondering if this was a veiled offer of partnership. He reappraised the man, who appeared to be almost of an age with himself. While at first, he had been concerned that the son would spoil the deal he had laboriously worked out with the father, he began to see Afren as an insightful and intelligent individual.

"Yes, I like that," he replied, "consider it done."

Afren smiled easily and gave a single nod – half acknowledgement and half thanks.

The other men were obviously in no hurry, and Vambolent was already learning the slower pace here in Mombasa, compared to the Western business climates. He waited patiently as Nbele reviewed the paperwork, and filled in forms. From his previous meetings with the elder Kwapili, he knew the deal was acceptable – it would make Mahmed comparatively wealthy – but these men seemed to want to know the character of the man they were dealing with.

"I am sorry for the loss of your friend," offered Mahmed quietly.

"Thank you. Cloy was a great man, and excellent with people. We will have trouble replacing him on this project."

"My son has a technical degree," Mahmed said slowly, looking steadily at Vambolent. "Perhaps he could be of some use?"

Vambolent looked again at Afren. No resume, no interview, but he was about to hire VSI's first employee.

"I would be glad of his help."

Mahmed smiled warmly, and clasped his hands contentedly over his belly. Nbele had finished looking busy and was watching the scene. Afren sat quietly, giving no indication of how he might feel. Parks suppressed a longing for the comfort and security of his own native culture, where somebody should be saying something right about now. Still, he noticed the earlier tension was gone, and everyone seemed pleased.

Nbele spoke up now that it was finally time to conclude their business. He pulled a contract from his lap drawer, and caressed the top page with the tips of his dry, cracked fingers.

"Well, well. Here is the agreement I have drafted. You have both seen copies?" Vambolent and the elder Kwapili each nodded. "Good, good. Now, before we sign, I will need some assurance that the agreed-upon funds are available and ready to transfer."

Vambolent handed a 3x5 index card to Nbele.

"This is the phone number of a bank in the Bahamas. The number beneath it is an account code. They will verify the existence of the funds, and their availability. When the paperwork is signed and notarized, I will call with an authorization number, and the money will be transferred."

"Excellent. I will just have my secretary handle this while we finish the other arrangements." Ignoring the intercom, he raised his voice slightly, and called Natala. The door opened almost immediately, and she walked over to the desk. Nbele handed her the card and gave her instructions. She turned to walk away, and he patted her behind. She seemed not to notice.

Afren Kwapili cleared his throat, and addressed Vambolent.

"In your proposal document, I see that you address the environmental consequences of this massive project," he began. "It seems that your project is ideally suited to our desert wasteland."

"Grazing land," his father interrupted.

"It was only grazing land when you were young, Father. Now, the goats and cattle have turned it into desert. With Mr. Vambolent's project, the land will be restored to usefulness, but it will no longer be ours."

"It is good land," the father insisted, weakly.

"It is good for this project. There are few migration paths overhead, there are no settlements to speak of, and there are no industrial effluents from the project itself. However, Parks, you do not seem to have anticipated all the potential outcomes."

"What do you mean?" Vambolent's brow creased with concern.

"The water you produce will transform the region."

"Isn't that what Kenya needs? There will be more food and more jobs."

"Of course, that is all true, and that is good for our country. But will this new prosperity not make us the target of jealousy from our neighbors? Already we have so-called 'political refugees' from Zimbabwe, where there is famine. With more food in Kenya, this problem will escalate. Will there not also be security concerns, as you are already so intimately familiar?"

"But aren't these problems the responsibility of your government?"

"That is too easy an answer, Parks. You are bringing a new technology to Kenya, and perhaps to all of the world. On the surface, it seems like a good thing all around. But I wonder what will be the final assessment of history. Many times in the past, inventors like yourself have sought to make peace, but their tools have become the implements of war. I wonder if you have truly thought through the downstream consequences."

"Your point is well-taken, Afren," Vambolent said, gaining confidence. Yet, when we compare this new technology to the devastation caused by today's methods for energy production, many serious threats to our survival will be completely eliminated."

"Perhaps we should all live simpler lives," interjected Mahmed Kwapili.

His son paused a respectful moment, then spoke carefully.

"I personally do not advocate the return to simpler technology. Such could not be done unilaterally. Whatever man, group, or nation still retained advanced technology would sooner or later use it to exploit others, as we have seen so poignantly in Africa's past. Is it not the history of technology that someone always uses it for their own selfish purposes?" He looked back to Vambolent. "Your project sounds almost too good to be true."

At this point, their discussion was interrupted by the entrance of Natala. She announced that the funds were available. She stayed behind Nbele's desk, standing to one side, her hip cocked at a sharp angle.

"So, we have everything in readiness, yes? Mahmed, you are satisfied with all arrangements?" Mahmed nodded. "Afren, you are satisfied also?" Afren nodded in turn. "Good, then. Mr. Vambolent, you will please sign where indicated? Natala is a notary public, and will witness all of our signatures."

They signed, and then each page was imprinted and countersigned by the notary. Nbele pushed his telephone towards Vambolent, who made the long distance call, and recited the ten-digit authorization code from memory. He passed the handset to Nbele, who received verbal confirmation of the transaction. Nbele wore a big smile, and shone it round to everyone in his office.

"Very good. *Very*, very good. Well, gentlemen, we are all finished here. Mr. Vambolent, I will send photocopies to your office in town. Now, let us go and have some supper. I know a place on the waterfront that serves an excellent *ugali*. Quite tasty! It will be *my* treat" He was clearly jubilant, and stood up in a surge of movement. "Now, Mr. Parks Vambolent, President of Vambolent Space Industries, your great project can begin!"

As the four men walked out of the building, two sets of eyes watched them from the dark-tinted windows of a non-descript panel truck. The laser interferometer trained on the window of Nbele & Associate's office shut down its invisible beam.

Mombasa

Winnimer Jenks shut off the recording of Vambolent's deal with the Kwapilis, and rubbed his chin. Moving over to the posterboards arrayed on his hotel bed, he made a new entry, then leaned back to scan them for connections. Scraps of magazine and newspaper pages, hand-written notes and pictures were glued down on paperboard. On each folio, related items were arrayed in an outward spiral from a central concept, written in his calligraphic hand.

The Wealthy Congregate, began the first, and oldest. *They will seek to preserve their wealth*, it continued. *Therefore, they will form coalitions to consolidate their power. Those with the greatest power and reach will be secretive. And they will be looking out for talented people to help them perpetuate their empires.*

While in High School, Jenks had first deduced the presence of secret societies of the extremely wealthy, and had dedicated his life to find one and be invited to join. His family was middle class; both parents professors at a small college in northern Ohio. They had taught him to "think hard, think through, and think beyond." It was a mantra he heard repeated many times until it became second nature. Any topic to which he turned his attention received the full treatment, often to the exclusion of all else. Long before he found the first evidence of plutocratic enclaves, he knew in his gut they must exist.

The research paper he prepared as a Junior on the topic took first prize in a state-wide competition. At the encouragement of his parents, he expanded his research and submitted it to scholarship competitions. While they awaited the results, he focused his powerful intellect on the test taking methods needed to ace the college entrance exams, practicing on every prior test and workbook he could find. Despite school grades which were either a C or an A+, his perfect 1600 score on the SAT together with his research paper, earned him a 4-year unlimited 'tuition, books, and fees' scholarship from an anyonymous philanthropic foundation. He earned admission to, and matriculated at Harvard University in Cambridge, Massachusetts. It was the perfect place, he thought, to begin his quest to find the right inroad to the plutocrats.

Campus secret societies were closed to him because of his amorphous pedigree. Instead, Jenks joined crew, took up fencing, and played on the lacrosse team in the spring. Every social contact he made, and every name he heard dropped, Jenks added to his relational database. He even volunteered as a tutor under the supposition that the less-than-brilliant offspring of the hyper-wealthy might have gained admission on the strength of family donations to the Alumni Fund. After a fruitless year, he gave it up, figuring that they probably employed private tutors, or hired the professors directly. *Or maybe*, he reasoned, *they just bought the grades they needed.*

By majoring in history, Jenks conducted considerable research into the origins and character of wealthy cabals, and the means of their rise and fall. In the Boston metropolitan area were certainly many rich people, and they certainly had their social circles. But these northeastern multi-millionaires were insular, and removed from the true power centers of the world, Jenks concluded.

He applied for, and received, a Rhodes scholarship to attend graduate school at Oxford College in England, studying the genesis and evolution of politico-economic systems of thought. It was during this three year period that Jenks formed his plans for global domination. Leery of becoming megalomaniacal, he strove to derive his strategy as an amalgamation of the most successful methods using throughout recorded human history. Thinking hard, thinking through, and thinking beyond brought him to a startling conclusion.

Throughout history, the conquest of lands and power has been motivated by at least one of: fear, greed, or need. In the case of need, Jenks broke this down further to be fueled by either natural resources or sheer manpower. In more recent times, as the importance of manpower declined, that need has been replaced by the need for energy. Then, with the advent of a global economy, the need to acquire land to obtain new raw materials was replaced by economic leverage, represented by money. Therefore, he reasoned, power and control, influence and wealth can all be boiled down to just two commodities: energy and money.

Of course, having energy and money would invite all the fearful and greedy people in the world to come relieve you of it. Therefore, a third factor becomes essential: protection. To protect your energy and money, you also needed weapons.

Following his graduate studies, Jenks took a job as a consultant in Los Angeles, working on contracts for the RAND corporation. In a series of increasingly complex projects, Jenks learned how the most powerful nation on earth managed money, energy and weaponry.

He found it woefully inadequate and appallingly wasteful.

Walking the Strand in the evenings, Jenks thought through the past, present, and future of world political and economic power. Since the turn of the twentieth century, the pace of change had been accelerating at a faster and faster rate. Extrapolating exponential growth could certainly lead to resource depletion. Without resources to survive, populations would be decimated back to sustainable levels. This had been postulated by Malthus in 1805. But Jenks saw another possible outcome. The science of physics is largely based on a law so fundamental, and so apparently obvious, that it had no name, no formalism. It is the principle that all systems eventually seek the lowest energy state, which is the most stable configuration. Many systems, far from this equilibrium, will undergo a phase change, or transformation, before finding a new, more stable arrangement. If he, Jenks, could deduce that future state, he might be able to place himself strategically at the tipping point.

Further, if he could find the plutocratic enclave he had been seeking for almost ten years, he could lead them to the same conclusion. They might then invite him into their ranks as a change agent who could help preserve their wealth and power in the new world order.

As a favor to his manager, Jenks had attended a business plan competition in his stead. He flew to San Jose, and took his place on the judges panel at Stanford University, the only judge there younger than 40. The event attracted a few fringe players, and a number of academics. But the real draw was Silicon Valley's beleaguered venture capitalist firms, as well as top CEOs in the electronics, biotech, nanotech, aerospace and defense industries. Jenks met as many of them as he could, dining with several of his manager's contacts. At night, he researched the key players, and worked them into his growing database. His judging duties he considered secondary.

Reading rapidly, he would quickly scan each submission assigned to him for key concepts and phrases. With a sharp eye for the essential, he could build a five-bullet summary of each business plan within as many minutes. That alone was enough to talk intelligently with the other judges the next day – most of whom had obviously not read past the Executive Summary section.

The submission by Parks Vambolent, Cloy Morgan, Franklin Juneau and Ned Austin gave him pause. His five-bullet summary of their work was still affixed to the center of his second poster board. He had developed his second tenet after thinking through and then beyond the inventors' business plan. They had already figured out most of it for him. But by the pleading, over-stressed verbiage, and the conference publications listed by the author, Jenks deduced that

Vambolent, et. al. had grown exasperated of not being heard, and were approaching desperation. The second tenet of his world domination plan had grown directly from Vambolent's business strategy.

In the New World Order, Jenks had written, *information will supplant goods, energy will be the basis of currency, and a single coordinating entity shall replace the nation-state system.*

The five Jenks-distilled bullets from Vambolent's plan were hand-written below it, although Vambolent would hardly have recognized them as his own:

1. Space is the High Ground, possessing height, speed, and visibility.
2. Resources are abundant - \$1T asteroids, 10^{14} watts of Sun on Earth.
3. Space Solar Power can be used for energy, or for weapons.
4. Leveraging space means launching factories, not finished goods.
5. US Patents for space factory architecture are held by one man.

Jenks had given Vambolent's proposal low marks, and pointed out enough of the flaws to make his recommendation triumph in the end. Certain judges liked the approach, and Jenks took careful note of their names and what arguments they forwarded. In their closed meeting, Jenks countered each key point: launch costs prevent space from ever being economically viable; in-space resources are far away and require great amounts of energy to make them useful; sunlight is diffuse and requires enormous installations far beyond current technology's ability to assemble and maneuver; and establishing a factory infrastructure in space was far enough in the future that Vambolent should be asked to re-submit in another 5 or 10 years. He did not mention his fifth point.

In the end, he had helped push the Vambolent plan far enough down into the ranks that it did not even receive an honorable mention.

On his flight back to LAX, Jenks formulated the plan that had carried him to the hotel room in Nairobi. Vambolent was clearly out of his league, did not have the skills or contacts to make good on his ideas, and would eventually abandon his patent. Even if Vambolent kept up the maintenance payments on his patents, they would eventually expire. And long before then, once Jenks was engaged with the plutocrats, together they would have started the research and development so that once the patents did expire, they could deploy the factories fast, free, and first. It would be an economic coup of nearly unparalleled precedent.

All it required was to give Vambolent enough stumbling blocks that he would give up. The appeal to Japan's MITI surprised him. The set-up in Kenya had forced Jenks to take more active steps, and faster than he had anticipated. A simple tip-off and a modest donation to the fringe raiders in Somalia had been eagerly accepted. To gain acceptance into the group of plutocrats, all he needed was control of the key patent.

But that had backfired when Vambolent came back from the dead.

Now Jenks was looking for his next move. He let his eyes drift across the spread of folios, letting his eye snag on cogent points, looking for opportunities and ideas. He remembered an article he had read in the local newspaper. The thrill of connection sparked between two places in his mind. He stilled himself with a breath, and shone the light of his intellect upon that bridge. A third piece appeared and gravitated in. Mentally grabbing the pieces, he pulled them into a whole, and began thinking through to projections of future consequences.

Yes.

He reviewed the concept from every angle. It validated as far forward as he could see. Now, he backed up to present actions that would lead to the desired outcome. It was so easy. He was so proud of himself. *Sometimes*, he thought, *the simplest solutions require the greatest genius*.

He would start with the Japanese, and they could be controlled in this through a simple tip to the right interest group.

Near Rei, Kenya

Two months after the deal closed, Afren Kwapili and Parks Vambolent overlooked the Kwapili lands. They stood on a hilltop above the broad plateau where the receiving antenna was now being constructed. Their view extended to the horizon where haze from the distant ocean melded with the scattered clouds high overhead. Beneath them, isolated crews of workers were pouring concrete footings in a regular grid laid down by the survey teams. Portland cement, made in the gigantic plant near Mombasa, arrived several times a day to prepare the foundations for the rectenna.

"In my grandfather's time this was valuable grazing land," Kwapili explained. "He was granted this land at Independence, and became one of the most prosperous farmers in the region. Sharecroppers and migrants tended the herds of cattle and goats, and spent their earnings in towns of Voo and Rei. In contrast to the nomads who first inhabited this region, the farmers became settlers. As more farmers moved into the region, they began to over-graze the land. My grandfather, blinded by the income, and perhaps ignorant of the consequences, allowed it. He never gave the land time to recover, and at the first serious drought, the entire economy here collapsed. Without enough grasses to hold the soil, the loess blew away, and the humus washed down the river beds until all that was left was this wasteland."

Vambolent looked over the site, trying to imagine the dun desert as green fields, dotted with grazing ruminants. "In my own country," he said, "a similar story occurred in the 1930s. They called it the Dust Bowl, and it was a major contributor to the Great Depression."

"Yes, I have read of it. It is also said that the Libyan Desert and the great Sahara were once lush and fertile lands. Through the advent of ranching, instead of nomadic herding, those vast regions were stripped of their fertility as well. They now form a desert region larger than the entire continental United States."

"When this project is complete," Parks said, " we can begin to restore this land to its original purpose. Perhaps one day, with Kenya as prototype, we can even begin to shrink the size of the Sahara."

"It is a beautiful dream, Parks. Yet, will we not then be dependent upon space solar power? The more complex the web of relationships, the more fragile it becomes. A simpler approach is to let the biosphere perform what it does naturally. The Earth created those grazing lands in the first place with no help from any man. She can do so again, if we will just cooperate, and not dominate. Information and communication are the answer. When we can teach Africans to farm sustainably, the yearly advance of desertification can be stalled. In time, it may even reverse."

"I'm not so sure of that Afren. Average temperatures are increasing fast, and the swings in bad weather seem to grow wilder every year. Tornados, typhoons, blizzards, and floods have killed more people in the last ten years than in the prior thirty. I'm afraid mankind's impact on the Earth has taken us past the point where it can heal itself."

The two men stood silent in their own thoughts, watching the work. Their vantage point on the hilltop approximated the view of a scale model which filled one half of Vambolent's dusty office in the Mombasa headquarters of Vambolent Space Industries. When completed, the antenna would be a flexible, woven wire mesh suspended two meters off the ground by steel I-beams embedded in concrete. The finished dimension would encompass eight square kilometers, making it large enough to be easily seen from orbit.

Beneath the grid, copper cables would connect the inverters on each grid sections to the massive central power converter, a large dull-grey box sheltered in a concrete bunker off the southern edge of the array. Tall shiny-brown insulators poked through the top of the bunker, forming the connection to the 100 kilometer transmission line that would carry AC power at750 kilovolts to Kilindin harbor in the southeast.

Along the northern edge of the antenna, where the land began to rise and could not be easily leveled, would be a large number of distillation vats. Each vat would be a shallow can, with ceramic-insulated steel pressure walls, and capped with a convex quartz plate. The deal Vambolent had arranged with Mtongwe and his cabinet was that the Japanese could use the array 100% during the daytime, to generate electricity for their hydrogen plant near Mombasa. Then, during the night, the satellite would make a tiny shift in alignment, and direct a portion of its power to the distillation vats to produce fresh water for eastern Kenya. Using the heat energy of the 2450 megahertz microwave radiation, they would boil seawater to produce steam, which would then be condensed in air-cooled copper tubing, and collected in a large cistern. The lands below the Kwapilis would then be converted from wasted grazing land to cultivated fields, bringing much-needed food to the nation. As an additional byproduct, the Japanese would cart off the remaining sea salt, where it would be mined for uranium and platinum-group metals.

"What's the news on the pipeline?" Vambolent asked.

"They are making good progress. But there is some of what you Yanks call 'good news and bad news'."

"What is that?"

"They have volunteers offering to help, which is good. This project is becoming very popular. However, the regular workers are grumbling. They are concerned that there will not be as much work for them."

"Can't we just re-assign the newcomers up here?"

"The array is coming along as fast as we can get cement." Kwapili thought for a moment. "We could start work on the distillation plants, but we would need the Japanese engineers to come early."

"I can talk with Yoshi."

"That would be very good for morale. Can I tell them of this possibility?"

"It may be premature, Afren."

Kwapili looked kindly at Vambolent. Vambolent felt like there was a slight mocking laughter in his eyes, as if he were missing something obvious.

"What?" Parks asked.

"Even if we do not resolve the situation at once, we can give them hope, which will be a great balm to them now."

"But if later, it does not work out?"

"We will find another way."

"Okay, then." Vambolent sighed. "I'm starting to feel like this project is moving out of my grasp, Afren."

"You are only feeling what is natural. All of us fear loss of control. Soon, your responsibilities will advance along your master plan, is this not so?"

"Yes, I suppose it is. But still, I feel an obligation to make this come out right."

"It will come out right, Parks. Each facet of the project is on or ahead of schedule. The people believe in you."

Vambolent harrumphed, and looked over at Kwapili.

"I don't know how I would have done all this without you, Afren. Your help has been invaluable. Even the workers, they group around you whenever you visit. What is your secret?"

"Oh, there is no secret, my friend. I just listen to them. The men working on the project know the problems, and often they know the solution. I simply relay their ideas to the crew bosses and the Japanese managers in terms that they can understand. That is all."

"I doubt it's really that simple," retorted Vambolent, smiling a crooked smile. "The supervisors tell me you drum with the workers in the barracks all night long."

"There are many good drummers. I am not such a good drummer. But they let me sing and chant, anyway. Swahili is a beautiful language for chanting."

Vambolent chuckled, and moved toward their Toyota Range Rover, with the new VSI logo painted next to the MITI insignia on the doors. "Come on, I need a ride to the train station. We can run down the line to see how the towers are doing."

Kwapili climbed up in the driver's seat, and keyed the ignition. The powerful V8 motor roared into life, and they drove down the hillside and turned onto the access road that led towards Mombasa. At each footing site they passed, work stopped, and the crews waved to the passing vehicle.

Along the right-of-way cleared for the high-voltage towers, pre-packaged kits of angle stock lay every 225 meters. The tower kits, cable and pulling trucks had arrived, along with pipes, pumps and antenna materials in Kilindin harbor on Independence Day, December 7th. The Japanese freighter *Sokoru* delivered the entire first wave of the project on-time and as-promised by Mifune. From the decks and hold of the *Sokoru*, tall cranes unloaded a heavy weight of crates, tube-stock, and vehicles, dropping her rusty waterline several meters in the process. A convoy of dissimilar trucks queued up at the wharf to carry the freight inland to the various worksites.

About half of the trucks, marked by magnetic VSI logos, chuffed over the New Nyali Bridge to take pumps and pipes to Kilifi. Seawater would be drawn from the bay there, and pumped uphill in stages to a tower on the Kwapili lands. The pumps used solar-powered electric motors, so they could fill the water tower above the distillation vats during the day. The motors were designed to be augmented with hydrogen fuel cells at a future date, so they could also pump during the night.

Reaching northward from Mombasa, another group of trucks carried the tower kits, giant spools of wire, and pallets of copper alloy lattice for the antenna. Now that the materials had all been delivered, the *Sokoru* loaded up with salt, soda ash, iron ore and limestone to sell at various ports on the voyage home. The next ship from Japan was due in the summer with the launch vehicle, which would be assembled for liftoff from the San Marco Equatorial Range. The rocket would deliver a solar array with a large-scale deployable antenna to geosynchronous orbit. From there, the solar power satellite would beam microwave power back to Kenya. The third and final ship was due to arrive as the rocket was being erected. A retrofit supertanker packed with MITI's porous silicon storage material would arrive, towing behind it the floating electrolysis plant which would fill the tanker with hydrogen for delivery back to Japan.

Driving back towards Mombasa, the land gave way from steppe to scrub land and then forest. The colors were bright green from the recent rains, and Kwapili negotiated the Range Rover through a few washouts in the rutted path. As they approached one of the jobsites, later in the afternoon, they found a mess. A pre-assembled tower now lay twisted and sprung on its side. Some workers were dismantling the angle stock and others were hammering the pieces straight. The crew boss trotted over to the VSI truck as they drove up.

"Hey, boss!" he called to Kwapili through his open window. He noticed Vambolent in the passenger seat. "Hey, big boss!"

"What happened here?" asked Vambolent.

"Oh, it was terrible, the water. The rain last night, it was west mostly, na much here. We build one tower, then wham! Out of nowhere come *big* water. We hear the sound, but too late. Two men, whoosh, and they are gone. We search, but couldna find a body!"

Vambolent and Kwapili got out to see the damage. The twisted tower was ruined, and despite the efforts of the crew, it would never hold its design weight after the structural members had buckled. Kwapili suggested they let the disassembly work proceed, but tell them the parts could be used only as spares. The hammering would give them something to vent their feelings. He would then arrange for a replacement tower to be brought up the line from the warehouse they rented in Mombasa.

The arroyo walls had been weakened by the excavation of the original footings, allowing them to be washed away in the flash flood. Two of the fat, concrete cylinders could now be seen jutting from the still-dark mud at the defile downhill of the worksite. Vambolent walked with the crew boss to pick a better site, while Kwapili squatted down with the rest of the workers to share their grief over the lost men.

"The climate here is more severe than it once was," Afren told Parks when they were back inside the vehicle. "We may need to factor that into the installation design." For the rest of the trip down, they talked about the continuing trend of global warming, climate shifts, and weather fluctuations, and how it might affect their project.

Night had fallen by the time the two reached Mombasa. Instead of dropping off Vambolent, as had become their habit after the Friday surveys, Kwapili parked the Rover and they both rushed into the train station. Kwapili had announced he too was going to Nairobi for a weekend visit with friends – a much-needed break after eight weeks of steady work on the project.

*   *   *

When they settled into their cabin in the sleeper car, Vambolent laid down gratefully on the soft bunk. His back was tender and sore from the bumpy ride. He flipped open a copy of the *Daily Nation* as they awaited the conductor's whistle to begin the journey. Inside the paper, an article caught his eye. Someone was accusing the Japanese of developing power beam weapons using discontinued technology from the USA. He wondered if this connected in any way with what the two of them were doing.

Parks had been spending long weekends with Elise and their kids in Nairobi, using the extra Monday there to help with her vain attempts to improve the education system of Kenya. At first a holiday, these weekly visits were becoming more of a burden for him. Elise more and more wanted to return to the States. Just as his project was beginning to take shape, she grew increasingly exasperated, and made no secret of her feelings. He felt himself starting to tense up.

"Afren, why don't you have dinner with Elise and me Sunday night?" Parks asked suddenly, raising up on one elbow. He was already thinking that he and Elise might get along better with company. Plus, she needed friends, and he had found Kwapili to be a fine gentleman.

"Are you sure I would not be imposing?"

"Of course not! Please, I insist!"

"Well, then, certainly. I have not met your wife, but I feel I already know her from all that you have told me. I would be honored."

"It's a date then!" said Vambolent, and he began to breathe easier at once.

The whistle blew, and moments later the train shuddered as the links engaged, and they were on their way.

Nairobi

Elise Vambolent stood tight-lipped and arms akimbo at the elevator door. The call button was dark, despite her having pressed the dirty thing several times. A uniformed security guard was explaining to her husband that the elevator had stopped working earlier in the week, and that the repair company was no longer in business, but an engineer was coming on Monday to fix it. He apologized profusely, but they would need to walk up the stairs to the restaurant. Parks had chosen the rotating restaurant with the panoramic view of Nairobi at night, high atop one of the newer office buildings in the city. But an ascent of thirty flights in dress shoes was not what Elise had planned for.

In her mind, this was typical Africa. There had been a few sweet weeks when she used the inevitable inconveniences and obstacles she encountered to boost her appreciation for her home in the USA. But that had passed. With each new impediment, she stacked another brick of frustration on her growing edifice of disgust.

Parks looked from the guard to his wife, and could read her emotions like a billboard. She was a beautiful woman, tall and graceful, with a face that would remain young-looking her whole life. Her hair was swept back, leaving curled wisps that framed her face, but there was no makeup in the world that could mask the fury in her eyes. The line of her eyebrows was straight and as rigid as an iron bar. Beneath them, her eyes were sharp-edged rectangles of flinty fire. Her sensuous lips had been pulled in, leaving a harsh line above her jutted chin. Parks shuddered involuntarily.

At that moment, Afren Kwapili entered the lobby, accompanied by a swirling wind of cool, fresh air. The three of them turned to see him silhouetted against the headlights and sign-glare of the street.

"Afren, how nice to see you," said Parks, gratefully reaching out his hand. "Please, I would like to introduce you to my wife, Elise." He stood next to Afren, still shaking his hand. "Elise, this is Afren Kwapili, our Site Director for the project."

Elise released the breath she was holding, in case she should need it to bark at someone, and effected a meager smile. The rest of her face started to relax, but slowly. "Hello, it's nice to meet you." She reached out her hand.

Kwapili released Parks' hand with a quick look, and took Elise's fingers. He kissed over the back of her hand, in the old-fashioned European way. Her smile started to become genuine. Parks breathed a sigh of relief.

"The pleasure is all mine. Parks has told me a great deal about you. You must be a remarkable woman."

Elise blushed.

"He has told me a lot about you, too. You must be a remarkable man," she said. Then, they both turned towards Parks. "I hope he has not over-rated us."

"Hmm," agreed Kwapili. He surveyed the situation, and turned to the guard, speaking rapidly in Swahili. The two men spoke frenetically, gesticulating quickly. It seemed they communicated more with their hands and facial expressions than their verbal exchange. Finally, they finished, nodding their heads vigorously, and shaking hands.

Kwapili turned back to the Vambolents.

"Well, this good man tells me the food upstairs is not so fresh today, because the bakery and the vegetable truck drivers would not walk up the stairs. I can recommend a very nice restaurant just two blocks away that serves excellent Indian cuisine. Would that be acceptable?"

Parks nodded at once, but looked to Elise for the verdict. She nodded also, and the trio walked out the front door.

Out on the street, a cool night wind was blowing the warmth of the day out of the city, bringing a clean scent from the hills. The breeze pushed them along, with Elise between the two men, walking with linked elbows. The neon sign for The Palanquin drew them halfway down the next block, and then up a flight of concrete stairs to a sumptuous waiting room with brocade wall hangings and bronze Hindu statuary. A polite Indian man seated them in one of the small rooms, designed like the litter mounted on the backs of elephants a hundred years ago. On the table were ornate silver plates and cutlery, with cut crystal goblets already filled with ice water. A subtle aroma of patchouli and the sound of gentle sitar music set a comfortable mood.

After ordering their drinks, and remarking on the decorations, Kwapili turned to Elise.

"Well, Elise, Parks tells me you are a vegetarian."

"Yes. Well, I was back in the States. Here, they seem to put meat into everything. I must admit I've had chicken a few times recently. So, it's nice you chose this restaurant."

"Are you a vegetarian for health reasons, or for ethical reasons?"

"I started out doing it for health, but the more I learned about where meat comes from, the less I missed it. It makes me feel good to know there are cows and pigs and chickens out there – at least some chickens – who are still alive because one person became a vegetarian."

"Where I attended college in England, there are many vegetarians," Afren said. "One friend in particular explained to me that each pound of beef consumed in the US uses the same resources as 40 pounds of soybeans."

"Wow," said Parks, who was feeling left out of the conversation, but didn't really mind.

"I've heard something similar," mused Elise. "Through the use of feedlots in America, they say 90% of our grain harvest is fed to livestock. If everyone on the planet stopped eating meat right now, by next year those same crops would produce enough to feed three times the world population!"

Kwapili smiled. "Of course, we could then grow less food, and ease the strain on the Earth's ecology."

"Right!" said Elise. "It seems incredibly selfish to me that people in this day and age should eat meat!"

Parks, who had been scanning the menu under the Lamb and Chicken sections, discretely turned the menu over, and began reading the entries under 'Vegetarian'. Elise did not usually mind him eating meat, but on this particular day, he did not want to press his luck.

"I think it is not so much selfishness as it is ignorance," said Kwapili. "If people were informed of the consequences of their choices, I believe they would change their habits. In fact, I think that is the only way for people to change. Like you, when you learned about meat production. It makes it easy to change if you know the whys and wherefores."

"That is so true! You can't change *anybody*." Elise stopped, and her eyes scanned back and forth across the wall hangings, but was not focused on them. "It's the same problem I'm having with this stupid education reform. Nobody wants to do anything they aren't already doing. It is so frustrating!"

"Yes, Parks mentioned you were having some trouble," Kwapili said, leading her.

Elise drew in a deep breath. Her brows knitted, and she began casting about for how to begin. Both men could tell there was a lot on her mind, and were willing to let her vent. Just then, the waiter entered their palanquin for their dinner orders. Elise ordered Malai Kofta for herself and garlic naan bread for their table. Kwapili chose the Chana Masala, and Parks ordered something called Alu Dumm, which seemed to be mostly potatoes and was probably pretty safe.

"You were saying?" Kwapili said gently, leaning towards Elise.

"Oh, I don't want to burden you with my troubles. We're here for a relaxing evening. And I know Parks doesn't want to hear this all over again."

"Perhaps I can help in some small way?"

Elise looked at him closely for the first time. He was a handsome man, and portrayed an image of sophistication with his erect bearing and neat clothing. His face was warm and open, and she sensed great vulnerability there, somewhere behind his self-confident exterior. His hair was short and curled, pulling back slightly from his forehead. His eyes were bright and alert, surrounded by fine wrinkles. His cheekbones were strong and high, divided by a broad nose with generous nostrils. A trim moustache with just a hint of gray arched above his ample and relaxed lips. His chin was split by a cleft hiding a short stubble that was probably hard for him to shave. She decided she liked him.

"Well,..." she said, working up trust in her new friend, "I don't know how much Parks has told you..." She looked at him again, but his face was completely attentive. She smiled despite herself, and wondered if he was popular with the ladies in Kenya. She decided he probably was, and started to let loose the million things bouncing around inside her head.

She told him about the Ministry of Education, and the incompetent bureaucrats who smiled at first but then did nothing. Now, they had stopped smiling, tiring of her constant badgering, and were still doing nothing. The Minister himself was a political appointee. His only qualifications seemed to be that he was an old friend of the late President Jemmat Mtongwe, and that he cleverly guarded his turf. She had made several friends amongst Westerners doing volunteer work, mostly missionaries, and they had warned her off from using her influence with President Jomo Mtongwe. It was an ace she could play only once, they advised her. Because after that, tolerance of her meddling would vanish.

As she had learned early on, no self-respecting man in Kenya took orders from a woman, and that was probably the worst obstacle she faced. She had resorted to every trick she could think of. Make them think it's their idea – that worked wonders in the States, but was almost useless for her here. She would lay journal articles on the desks of ministry officials detailing the latest in education theory, only to notice them in the dust bins the very next day. She even tried a pilot school, but the Minister had insisted on hand-picking the children. He chose the worst trouble-makers the Nairobi school system had to offer, and challenged her that if she could make them successful, he would let her expand her program. But Minister Buisi had picked well. These incorrigible street rats had absolutely no interest in education, and she spent her entire class time in a vain attempt to enforce discipline.

Twice she had gone out with the missionaries to the bush schools, to see what might be done from that angle. While supportive in the capital, her supposed friends from the Christian churches became politely uncooperative in the field. It seemed they did not want her help, and they were doing a fine job converting the locals to the true gospel, thank you very much. Good luck back in Nairobi.

Afren Kwapili listened attentively, nodding easily, and following her every word. Elise sighed, spent from unloading her burden of grief.

Parks cleared his throat.

"Afren, you've done so well on the Energy Project, we can probably manage for a week or so. Is there anything you might be able to do here for Elise?"

"Oh, no, Parks!" Elise interjected. "You're needed out there, Afren. I can manage here just fine," she finished weakly. But she convinced nobody – not even herself.

"I don't know exactly how I *could* help," started Kwapili. "But maybe if you tell me more about what it is you would like to do, I might be able to offer some suggestions?"

"Okay! Whew, it's been so long since anyone has asked, I hope I can still remember it all." She looked at Parks and smiled.

The bread came, and she took a piece and bit into it, talking around her mouthful. "It all started when Parks called me last autumn when he first came over here. The basic idea was mine, but Parks added most of the details."

Parks smiled back, and some of the warmth that had slipped out of their relationship in the past several weeks rekindled in that moment. He sat back, munching on the garlic flatbread, and listened to her story.

"Well, okay, um, the basis for the whole program is called Thematic Learning. For each grade level, you have some kind of basic topic, or theme, that connects together lessons in all the basics. For example, your theme might be food production. A second grader might learn the names of foods and how to spell those words for language arts; then learn about soil types and the plant growth cycle for natural sciences; bring in coffee beans to study math skills like addition and subtraction; and finally explore how crops like cotton and sisal connect Kenya to the outside world for social studies. Higher grades might focus on research reports, biology experiments, economic analyses, and so forth."

"Excellent!" exclaimed Kwapili. "I like it already!"

Encouraged, Elise went on, speaking faster and becoming more animated as she went.

"Now, as Parks explained, the basic problem in boosting Kenya's literacy rate is the shortage of trained teachers and the lack of teaching materials. A traditional approach would be too costly, so the idea I gave to him was to use computers. Computers are so cheap now that we can get large companies to donate their older models. I've called a few places, and have a couple which are interested, but so far, I don't have the infrastructure in place to *do* anything with them. There is a warehouse near the airport with a few hundred units donated already, but I've been reluctant to press this until we have the lessons ready.

"The lessons have been the only part that is going well. I've got a couple of university professors developing an on-line, interactive Montessori-like approach. It's fascinating, and until they showed me some early versions, I had no idea what they were capable of. This part was initially Parks' idea, but they've taken it even further. You see, if we are to implement this program across the whole country, without qualified teachers and in some cases without school buildings, we need the kids to guide their own learning. These professors, Gerold Land and Euren Belew of Georgetown University, have created a high-level programming language they call METAXA. With their operating system and foundation classes – these are their words I'm using, but don't totally understand – a handful of teachers located anywhere in the world can develop modular lesson plans for any number of students, all based on themes, with multiple levels of entry.

"This way, students can start right away with graphical lessons, and gradually move into more language-based and mathematics-based modules as they progress through the grade levels. Each student would have a computer, ideally, so they could learn at their own pace. But the really exciting thing is that METAXA even allows the students to change their own stuff. It is really easy for them to change graphics, interactive personas, and the look and feel of their lessons. And for those who want to go further, the METAXA code is open source, so they can even modify existing lessons, or create new ones!"

Their food arrived, but Elise was on a roll, and did not stop talking. Her audience of two sat politely, not starting their meal. The steamy fragrance of Parks' Alu Dumm was making his mouth water.

"And that brings us to the neatest part of this whole idea," she said. "You see, younger kids need manipulatives – blocks and paper, toys and fabrics – to develop fine motor skills and spatial reasoning. Well, the older kids will have modules that are geared towards product design and manufacturing, sort of like vocational education. What we want to set up is where the older children actually start their own businesses, and begin making things that can be sold and transported to schools that need their products."

Elise paused, and spooned her entrée over a heap of rice. She poked around it with her fork, and lifted a hefty bite to her mouth. Parks and Afren followed her lead, and they all chewed thoughtfully for a moment.

"Mmm," Parks mumbled. His food was delicious. He raised his eyebrows and nodded his head to express his appreciation.

Kwapili swallowed, and directed a question to Elise.

"I think I am following the basic idea. I like the concept of teaching the older children about how to start a business. With the economic growth that your husband is helping to generate, we will need energetic young entrepreneurs to carry on. But I am curious.

How will the older children 'sell' their products to the younger children?"

"Yes, that's something we haven't quite worked out. It could be an exchange, but then, I don't think the younger kids will have anything the older ones will want. We could encourage the communities to pay the bills, but I despair of *that* ever happening. Do you have any ideas, Afren?"

"Actually, I was thinking that the older children, who are learning manufacturing, could actually *export* their products. I would think there is a market around the world for such items, and I am sure your students could produce them at competitive prices. Any profits they generate would go into the school treasury. Then the teacher, or whomever you have administering this on a local level, would use these proceeds to purchase the toys and things for their young ones."

"That's a *great* idea!" Elise exclaimed. "And you're right, we can't do away with teachers entirely. There would definitely need to be an adult in charge, to enforce discipline, and make sure everything ran smoothly."

"Would there be any issues with exporting products like that?" asked Parks.

"Perhaps," said Afren. "But such a market has no competition in Kenya at this time, which is good. As for export licenses, we could approach President Mtongwe. I'm sure he could help make this a workable idea."

They ate for a while musing over the possibilities. Each commented favorably on the quality of the food. Parks' phone rang. He opened the flap and answered it quietly. The call was from Japan. Sugeto Mifune said he had some important items to discuss which could not wait. Parks explained this to Elise and Afren, and excused himself, taking one last bite of his Alu Dumm. He walked out to the stairwell where he would not disturb anyone.

In the palanquin, Elise and Afren talked in more detail about her education reformation plan. Kwapili made suggestions and demonstrated an astute knowledge of the power layout of the Kenyan government. He quoted names and showed a familiarity of numerous intricacies of which she had only glimmerings of awareness. As he talked more and more, she started thinking of a plan. He noticed her watching him, and stopped abruptly.

"Yes?"

"I was thinking. You seem to know many of the people I deal with. Maybe you could meet them with me, and help me out?"

"Oh, I know of them, but I am not exactly friends with any of these people."

"But do they know you?"

"Well, actually, I was something of a celebrity a few years back."

"Oh, really, what did you do?"

"I ran marathons."

"Did you win any?"

"Yes, in fact I won quite a few. I ran for my country in the Summer Olympics eight years ago and took a gold medal."

"Oh, my goodness! That's so amazing! I'm sorry, I didn't know. I should have!"

"It was a long time ago, and I was famous for just 15 minutes, as Andy Warhol once said. It is only in Kenya that people still remember my name."

"Well, that is quite an accomplishment. Do you still run?"

"Oh, I run to keep fit, but I do not race any longer."

"But you are still well-known…"

"People know my name, Elise. That does not mean I have any influence."

"But you are a man. You could do so much more than I have done."

Kwapili kept his mouth shut. The look on his face was pained humor, as if he already knew where she wanted to take him, and had grave misgivings.

But Elise was clever. She ran back through her mind the things he had said, looking for a way to enlist his help.

"You obviously love your country," she said.

He nodded slowly, non-committal still.

"And you have been very supportive of my husband. In fact, he tells me that your help has been instrumental in the successes so far."

Kwapili tucked in the corner of his lips, wary of flattery.

"I think you have a way with people," she said. "I think you want very much to help Kenya. That is what I'm trying to do also. But I cannot change people. They have to change themselves. But the only way they can do that is with information. If they can see the reasons and the benefits, the 'whys and wherefores', then they just might change. But me, I'm a woman. A white woman from America. They don't listen to me. What we need is a messenger. Someone they will listen to. Afren, we need you to help show them the vision. I need you to put the ideas in their heads, so I can make it happen. If they think you are behind the whole plan, then they will do it for you. I'll play the role of your functionary. But you can be the visionary, the messenger, the fleet-footed warrior out to raise the awareness of all Kenyans, Afren. You could change the whole country!"

Kwapili breathed in heavily through his nose, and then back out again. Still, he did not say anything.

"You'll be like Mercury. You can run across Kenya, to each of the school districts, wearing your gold medal. You can talk to the parents and the children, bringing the message of this new way of learning. Describe for them their future, and they will all want to be a part of it. C'mon Afren. This is the opportunity of a lifetime!"

He stared stoically at her for a moment, but from deep inside him, a happiness swelled that he could not hold back. Like a tide it rose, gently but inexorably, until it crested over his neck and emerged as a smile that formed on his lips. He began to chuckle silently at Elise, who looked at him with eyebrows so high on her head, the very image of expectancy. But before he could answer, Parks returned.

"I have some… news," he announced, sitting back at his place at the table. "Director-General Mifune is now former-DG Mifune. He was fired."

"Oh my God, Parks," Elise said. "What does that mean for your project?"

"Probably, it's over," Parks said. His voice cracked.

"Parks, that's awful. After all we've done here. And it was just starting to take off." She looked over at Kwapili, aware that he had not yet consented to her proposal.

"It will all work out, my friend," Afren said to Parks. "You have a grand idea, and it will come to pass."

Parks grunted and ground his knuckles into his eye sockets.

They heard a loud commotion in the restaurant. Suddenly, there was yelling and strident voices. A woman's wailing cry could be heard. All three of them rose quickly from their seats and rushed into the restaurant's main room.

"What's happening?" demanded Vambolent, catching the arm of a waiter hurrying past.

"It is war. A nuclear war. Pakistan has bombed India."

The waiter pulled his arm from Vambolent's grip, and rushed towards the kitchen where the noise was loudest. Parks stood there in shock. Elise muttered "Oh, my God." Kwapili bowed his head. The three of them stood there, unsure what to do next. The wailing from the kitchen was grievous.

Kwapili slowly unwrapped his money clip, and laid a stack of bills on the maître d''s table, in excess of what their check would amount to. The three of them walked quietly out the door and down the stairs, into the night of a world that seemed to have no end of bad surprises.

Tokyo

There were public calls for him to take his own life. Being fired as MITI Director-General was considered too light a punishment for Sugeto Mifune. His actions had shamed Japan in the eyes of the world.

Strange lights reported in the skies over the Russian island of Sakhalin caught the attention of the growing international network of environmentalist organizations. These auroras lacked the sinuous, curtain-like quality of a natural formation. Instead, they appeared as bands of straight lines, sometimes with a bend in the middle, like two legs of a very obtuse triangle.

Similar phenomenon had been reported in Alaska a decade earlier, as the USA experimented with its HAARP device. Once atmospheric scientists had proven that beaming enormous amounts of energy into the upper atmosphere accelerated depletion of the ozone-layer, the US had halted its experiments.

Using the GreenWeb, developed by Willow Masters, environmentalists from around the world collaborated to identify and expose crimes against the planet. Careful sketches, taken from observers on Sakhalin, were uploaded and shared with green-friendly physicists who then deduced the source of the electromagnetic beams. Three intrepid individuals traveled to Japan's northernmost island, Hokkaido, and hid in the scrub until the next appearance of the anomalous auroras.

On receiving a signal from a Greenpeace ship in the Sea of Okhotsk, they launched a flight of aluminized balloons filled with hydrogen gas. As the balloons rose into the power beam, emitting from a heavily-guarded facility thought to be the origin, the three captured telephoto video evidence of a Tesla power beam. Their balloons sparked and sizzled, then burst into flames. One by one, the balloons exploded as they hit the invisible wall of beamed electricity. Within minutes, their digital images were uploaded to GreenWeb so that environmentalists around the globe could take action.

Outrage avalanched around the globe overnight and slammed into Japan with the rising of the sun. Shamed by this news, the newly-elected Japanese government ordered the experiments shut down at once, and the responsible parties brought to justice.

Mifune was escorted from his office before ten o'clock in the morning.

By afternoon, MITI had a new director, Dr. Makoto Sawahata. As his first act in office, he ordered the Tesla facility dismantled, and the scrap metal used to build underwater power cables instead. His next act was to denounce Mifune's plan to beam microwave power from space. Japan would rely instead on their established nuclear power infrastructure to supply the electricity for hydrogen electrolysis.

Funding for Mifune's Energy Project with Vambolent vanished from the accounts, and work was stopped at the aerospace contractor building the satellite assemblies. Word was leaked to the press that the patents on that project were held by an American. This man had even patented the technology in Japan, right under Mifune's nose. There were even rumors that the awful American brought the Tesla technology to Mifune and charged him an appalling sum of money.

Under pressure from her family, Mifune's wife left him. He drove aimlessly through the city that day, listening to the radio broadcasts and op-ed pieces. He was shunned by friends, denied access to his golf club, and made unwelcome at his favorite restaurant. Everyone expected him to do the right thing, and remove himself from this world.

It was late when Mifune drove back home and began to settle his affairs. He called Vambolent, trying not to sound too final. Then, found he had nothing left to do.

His long-term dream for energy independence in Japan no longer belonged to him. The term Hydrogen Economy had been coined by a General Motors engineer in the 1970's. In the early 2000's, then-President George W. Bush initiated the Freedom Car program in the US to reduce dependence on foreign oil. The US Department of Energy issued a Grand Challenge, setting ambitious and unrealistic goals for the cost and efficiency of hydrogen storage methods. After the public hype brought greater awareness of the entire energy cycle for the Freedom Car, two significant drawbacks became clear. Fuel cell cars still relied on methane gas reformation for their hydrogen, and while methane is more abundant than crude oil, much of it is still imported. Secondly, methane reformation produces carbon dioxide as a byproduct. Because methane and carbon dioxide are both greenhouse gasses, leakage in the supply chain, and effluents from the vehicles both contribute to global warming. With global climate changes finally becoming noticeable to the average citizen through increasingly violent storms and weather fluctuations, the Freedom Car initiative faltered.

Watching this dynamic from overseas, Sugeto Mifune had been well-prepared to receive Parks Vambolent's proposal. By converting sunlight to microwaves, and rectifying microwaves to a DC current, they could generate hydrogen gas directly from water in a process called electrolysis. This methane-free and carbon dioxide-free approach found enormous appeal with many technologists and environmentalists. When Mifune combined the SPS-based hydrogen generation with his own silicon-based hydrogen storage research, Japan found itself holding the keys to a true Hydrogen Economy.

Now, on the outside, he was powerless. Nuclear-sourced electrolysis, the option sure to be pursued by the MITI, was not a long term solution. But that was no longer his concern.

Mifune took the elevator to the parking garage under his apartment building. In his hand, he carried a hose and a roll of duct tape. Backing his octane burner into a far corner of the underground parking lot, he then fed the car's exhaust into a window gap. He mused with wry humor that this method of release would be denied an owner of the Freedom Car.

Inside, waiting for sleep, he turned on the radio.

Excited voices told of the start of a nuclear war.

He sat up straight, anxious, heart beating rapidly. He was only one generation away from a similar tragedy in his own country. He knew people who knew people who had been killed by the American atomic bombs. But then he sat back, tired and sleepy, remembering that these events no longer concerned him.

As Mifune waited to die, a small part of his brain grew restless. True, Japan was far from Pakistan and India – they might not get any of the fallout. Yet his fading mind began to see the possibility for a new dynamic, a new twist from which he might benefit. A plan formed, slowly, in his partially poisoned brain. He blinked his eyes hard, then slapped his face – twice – to bring alertness. He put the car in gear, opened all the windows, and sped out onto the streets of the city.

New York City

In the United Nations building, under seige from maruading insects like the rest of the City, the General Assembly prepared to vote on a hastily-drafted resolution to impose tighter limits on the use of nuclear power. Representative L. Marshall Washington, from the United States of America, was narrating a video summarizing the India-Pakistan limited nuclear exchange. The presentation had started with charts showing the build-up of troops on the border between the two countries, indicating over one million armed soldiers total. He showed media footage of the periodic skirmishes in the Kashmir region, as well as the results of terrorist attacks on Hindus or Muslims in the cities. Next, he showed graphically the build-up of their arms race; how technology acquired by one country would be rapidly matched by the other. Whether it was gyro-stabilized automatic rifles, laser-guided shoulder rockets, Gen IV night vision goggles, airborne radar, uninterruptible battlefield communications, or even mercenary generals hired to bring them the latest tactics, each side had spent progressively more and more on their military. Each country had their own nuclear power plants, as well as secret processing facilities to manufacture weapons-grade plutonium from the byproducts, and had been stockpiling fissile material for decades. Both had ballistic missiles, with overlapping circles of coverage that included the entire country of their enemy, as well as their neighbors for thousands of miles in every direction. Half of the world's population was within reach of the nuclear bombs of India and Pakistan.

Ambassador Washington explained to them how the open tests of nuclear weapons from both sides before the turn of the millennium had failed to evoke sufficient response from the world community.

Now, their nuclear arsenals were approaching the limit of MAD, or mutually-assured destruction. Yet, from the political rhetoric of the Prime Ministers, and the hawks of each military, they did not appear to have the same restraint as the US and USSR had exhibited during most of the Cold War. Washington played sound bites of the invective hurled by both Pakistan and India, and explained that the world had sat idly by as they escalated to an actual launch of nuclear missiles.

Now, the images he showed switched to satellite views of the exchange itself. From the northern mountains of Pakistan, a grainy orbital image showed the bright yellow-orange lines of boost-phase rocket exhaust. In just seconds, the condensation trails of the accelerating missiles began tracing paths that fanned out southeast. The arcs of three missile trails reached skyward, and then began curving back towards Earth. They did not have far to travel, and in less than a minute were directly above the greatest concentrations of Indian troops. When the first nuke detonated, the satellite image saturated into bright white and tortured gray chaos as the contrast exceeded the capabilities of the camera. Horizontal slashes of static ripped across the picture, and the visual went dead.

He then recounted the events after the first launch. India had responded within twenty minutes, sending a salvo of nukes to incinerate Pakistani troops in retaliation. World leaders had called at once to demand a cease-fire, but the damage had already been done. Clouds of radioactive dust and debris had been lifted high into the skies over Kashmir, and were drifting eastward, raining fallout over the beleaguered region. He then cut to a static view of a US aircraft with a conspicuous black circle in the side of its white hull.

"This is one of the anti-missile planes we have been flying continuously over the region. With this airborne laser, we can destroy most missiles during their boost phase. Yet, with the airborne radar capabilities of both Pakistan and India, against the small number of our jets equipped with this capability, it will be easy for them to evade us. Esteemed representatives, diplomacy has failed us and our military might is inadequate to prevent a repeat of yesterday's heinous acts. Both countries have sacrificed the welfare of their own citizens to build up their military, and now they have put the entire world at risk from nuclear attack and radioactive fallout.

"I strongly urge you to ratify the Banzhof Resolution to limit the ability of these two nations, and others who present similar risks to the entire globe, from developing and deploying nuclear weapons. Thank you for your consideration."

L. Marshall Washington sat down, and Ludger Banzhof of the International Atomic Energy Commission approached the podium to explain the resolution. In German, he explained the basic tenets of the proposed legislation. The four principles could be elaborated further, he explained, but in essence, the plan was very simple. First, any nuclear electric-generating plant would need to meet strict IAEC guidelines governing the entire fuel cycle from tailings to tomb. Every milligram of actinide material would need to be accounted for, and the final repository would need to be geologically stable and inaccessible, so that it did not need to be guarded. Secondly, pelletized reactors with high-pressure gas working fluid would be required on all new plant constructions, to minimize the damage in the event of sabotage or attack. At this point in his delivery, he gripped both sides of the podium and leaned forward towards his audience. The third tenet was that the United Nations would act swiftly and decisively against any country shown to be diverting, developing, building, testing, or deploying any nuclear bomb technology. There were gasps of shock and alarm, delayed by the translation time.

"Now, hear me," he said quickly, expecting this response. "Those nations with existing stockpiles, and who have managed them wisely, would be grandfathered."

A delegate stood up suddenly and raised her voice over the growing din.

"Neither Pakistan or India signed the NNT!" she shouted, meaning the Nuclear Nonproliferation Treaty.

At that outburst, the collective outrage and indignation went out of control. The buzz of noise in the giant amphitheater grew more heated and clamorous. Delegates were standing up, some shouting at Banzhof, some shouting at each other. The Secretary General, sitting to one side, banged his gavel for order, but to no avail. After banging until his wrist tired, he looked about, and then reached for the room lights. Using a slide bar, he dimmed the lights until it was nearly dark. The delegates finally started to settle down. He strode over to the podium, and politely hustled Banzhof off, ignoring his complaint that he never got to explain the fourth tenet. Taking the microphone in one hand, and raising the other, he dismissed the assembly, and asked that they reconvene the next day to consider other proposals.

CHAPTER 4

Nairobi

Parks stayed on at their house in Nairobi. While the Energy Project languished, he sank into despair, and Afren Kwapili lost his only excuse not to help Elise. She put him to work right away, and within the first two weeks, they had begun to turn around the entire Ministry of Education.

Elise would return home of an evening, full of energy.

"You should have seen him, Parks. He ran all the way from Nakuru to Naivasha. I followed him in a Jeep, handing out brochures and candy to the children. There were hundreds of them, a thousand bare little feet trampling the road like a herd of elephants. At every small town we passed, more would join, and run with him for a mile or so. He wears his gold medal, and they treat him like a mythical hero."

"That's great," Parks said.

"He jogs slowly so they can keep up. He talks to them about the importance of education, literacy, and learning about the world. They are all ears, Parks! They are hungry to learn. And get this: village leaders have been swamping the Ministry with requests for our program. It's working better than I ever thought possible."

"Congratulations."

"Parks," she said, annoyed, "this is your project, too. Your idea, your vision. We've got the support of the President of the country, and soon, I hope, the Minister of Education. I've got new leads on donated computers. It's amazing. A month ago, getting donations of old machines was like pulling teeth. Now, when I tell them up-front that Afren Kwapili, the *gold medal Olympian*, is the champion of our program, they open right up and ask me how much I want."

"But, Elise, it doesn't matter anymore. The project is dead."

"How can you sit there and *say* that? Are you saying education doesn't matter? That just because your project hits a bump in the road, just when you're on the verge of success, you want to quit? I can't believe you. It just makes me feel sick. Snap *out* of it," she said, and turned on her heel, walking away.

"It's more than a bump in the road," he called to her back.

She whirled, fury in her eyes.

"Wake up, Parks. The bombings in India and Pakistan change *everything*! If you'd watch the news, you'd know that the United Nations is *right now* preparing to clamp down on nuclear power all around the world. You think the Japanese will continue going gung-ho on a nuclear Hydrogen Economy with all *that* going on?"

"But the environmentalists have everyone convinced that SPS microwaves will harm the ozone layer just like their Tesla beams."

"No they haven't, Parks. The Japanese just did a knee-jerk retreat from *all* power beaming after their wrists got slapped on the world stage. You already researched that. I remember you telling me that your power beams won't harm the ozone layer. Is that still true?"

"Yes."

"Well, then," she said, staring at him.

Parks had no response.

Elise shook her head slowly from side to side. She harrumphed and stalked out of the room.

Olduvai Gorge, Kenya

Parks used the respite to spend time with Mindy and Eric. They took a long-delayed bus trip to the Olduvai Gorge, where some of the oldest human fossils had been discovered. The archeological tour would allow them to pick and brush away compacted dirt and stone from a cliff wall that had once been topsoil a million years before.

Mindy, an intensely curious child of six, adored her father, and spent most of the trip on his lap. Eric played a video game as Parks spun them tales of civilizations come and gone, mythical societies from before the dawn of recorded history, and mighty continents that were now little more than archipelagos of islands in a wide, wide ocean. He told them of extinct mammoths found frozen for ten thousand years whose stomachs were filled with green tropical plants. And he told them of the fossilized human footprint found within the much deeper footprint of a dinosaur, and how legends of dragons and dragon slayers could be found the world over.

Eric put down his game, and looked up at his father.

"Perhaps," Parks told them, "Earth's history is fresher than we think, and human history is much, much older than we imagine."

"Where's the cities?" Eric asked.

"Buried."

"How?" asked Eric.

"I was in Krakow, Poland once," said Parks. "In their town square is a church almost a thousand years old. Through all that history, the parsons of the church kept it clean by sweeping dust and dirt from the floor and brushing it outside. Over time, the dirt outside got taller, while the floor inside stayed the same. Nowadays, you have to walk down three steps to go inside."

"Wouldn't that just make a hill outside the door?" asked Eric.

"Yes, but the rest of the town square has also been gathering dust and dirt for all that time. Except nobody swept it away. Gradually, so slowly that you can't even notice it during one human lifetime, the ground builds up, and buries the past."

Mindy asked: "Where does all the dust and dirt come from?"

"Most of it was tramped in by the people going to church, or coming into the town square from the fields. A lot of it came from the people themselves, whether it was hair, flakes of dead skin, or lint from their clothing. Some of it comes from dust storms, as winds blow in tiny specs of rock weathered from the distant mountains. And some of it falls from space. The Earth is constantly bombarded by tiny rocks from space, meteorites. I don't know how many tons a year, but those meteorites drift down through the air, and settle as dust. Dust which builds up and buries our cities."

"Cool," said Eric.

Mindy's brow furrowed in concentration. "So, the deeper we dig, the older the cities, right Daddy?"

"Right, Pumpkin." Parks smiled. "Except sometimes, cities get buried even faster. The ancient city of Pompeii was buried by an eruption from the volcano named Mount Vesuvius. In just hours, the entire city disappeared under a blanket of volcanic ash. Archeologist in modern times have uncovered whole households, with food still in their pots, and the unfortunate families curled up on the floor."

"Dead?" asked Eric. "Drowned in dirt?"

"Yes," said Parks.

"Way cool."

Parks did not notice the concern that was creeping into Mindy's face. He was turned towards Eric, pleased to be having an actual two-way conversation with his boy.

"And that's not all," Parks continued. "Recorded history goes back only about five or six thousand years. Yet in nearly every culture around the globe from that time, there are stories of incredible floods. Some people think there was a world-wide flood. Maybe the Earth turned on its axis, so the polar ice caps moved to the warm equator. When all that ice melted, everybody living on a coastline was swamped as the sea level raised up a hundred feet. Huge fog banks from the melting ice caused horrendous rains that washed down millions of tons of mud from the mountains. Cities on rivers would have been buried in mud, so instead of taking a thousand years for the ground to rise around the church, it could happen in days."

Eric squirmed in excitement. Mindy made a noise of despair, and spoke in a small voice.

"How did the Earth turn on its axis?"

"Well, there are a few theories. One, proposed by a guy named Boardman, and backed up by Albert Einstein, is that polar ice caps grow and grow until they become so heavy that at some point, they suddenly 'fall' towards the equator. Nature always seeks the lowest energy, and having the heaviest part of the Earth at the equator is lower energy than having it at the poles."

Mindy looked shocked. "And the other one?" she asked in a whisper.

"Another guy, Velikovsky, says that the planet Venus is a newcomer to our solar system. He believed that Venus was a rogue planet, traveling thousands of light years until it came close enough for the Sun's gravity to capture it. On the way to its current orbit, it passed close by the Earth. The gravity from Venus tugged at the Earth and caused our whole planet to tilt over."

"I don't believe that one," she said.

"I tend to agree, Mindy, but there is some interesting evidence. For example, Venus spins on its axis in a direction opposite to every other planet in our solar system. It's the only odd ball, and can't be explained by our normal theories of planet development So, it's a mystery. Ancient calendars from the Mayans of Central America were very precise and showed the length of the year as 360 days and the cycle of the Moon as 30 days, forming the basis for our studies of geometry and time. Nowadays, that has been shifted to three hundred sixty five and a quarter days for each year and 28 days for the Moon's orbit, so nothing matches evenly up anymore. Maybe Venus came by and gave the Earth and Luna a shove, knocking them out of their perfect harmony.

"Then, in the Bible, in the Old Testament, there are references to a time when the Sun stood still in the sky. Maybe that was when

Venus came closest and its gravity yanked on our planet, stopping its rotation for a while. There are other stories of 'manna from Heaven', a sugary bread that fell from the sky for years while Moses led his people out of Egypt in search of the Promised Land. It kept them from starving to death. Some say that the atmosphere of Venus contains complex hydrocarbons which were sucked into Earth's atmosphere, by gravity, as it passed. The gas from Venus may have combined with Earth's air, and this stuff fell to the ground as sticky sugar."

"I'm hungry," said Eric.

"We're almost there," Parks said, but fumbled in his shoulder bag for a granola bar.

"What will happen to *our* culture, Daddy?" Mindy asked. "Will we get buried, too?"

He handed the snack to Eric, and put his arms around his daughter.

"No, dear. Our society doesn't have the patience to wait around to get buried."

Their bus turned a tight corner, and the driver downshifted. The gears rattled and the engine whined as they rumbled down into the riverbed. Eric craned his neck to see the drop-off just beyond his window.

"Then, how *will* we end?" she asked. "All cultures do, right?"

"All the cultures which are gone have ended. That's true," Parks told her. "But there are many cultures and societies around today which have not ended."

"But what about the nuclear war?" she asked. "What about terrorists? And you're always telling us about global warming and energy shortages," she said, working up into a panic.

"Mindy, it's okay." Parks said, stroking her hair. "You're much too young to worry about things like this. Besides, there are a lot of smart people working on those problems. We'll find a way."

"But your project, Daddy. You were working on a way, too." she said.

The bus lurched to a halt. The cloud of dust which had been trailing them caught up and engulfed the bus. Inside, the light turned dim and brown. Parks was at a loss for words. The bus of tourists started to empty.

"That city under the volcano," Mindy said. "Did they find any children?"

"Well… yes."

"And this dig today, will we find any…any children?"

"I don't know, Mindy. Probably not."

"Do you promise?"

An impatient Eric climbed over the seat in front of them, and ambled up the aisle.

"I seriously doubt it, but I can't promise," he told her.

She climbed off his lap, sat firmly against the seat back, and folded her skinny arms across her chest.

"If you can't promise, then I'm staying *right here*."

She looked just like her mother.

Parks watched her for a moment, looking for any signs of uncertainty. Finding none, he left her a water and a granola bar. Then, he chased after Eric.

Nairobi

The following night, Parks sat with Elise in the reading room of their rented home. Afren Kwapili had joined the Vambolent family for dinner, and stayed after to help the parents work out the details of their departure.

"I understand we don't have any income right now," Elise said. "But I'm not ready to leave. The kids love it here. And we're doing so well on the project. Aren't we, Afren."

Kwapili gave a pleasant, but noncommittal, smile. He sipped from his coffee cup.

Parks turned to his friend.

"Afren, how about you? You're in the same boat I am. Since the Japanese cancelled the project, we're both out of a job. How do you make ends meet?"

"Parks," Elise said, shooting him a warning look. "That's none of our business. Afren, that was rude of him to ask. Don't feel obliged to answer."

"It's all right, Elise," Afren said. "I have very few expenses, and since the Japanese did pay well, I have some little savings to fall back on. When in Mombasa, I lived at my father's house. In Nairobi, I stay with old friends. For me, it is easy. For the two of you, it is not so easy. I wish there was something else I could do."

"I know you won't spend your patent savings," Elise said to Parks in a voice that belied the many arguments this point had caused, "but what if we cash in on your 401(k) from when you worked for Uncle Bryce?"

"We'd get clobbered by early withdrawal fees," Parks said. "We could stay here maybe another two or three months. Then, we would *still* have to go home. But we would have nothing, Elise. We'd be broke and unemployed. I need to get back ASAP and get a job."

"What if I stay on for a while. Just a little bit, until…"

"No, Elise," Parks said. "We stick together as a family. You'll have to run your end of it from the States."

"Fine, then," she said, crossing her arms. "It will all just fall apart, Parks. It will be just another failure feather to stick in your cap of catastrophe."

"Elise," Afren said, with a pained expression on his face.

"I'm sorry. But it's not fair. I'm needed here. If I leave now, before things really catch on, everyone will just go back to what they've always done forever." Her jaw jutted and worked side to side. Both men could hear the grating sound of her molars.

"Please excuse, Miz Elise." It was Bakuu, the house servant included in their monthly lease. She stood just outside the doorway to the reading room. "There is a man to see Mister Parks."

Parks rose and walked over to the front door.

"Hello?" he said, and stepped out onto the porch, closing the door against the mosquitoes.

"Parks Vambolent," said the man, a bit breathlessly. He reached out his hand. "It is an honor to meet you, sir. A very great honor indeed. I have traveled a long way to meet you. I can hardly believe that I am seeing you now. It is indeed a great, great honor."

"And you are?" Parks asked, still being shook by the hand.

"Ah, me? I am Dajwal Chellapilla, sir. At your service."

"Dajwal?"

"Oh, call me Dodge. Everyone at NASA did." A sad look came across his face.

"You were at NASA?"

"Yes," he said, brightening quickly. "At Headquarters." He chuckled. "That was before I was 'invited' back to India. Oh, it was not a happy thing, no, not happy at all." His face turned down again, as he wagged his head back and forth.

"Won't you come in, please?"

"Oh, I would be delighted. Are you sure? I could come again tomorrow if that would be better."

"No, please. We are just having tea and coffee."

Parks let them into the house, and introduced Dajwal to Elise and Afren. The four sat down. Eric came running into the room, chased close behind by Mindy. The two children came to a dead halt at the sight of the stranger, and stood still as trees.

"Hello, children," the newcomer said. "Please, call me Dodge." He reached into a pocket of his jacket and pulled out two small boxes, one square, the other rounded. Each was of dark wood, inlaid with silvery metal and closed by a silver clasp. With one box on each palm, he held his hands out towards them. "Is it okay?" he asked Parks.

Parks looked at Elise, who nodded. Mindy and Eric craned their neck to see the response. Before Parks even gave *his* permission, they reached for the gifts.

Eric accepted his with both hands cupped. He opened it at once. Inside, a large shiny beetle seemed to run in place with lots of twitching black legs.

"Cool!" he said, and took it over to his mother, who shied away from it. Having grown up in Indiana, she had still not adjusted to the many large and ferocious bugs of East Africa.

Mindy looked dubiously at her own box, held out at arms length.

"No beetle in yours, my dear child," Dajwal said. "No beetle a' tall."

She sized him up, then appeared ready to trust him. She opened her clasp. Inside was an iridescent butterfly, shimmering in the room lights.

"Oh," she gasped. "It's beautiful. Thank you, Uncle Dodge."

Dajwal gave her a huge smile, nodded his head, but said nothing.

"Say 'thank you', Eric," Elise said.

"Thank you," he said.

"You are most welcome, young sir."

"You two go off to bed now," Elise scolded gently. "I'll come tuck you in a few minutes."

"Okay, Mommy," Mindy said, kissing her cheek. She flew over to her father, and gave him a hug, then ran out of the room, followed by Eric.

"Such beautiful children," Dajwal said.

"Thank you, Dodge," Elise said, enjoying the compliment.

"It is my pleasure. You see, I knew you had two children. I have been studying your husband for some time."

She gave him a decidedly suspicious look.

"Oh, no, it is nothing like that. I am not some stalker, some weirdo." He laughed easily. "No, I was working for NASA until just a short time ago. You see, it was my job to find out about your husband, Mr. Parks Vambolent. He has caused quite a stir up at HQ. Yes, ma'am, quite a stir indeed."

"I'm afraid I don't know what you're talking about," Parks said.

Dajwal looked around at the three of them.

"I'm sure this must seem highly unusual to you. I have been a most ungrateful guest. You were obviously enjoying your evening, and here I am, interrupting."

Elise rose to her feet. "Dodge, it is we who have been remiss as hosts. May I offer you some tea?"

"That would be splendid, absolutely splendid. Thank you very much."

While Elise went to give her orders to Bakuu, Dajwal turned to Parks.

"What do you think of the Chinese landing on the moon?"

"They did already? Oh, my God. I knew they were planning to," Parks said. "I guess I've been preoccupied."

"Yes, they spent four days there, setting a new record, and returned with lots of rock samples. They left behind what they call a 'base'. It is really nothing more than a pressurized coffin, in my opinion. But this, too, has really stirred the pot at NASA."

"Did they return safely?" asked Kwapili.

"Oh, they are still on their way back, I think. The Chinese people are very proud."

"I feel embarrassed," Parks said. "I'm a bit isolated here. And frankly, for the last several weeks, I've not really had much interest in the larger world."

"Oh, there is so much happening. Perhaps you don't know yet about the new International Atomic Energy Commission? No? Well, the United Nations, under great pressure after my country and Pakistan performed their atrocious acts, has strengthened the IAEC considerably. This new body will oversee all uranium mining, enrichment, processing and disposal operations. Costs are to be charged to the users of the radioactive materials. Nuclear energy is about to become much more expensive."

"How are people reacting?" Parks asked.

"The people, I think, are very happy. But some of the governments, understandably, are very upset. Japan and France protest bitterly."

"And the US?" Afren asked.

"The US has made no real comment. I suspect the US will ignore the IAEC. But this is only my opinion. The UN wants to prevent any country from being able to use nuclear materials to make weapons of mass destruction. The US, I think, wants to be the only exception."

Elise returned with tea, and set a tray with cream and sugar before her guest.

"So, Dajwal, what brings you to our home."

"Yes, well, I will tell you the short version. There are a great many details, but I do not want to keep you up all night."

"Oh, my gosh, the kids," Elise said. "Go on, Dodge. I'll be right back."

"Well, Parks Vambolent, let me say again what an honor it is to meet you at last. You see, the State Department learned of your work with the MITI in Japan. Well, the State Department talked to the White House, and the White House talked to Mr. Lewis Rinker, our Administrator at NASA. Mr. Rinker talked to the head of my department, Technology Assessment, and asked her to look into the matter and report back to him. My department head pulled me off the work with the Russian company Energia – me plus several of my co-workers. She told us to determine exactly what arrangement you had with the Japanese, where your funding was coming from, what was your intellectual property position – and a whole host of other issues. As you might imagine, with the Chinese about to land on the Moon, they did not want the Japanese also to put egg onto their faces."

"So, you must have found my patents. I thought nobody at NASA would ever look at them," Parks said.

"We looked at them very closely," Dajwal said, using a dramatic voice that still seemed sincere. "And I must say, they are very good. Very good, indeed. Your concepts on isotope separation in micro-gravity are well ahead of your time. But as far as I could tell, your work with MITI did not involve those patents. Not yet, at least."

"That is true," Parks confirmed.

Dajwal nodded. Elise came back in, and sat next to Afren.

"Of greatest interest were the many presentations you made at space conferences over the last five years or so. I wonder if your audience even appreciated the technology you described. Very advanced. Very unconventional. Most impressive."

"Thank you," Parks said. He tried to suppress the smile that threatened to spread across his face.

"In fact, one of your ideas, on ozone scanning using refracted sunlight, is being pursued by an Israeli university. Did you know about that?"

"No, I did not."

"I was not sure. There was no patent on that particular idea. It was one of your earlier breakthroughs, I think. Perhaps you published it before you realized that you lose patent rights by public disclosure."

"That's exactly right," said Parks. "Geez, this is spooky."

"Oh, it gets downright frightening in just a minute."

"What do you mean?" Elise asked.

Dajwal turned to her. "Well, after our initial report was submitted, nothing happened for a week or so. Then, the three of us who worked on the report were called into the Administrator's office. He gave us a discretionary budget, and told us to locate people who could beat the Vambolent patents."

"What!" Parks exclaimed.

"Yes, it is true! I was told to find a way around your claims. I am ashamed to admit it, but those were my orders from Mr. Rinker himself."

"Oh, my God. I had no idea."

"It gets even worse, I am sorry to say."

"What do you mean?" Parks said, his pulse starting to pound inside his head.

"Well, the three of us were flown to St. Louis, to Seattle, and to Los Angeles to meet with senior executives at the primary aerospace firms. They, in turn, brought in partners in mining and materials. The NSF got involved, and DARPA stepped in. NASA called in key researchers at the top East Coast technical universities, UMR, and the Colorado School of Mines. You see, Parks, they finally read your papers on cislunar architecture for a sustainable space economy. They didn't want you to do it with the Japanese. So, they decided to do it themselves."

Parks fell back in his chair, putting a hand to his forehead. Mixed emotions of pride and outrage battled in his mind. Elise looked at him, and a new measure of respect showed in her face.

"They're calling it the Seven Dwarfs mission," Dajwal said.

"What do you mean?" Parks asked.

"I mean that NASA is partnering with six corporations to capture an asteroid, bring it back to Earth orbit, and harvest the minerals."

Parks was shocked into silence.

Afren spoke gently. "I have not read Parks' papers, but he speaks always of building solar power satellites to help ease the global warming problem. Is this their intent?"

"Indeed, Mr. Kwapili," Dajwal said. "That is their long-term plan, especially if the new IAEC starts to play rough. But of more immediate interest is the extraction of precious metals."

"Gold?" Elise asked.

"Not gold so much, but platinum," Dajwal said. "Part of the failure of the old Freedom Car initiative was the high cost of this metal. It is needed as a catalyst for fuel cells. Even with a layer of platinum only one atom thick, a fuel cell vehicle costs too much money."

"And with a good sized chondritic asteroid," Parks continued smoothly, "you've got a trillion dollars worth of platinum."

"A trillion dollars!" Elise exclaimed. "No wonder they're going after it!"

"And," Parks added, "there will be enough silicon to provide every person on Earth all the pollution-free power they need forever."

"It is right out of your ISDC paper," Dajwal said.

"I need a drink," Parks said, rising from the sofa. He moved to the wet bar on unsteady legs. "Anybody else?" he asked over his shoulder, and poured himself a scotch. The bottle chattered spastically against the edge of his tumbler. He gulped from the top of the glass, then tilted the bottle again to fill it back up.

Sitting back down, Parks took a deep breath and steadied himself. He looked closely at Dajwal Chellapilla, who was stirring his tea.

"So, they are trying to beat me at my own ideas. Without me."

Dajwal watched Vambolent closely, allowing him time to deal with his emotions. Elise and Afren were also watching Parks.

"I must have a million questions. But perhaps the first one is why are you telling me all this? Did they send you here? Why exactly did you come to my house tonight?"

"A fair question. My motives are my own, I assure you. Perhaps it would help if I told you my story. At least the last month. Is that okay?"

"Will it answer my questions?"

"Oh, most assuredly."

"Then, please continue," Parks said, and slurped his drink.

"Well…so much has happened. Who could know that things could change so fast. You certainly remember the night when Pakistan launched its missiles against India?"

"Yes," said Elise. "We were having dinner at an Indian restaurant. The three of us."

"Is that right?" said Dajwal. "I was flying back to D.C. when it happened. By the time I landed, and learned of the news, India had already retaliated. I took the day off from work. Many of my Indian friends came over, and we wept together. It is a terrible shame, a terrible outrage. Then the phone rang. It was a call from the ISRO. That is the space agency for India – their NASA. Apparently they had a list of all people with Indian heritage working in aerospace in the US. They begged me to return and join their 'peacekeeping' force.

"Now, I was born in Bangalore where the ISRO is headquartered, but my family emigrated to the States when I was a boy. I am a naturalized US citizen, and carry a US passport. I told them no. But after I hung up, when I explained to my friends about the call, they became very excited. They criticized my choice. Then, they all left to return to their own homes, hoping for their own phone call. Naturally, I reconsidered my decision."

"That must have been a difficult day for you," Afren said.

"Oh, it was. You cannot possibly know. Well, perhaps you can, but I certainly had nothing to compare it to. Changing my decision was agony, but I did it. I called the ISRO back and told them I would return, at least for a while. The next day I called my department head and asked for a leave of absence. But by then, so many had already asked, that she said my only choices were to stay or quit. So I quit."

"You quit NASA to go back to India?" Parks asked.

"Yes. It was a foolish decision now that I look back. I did not know what awaited me in India. So, I went, full of hopeful expectations. Instead, what they had told me was a peacekeeping force was in truth a crash program to develop ground-following cruise missiles. As soon as I learned their plans for me, I complained. That got them really mad. But they put me on another program. This one was to develop anti-missile satellites using nuclear-powered laser beams. This time, when I complained again, they got even more mad. They called me a traitor, and threatened to place me in a jail cell until my thinking got straight. I could not believe it!"

"Oh, that's terrible," said Elise.

"So, I promised them I would work on whatever program they wanted. I told them what work I did for NASA, exaggerating ever so slightly, and they settled back down. That night, I left by train for Bombay. I found a freighter leaving for Mombasa and begged the captain for work. He accepted me without question, and I worked my way across the sea to Kenya. You see, I was coming to find you, Parks Vambolent. We arrived in port just two days ago, and now I am here with you."

"I see," said Parks, digesting this whirlwind of information. "But why?"

"The reason why is that I want to work for you!"

"What?" Parks blustered. "Um, Dodge, I hate to give you more bad news, but when you knocked on the door, we were just planning to leave Kenya. I'm broke. I'm flattered that you came all this way, but I can't pay you a single day's wage."

"That is okay," Dajwal said. "I will work for just one dollar. You can pay me that dollar today, if you will hire me. Then, when you become a wealthy man, you can pay me whatever wage you think fair."

Parks looked at him as if he escaped from a mental institution, instead of from India. Afren stirred, and moved in a precursor to excusing himself. Elise leaned towards the stranger.

"Let's get this straight, Dodge," she said, "you came here and explained to my husband that you were part of a group trying to bypass his patents. Then, you told us that a group of seven huge American corporations are pursuing his ideas without including him, and presumably without giving him any credit. Then, you have the nerve to ask him for a job, and you persist even when he tells you that he has no money to pay you. Forgive us if this seems a bit, well…"

"Crazy," Dajwal finished, with a huge smile on his face.

"Yes, crazy. And the way you're smiling makes me even more uncomfortable."

Dajwal's face immediately fell into a very serious expression. His eyes blazed with intensity.

"I'm sorry. I rehearsed this meeting so many times, I had no idea how I would come across. For you see, more than anyone else alive on the planet, I have become an expert in the ideas of Parks Vambolent. Parks, I know everything you have ever published or patented inside and out. I am a highly-trained aerospace engineer, and I have seen the spark of genius where others have not."

He paused. Elise sat back. Afren settled into his seat. Parks got up to re-fill his tumbler.

"But Dodge," Parks said, slurring his words slightly. "You must also know that the Japanese have cut me loose. They dropped the Energy Project like a hot potato. It's over."

"Yes, and that is what is so perfect. It makes what I have to propose so much more agreeable."

"Please, then, explain your proposal," Parks said, throwing out a loose hand of invitation.

"Everything comes back to your first patent, the microgravity isotope separator. You have written many times that is the 'foundation of space-based manufacturing', have you not? Well, I believe it is true. It is a good patent – very broad and full of good claims. Because your patent is so defensible, the consortium of the Seven Dwarfs mission has chosen to use conventional materials processing and metallurgy. It is brute force to be sure, not elegant and self-sustaining like your isotope separator. But they are relying on the economic payoffs of platinum to finance launch costs for consumables. This leaves them economically free to produce iron, aluminum and silicon very cheaply. Once users of these materials begin to buy them, the consortium will become financially viable, and a cislunar economy will have been born."

"You make it sound very convincing," Parks said. "So, I'm still wondering why you are here, asking for a job from me, instead of over there, asking for a job from them!"

"Ah, but the seeds of their own failure are planted in their very business plan."

"I must admit, I don't see the flaw."

"The first flaw is that they ignore the law of supply and demand. The 'trillion dollars of platinum' assumes current prices and known sources. But with a whole asteroid packed with the metal, prices will plummet before they sell the first troy ounce. Even if the prices drop by only 50%, surely someone else will go get the second asteroid. The Chinese could do this, the Russians maybe. Within a decade, platinum will cost no more than silver."

"Wow." It was all Parks could think to say.

"The second flaw is in the customers for their other materials," Dajwal continued. "Transport in cis-lunar space dominates all other costs. For customers to pick up their iron, aluminum or silicon, they must either fly up to space, or pay for the goods to be dropped from orbit. Don't you see now? They will have low-cost materials, yes. But nobody can afford to come get them!"

Afren cleared his throat. "Just one question please, Dajwal. Won't the materials from the asteroid still be far less expensive than launching them from Earth? Combined with the advantages of being in orbit, and the benefits tip back to the customer, does it not?"

"A most astute observation, Afren," Dajwal said. "But what can anyone do with raw aluminum, pig iron, and amorphous silicon? Those customers will need to launch their own factories. Factories which will need supplies like acids and bases, gasses and spare parts, and of course power supplies. The economic pay-back keeps jumping ahead, no matter how far along you go."

"Doesn't the consortium realize this?" Parks asked.

"They are very smart people, indeed. But like so many Americans, they don't think ahead. I mentioned this logic to them once or twice, but they are not ready to listen. They are too eager to show up the Chinese, and to beat the Japanese to a Hydrogen Economy."

"What about their powersat concept?" Parks asked.

"The asteroid will be parked at L5," Dajwal said, then turned to Elise and Afren. "L5 is a libration point, one of five places where the Earth's gravity cancels that from the Moon. L5 is located along the Moon's orbit path, but 60 degrees further along." He turned back towards Parks. "Their plans are not yet finalized for power satellites. DARPA definitely wants a Moon base for security, even though it makes no sense for asteroid processing. The commercial people want to beam power from at L5 to Earth via relay sats in LEO (*Low Earth Orbit*)," he whispered to the other two. "And the academics want to transport solar arrays to Geostationary Earth Orbit (*GEO*)," he whispered. "Plus, they've got diplomats braying about 'military presence', and the press corps is parlaying the whole 'death ray' thing, after the debacle in Japan. So, what did they do? They formed a blue ribbon committee to study the problem. They will almost surely issue a Broad Area Announcement on this issue, and then make a final recommendation to the Seven Dwarfs Directorate. "

"Arrays at GEO is the best answer," Parks said.

"I agree," said Dajwal, "and that will hopefully be the same conclusion reached by the committee. And to think, you have envisioned all this well before them!"

Parks beamed, and looked at his wife. She got up from the couch, and plopped down on the arm of his chair, throwing her arm across his shoulders. With a wet smack, she planted a loud kiss on his cheek.

"My handsome genius," she said. "And you were just about to quit! Now, we can stay here in Kenya and finish what we started!"

Afren cleared his throat.

"What?" she said, annoyed. "We can stay now, can't we, Parks?"

He shook his head in the negative, but did not verbalize the answer. Elise conveniently forgot that the Kenya Energy Project was Japan's, and that they had cancelled it. That the US also did not need her husband sank in. Her hand came back into her lap, but she remained on Parks' chair.

"So, you believe the consortium will fail eventually?" Afren asked Dajwal. "How soon do you estimate that occurring?"

"Well, there's no telling how soon they might look ahead and figure this out," Dajwal answered. "There is powerful backing for this on Capitol Hill. It will create much-needed jobs on both coasts. The momentum will probably not slow down until the answer is obvious to everyone."

"How many years might that be?" Afren asked.

"Who knows. The asteroid they have selected can be brought back in four years – maybe five."

Elise went limp as the full realization swept over her. NASA was charging ahead without her husband, these Seven Dwarfs people were not using his patents, and by the time they figured out they might need those patents, they would be nearly expired – useless. She took Parks' empty tumbler from his hand.

"Anyone want a drink?" she asked.

"Na more for me, thanks," Parks said.

Without waiting for other answers, Elise went to the bar and re-filled Parks' glass. She took a deep sip from it, then stood there, cradling the arm with the drink.

Looking from Elise to Parks to Afren, Dajwal waited. He was on the edge of his seat, and appeared ready to leave.

"It sounds like my luck has run out," Parks said softly.

"I don't see it that way at all," Dajwal said.

A cross expression formed on Parks' face. "I appreciate you coming here to tell me all this, Dodge. I'm sure it's exciting to you. But to me, it's a death knell. Japan has dropped my project. NASA is beating my patents. I've lost my best friend trying to save the world. How can you *not* see it that way?"

Dajwal asked politely about the loss of Parks' best friend. Afren answered, since Parks had withdrawn with an inward stare. Dajwal expressed condolences after the explanation, and they all were quiet.

"You've all been very generous, and I've taken up too much of your time. I'm sorry," Dajwal said. "But there is just one more thing I came here to tell you. I hope it's not too late."

"What is that?" Elise asked.

"Parks, I know you tried very hard to get funding and support at home," Dajwal began. "It must have seemed impossible."

"Why do you say it that way?" Parks asked.

Dajwal's face brightened with enthusiasm. "You probably know of Senator Mason Barnes, yes? He represents Indiana."

"We met him once," Elise said. "Parks dragged me to the Moon-Mars Blitz one year, and we lobbied in his office. Senator Barnes came out, shook hands and had a photo taken with us. We sent him a print, which he signed and sent back to us."

"Excellent!" Dajwal said. "Then you must know that he was almost an astronaut once. Now, he is a major advocate of space tourism. And, he serves on the Subcommittee on Science, Technology and Space. This year, he received pork barrel allocations for space development through the Indiana Space Grant Consortium. Yet, I happen to know, confidentially, that he is very frustrated with the way universities in his state want to spend that money. He is sitting on the funds, waiting for the right project."

"With the Seven Dwarfs mission, what could we possibly do?" Parks asked.

Dajwal looked surprised, just for a moment. "Do you remember your second article in *Ad Astra*? Did you not suggest that the slag from your device could be used to build rotating habitats with artificial gravity?"

"Why yes. You really did your research, didn't you?"

"Thank you, Parks. But here is the kicker. Once the asteroid is parked at L5, even if you do nothing with platinum or silicon, you can build space hotels for Senator Barnes to come visit!"

"I see," Parks said, frowning in concentration. "So, we get funding to build an isotope separator to use for habitats. Then,..."

Afren caught the idea, and finished the sentence. "Then, when the consortium fails, as Dajwal predicts, you will be free to use your invention to make solar arrays."

"Oh, my God, it just might work!" Parks said, rising from the chair. "Honey, are you ready to go back home?"

Elise sipped her scotch. "Afren and I have things to do here in Kenya, Parks. You and I need to talk."

Dajwal stood up also. "It is late, and I'm sure you are all as tired as I am. Thank you so much for your hospitality, Elise, Parks. Afren, it was a very great pleasure to meet you."

"The pleasure was all mine," Kwapili said, rising to shake hands.

"And Parks, I will be working my way back to Georgetown. This card has my e-mail and phone. Please call me when you are ready, and we can discuss more details."

Parks showed their guest to the door.

"Thank you, Dodge. I hope to contact you very soon."

Dajwal looked over Parks' shoulder, as if embarrassed that anyone should hear. "One more thing, sir. I am currently without employment. It would be a great honor for me, sir, if you were to hire me into your corporation."

Parks barked a rough laugh. "Dodge, I have no income. Nothing to pay you. Why don't we wait until something more concrete comes up."

"For many reasons, sir, I would like to be in your employ starting immediately. A single dollar would hold my services indefinitely. You can pay me more when you become rich and famous."

Parks stared incredulously for a moment. With a wry grin, he pulled out his wallet and fished out a single US dollar. "Here, Dodge. You're hired."

"Thank you very, very much, Boss."

They shook hands vigorously. Dajwal turned, and walked down the path, bouncing on the balls of his feet. He turned to look back at Parks. There was a great smile on his face.

Parks sighed contentedly. He looked up to see the Pleiades, his favorite constellation, glowing fuzzily through the city haze. Its half-dozen visible stars twinkled in the shifting layers of air. He couldn't suppress a smile. A mosquito buzzed in his ear, but he hardly noticed.

Afren came out, and joined him on the porch.

"A nice night," he observed.

"Beautiful," said Parks, and slapped his neck where the mosquito pierced.

"She can stay here for a while longer, you know," Afren said, putting his hand on Parks' shoulder. "I could help her find a cheap apartment downtown. Or there may be ways to keep this house. It means so much to her."

Parks shook off the reverie of his own dreams to think of Elise. "You're right," he said.

"She has dreams which are just as big as yours, Parks. And hers are right within reach. What she is doing here is a new thing for Kenya. But it is also a new thing in all the world. Your dreams may be years away, but hers are close enough to touch. Think about it. I will help you in any way I can, my friend."

Parks turned towards his friend, moved, and they embraced.

"Thank you, Afren. You're a great friend."

Afren went back inside. Parks stood a moment longer, rubbing his bite, and thinking that the only way he would get an embrace from his wife would be to agree for her to stay.

Near Mt. Fuji, Japan

Sugeto Mifune stood in a soft rain outside an estate in the foothills of Japan's revered volcano. It was after midnight. Carefully tended cherry trees, blossoming earlier than ever, dripped water onto suspended stones, and the trickles joined briefly before cascading down a series of rounded black rocks. The sounds enchanted him, and helped him not notice the time passing as he waited to be acknowledged. There was no guard house, no call box, no knocker or buzzer. The heavy doors of a red wooden gate were shut, but unobtrusive cameras, and probably other sensors told those inside of his presence. Of this, he felt certain.

This was the residence of Hiyate Kindoh, a reclusive industrialist whom Mifune had never met. During Mifune's tenure at MITI, his role had been to shepherd research projects from development lab prototypes to fully commercialized products. As head of the organization, he became responsible for all of them. MITI's industrial partners were many, and Mifune had been privy to their documentation. Kindoh's name appeared as director in more than a reasonable share of those companies.

For many years, Mifune had sought an audience with Kindoh. Wary of the appearance of a monopoly, eager to work with an influential player, and curious about the man, several times he had sent indirect requests for a meeting. Efficient executive managers, polite but terse, told him a meeting was not necessary, but that any matters requiring Kindoh-san's personal involvement would surely be brought to his attention. Not wanting to create a fuss, and without anything more concrete, Mifune had backed off. Yet, he remained curious, and continued looking for the Kindoh name – finding it often.

Now, disgraced and disconnected for the first time in his life, he had no one to tell of his idea. He possessed too much pride to bend the ear of Sawahata, his replacement. He would not enter the campus of the MITI ever again. But while he yet lived, there was one more thing he must try. If this did not work, he still had the hose and the duct tape.

The massive gate creaked, and swung slowly, opening just part way. Mifune watched, standing erect, hands at his sides. From the shadows of the drive, an umbrella appeared in the dim light that illuminated the gateway. Mifune watched the body of a man materialize below the umbrella, which drizzled rainwater from its tips.

The umbrella tipped upward, but the man's face remained in its shadow. From the shuffling gait, Mifune guessed the bearer was old. Was this Kindoh? Or just a servant.

"It is a cold night," said a soft voice. "Won't you come in for some tea?"

Mifune bowed.

"Thank you very much."

"You are a long way from home, Director-General," the old man said, turning around with small, scraping steps.

Mifune followed, a pace behind.

"I am no longer the Director-General," he said.

"Ah, then we must find something else for you to do."

Nairobi

Two days after Dajwal Chellapilla's unexpected visit, Parks had flight reservations to return to Indiana. Together, he and Dodge planned to get research funding through Senator Barnes. Into his two suitcases, he tried to pack as much of their household goods as would fit, so that Elise's burden would be a little lighter when it came time for her to join him. His wife, who had been working with Afren in their living room, came in with a cordless phone.

"Phone call for you," she said. "It's Yoshi."

When he had accepted the handset from her, she began immediately to repack his clothing and the boxes, taking everything out in preparation for fitting it back in to suit her. Parks was distracted as he answered.

"Hello? Yoshi?"

"Hello Parks. How are you?"

"Fine. Look, I haven't heard from you in a month. I'm kind of busy right now."

"I am sorry I have not called, Parks-san. Closing down our office in Mombasa has been very tiring. But now I have stopped."

"Are you finished? Going back to Japan, now?" Parks asked.

"No, Parks-san, I am not returning to Japan. That is what I am calling about. The Energy Project is starting again."

"What? That can't be. Can it?"

Elise stopped what she was doing and looked up.

"What?" she asked.

Parks shushed her. "What do you mean? I mean… Why are you calling me?"

"I would like to invite you and Afren to return to work," Yoshi said, with a joyous lilt to his voice.

"It's a little late for that," Parks said. "I'm leaving the country tomorrow."

"Oh. Then, I must see you tonight."

"Look, Yoshi, I've got a better offer, back in the US."

"Please, Parks-san, please wait. You must hear what I have to tell you before you leave. May I come over tonight?"

"Tonight?" Parks echoed. Then, he turned to Elise, placing the pickup against his shirt. "Elise, he wants to come over tonight."

Elise shrugged. "Sure. I'll ask Bakuu to set an extra place."

Parks thanked her, then returned the phone to his face. "Sure, Yoshi. You can join us for dinner – wait, where are you calling from."

"I am in the airplane now. We land in Nairobi in about an hour."

Parks felt dizzy. "Fine. Fine. We'll expect you about seven, then?"

"Seven is good. Will Afren be there also?"

"Yes, he's here. We'll talk. But I don't know if there's anything you can do. We're pretty much decided on a course of action. As long as you understand that from the start."

"I understand, Parks-san. Thank you."

Yoshi ended the call, and Parks dropped his arms to his side. Elise had tucked rolled clothing into odd spaces, and was pressing down so hard the seams of the suitcase bulged and stretched.

"There!" she announced. "Now, you have room for some of some of the carvings I've been given by the people. They are calling me Aunty Lise."

* * *

Yoshi Hanazawa arrived with a wrapped gift box. Parks received it using two hands, and opened it to find a sake warmer and a four-pack of small white bottles. He loaded the unit, plugged it in, and then brought Yoshi to the back of the house to meet Elise. They came upon her and Afren, heads tilted together, talking in hushed tones like co-conspirators. Elise looked up guiltily, then eyed Yoshi suspiciously. His height equaled Parks', but he appeared to be more large-boned. His short arms look powerful, with fat, rounded hands. The skin of his face is medium-dark and ruddy with moles, freckles and acne scars. When he smiles at her, folds of skin drown his eyes, so she wondered if he could still see. Parks introduced them.

Yoshi bowed. "Nice to meet you," he said.

She inclined her head in return, “Nice to meet you, too.”

Afren stood, and greeted Yoshi warmly. The two men gave a playful lunge towards each other and grasped hands up to the elbows, grappling their arms, pounding each others’ back, and exchanging warm greetings.

Elise watched with a face equal parts surprise and confusion.

“We have both studied judo,” Afren explained to her. “Yoshi was a national champion when he was younger.”

Yoshi bowed his head humbly, and Afren used that moment to stage a mock attack. The two tussled like rowdy boys as the Vambolents looked on. Neither had ever seen Afren like this.

Bakuu called them to dinner. Elise had planned a simple meal starting with miso soup and a radish dish. The next course was soba noodles and shiitake mushrooms topped with green onions. For the entrée, Afren and Yoshi ate steam-baked tilapia alongside a baked yam. Parks had a small piece of fish on his plate also, next to a serving of the red beans and rice dish that Elise was having.

After the meal, a heavy rain began pelting the house. Parks led the men into the reading room, where they settled into chairs. Mindy Vambolent wandered over from the children’s wing and stood just inside the doorway, her eyes down, waiting to be acknowledged. Elise brought a tray of cookies into the room, noticed her daughter, and introduced her to Yoshi.

He smiled to her.

Mindy ran over and stood before him, her hands clasped behind her back.

“Are you a real Japanese?” she asked.

“Yes,” Yoshi answered, laughing.

“I mean a real native,” Mindy stressed. “Not just somebody born there, or who moved there, right?”

Yoshi’s smile drooped a little as he answered. “That’s correct. I was born and raised in Japan. However,” he said, leaning toward her, “when I was at university, I was teased for my dark skin and my thick fingers. I was called a ‘native’ – actually a similar word in Japanese – and made to feel bad about my ancestry.”

“That’s horrible,” Mindy said, aghast.

“Oh, I think all kids make fun of what is different in others. This is how a culture survives from one generation to the next. If you are out of the ordinary, you will be harassed until you change to become the same as everyone else, or you are made into an outcast. Even outcasts help continue a culture, by defining what it is *not*."

A look of concern furrowed Mindy's petite forehead.

"But I am no longer troubled by this," Yoshi said, reading her expression. "I have traveled many places in the world, and found people to be much the same everywhere. Only cultures are different. To better understand the *difference* between cultures, and also to understand what is the *same* in each one of us, I have made myself a student of other cultures."

"Me, too," said Mindy, breathlessly. "My father said that many ancient cultures have myths of a big flood. I was wondering if your culture had legends like that?"

"No," he said softly. "None that I was told. My heritage is from the small islands at the very southern edge of what is now called Japan. My father was a fisherman, and my mother worked at a cannery. They remembered a few old myths *they* were told as children, but nothing of a flood."

Mindy looked crestfallen.

"But there *is* something interesting," Yoshi said, remembering. "Off the island of Yonaguni, which is almost the southernmost island of Japan – nearly all the way to China – they have found great slabs of carved stone. What is unusual is that these stones lie deep under the ocean, as far down as a seven story building is tall!"

"Gosh," Mindy breathed, eager for more. The rain outside grew heavier, and they had to raise their voices over the din.

"You have to wear scuba gear to reach these old ruins," Yoshi said, "so clearly they must have been carved when that part of the land was much higher. Maybe when the original island sank down into the ocean, my ancient ancestors would have called it a flood."

Mindy's eyes grew wide.

"Did it happen fast?" she asked, raising her little voice above the rain.

"Oh, I do not know. I don't think anyone knows. It must have been very long ago."

"Were people killed?" she asked, then added in a frightened voice: "Were any *children* killed?"

Yoshi held out a large hand and rested it gently on her thin shoulder. His face carried more compassion than she had ever seen in an adult.

"I don't think any children were hurt, Mindy," he said to her. "Land usually does not rise or fall so fast. They probably had plenty of time to pack up their belongings, get into a boat, and take their children to another island. Every family has a boat, and there are plenty of other islands around."

Mindy absorbed all this, and then opened her mouth for another question. Elise spoke first, announcing her bedtime.

"It was nice meeting a true Japanese," she said to Yoshi.

"Thank you," Yoshi said, laughing. "It was nice meeting another student of human culture."

"Good night," she said to him; then ran across the room to hug her father. She hugged Afren also, and then followed her mother out into the hallway. Her head turned far around to watch Yoshi as she left. The rain abated, and the room seemed to grow still, the mood expectant.

Yoshi looked at Afren, then at Parks. A look of equal parts seriousness and sadness came over his face. He waited patiently for Parks to speak first.

"How is Mifune-san?"

"He is good. He is a changed man after what happened. But he is good, Parks. Thank you for asking. He asked that I extend his warmest regards to both you and to Afren as well."

"I related to Afren what you explained on the phone," Parks said. "We both feel the same way. It was great working with you, but we've each got other opportunities to pursue now."

Yoshi sucked in a breath, deciding on an approach. Several moments passed. Sensing that the traditional Japanese approach to such a subject might not be the most effective, Yoshi went against his nature and dropped them a bombshell.

"First, I should tell you that the power satellite has been completed," he began.

Parks stood up involuntarily. "Really?" was all he managed to say.

"After all the news, we thought the whole project dismantled," Afren said.

Knowing that he had their attention, Yoshi continued. Parks slowly reseated himself.

"There are many factions within the Japanese Aerospace Exploration Agency," Yoshi explained. "When JAXA was formed in October of 2003, three agencies were combined into one. Yet, even now, there still remain rivalries between groups. So, you can imagine their collective anger that MITI was leading the development of space solar power. Then, when the public learned of power beaming experiments, and the Prime Minister denounced all such projects, MITI dropped our Energy Project. Naturally, JAXA could not publicly re-instate the project. However, Mifune-san has found a way to keep the project alive."

"Why didn't he call here? We could have continued working on the rectenna," Parks said.

"But all this is being done in secret," Yoshi told him. "It is hard to explain."

"Try us," Parks said.

"You have to understand our culture," Yoshi said. Elise came back in and sat down. She motioned for Yoshi to continue. "When Mifune-san was removed from his office, he was disgraced. The shame he bears is more than the shame of one man. He also wears the face of shame for our whole country."

"What do you mean?" Elise asked.

"The whole story did not appear at once, but came out bit by bit like someone was planning it," Yoshi said. "And by coming out in this way, the dishonor upon MITI and on Mifune-san grew greater and greater with each revelation." He paused, gauging his audience. They were all eager to hear more.

"The beginning you know already. The public discovered his institute testing beam weapons for a next-generation missile shield. Ever since the surrender terms imposed by the USA in 1945, Japan has built only defensive weapons. Some of our defensive developments could be construed as offensive, but a Tesla beam falls clearly to the other side of that line. To have this fact revealed so blatantly in the worldwide press brought many strong reactions. The entire power balance in the Asia-Pacific region has shifted. An arms escalation between Japan and China is certain. Mifune-san's name is now associated with this new development.

"Next, investigative reporters probed more deeply into the beam technology. Their news programs described how both the US and the former USSR had developed Tesla technology throughout the Cold War, but had now largely dropped it. For decades, we have prided ourselves with being innovators and leaders in science and engineering. The Japanese people have great pride. To be portrayed as being a follower of out-dated technology is a considerable embarrassment.

"Finally, you should know that for years, the Japanese people have worked very hard at energy conservation, resource management, and reducing solid waste streams. Now, the environmentalists are shouting that we are destroying the last shreds of the ozone layer, and ruining the planet for all generations to come."

"I had no idea," Elise said.

"Oh, I hate to say this, Elise-san, but the story gets much worse."

"Worse?" she asked.

"When it was discovered that MITI had not one, but two energy beaming projects, the blame became much more focused. Mifune-san would have taken the blame for the Tesla beams; that is for sure. But when the Diet and the Prime Minister learned that he had two such projects, it became convenient for them to place all the blame on a 'rogue' ministry, one that did not represent the true face of Japan. Mifune-san became the scapegoat. There were calls for him to take his own life."

"Oh, my God," Elise cried. "That's horrible."

"There is more," Yoshi said. "Once Mifune-san was vilified, more information surfaced to disgrace him even further. An hour-long documentary was aired on national television. They portrayed Mifune-san as a puppet to the Americans. The secretary who served him for eight years gave an interview saying that an American – you Parks – brought this whole idea to Mifune-san. She said you were given money and flown to Kenya to work in secrecy. The reporter uncovered what he calls a 'front company' called Vambolent Space Industries. He claimed that VSI is part of a CIA plot to disrupt the Japanese plans for a Hydrogen Economy."

"What!" Parks yelled. "How?"

"By leading Mifune-san into yet another old technology, solar power satellites. The documentary showed the results of NASA studies – many of them, over many years – all saying that the technology is a dead end. The economics never make the energy cost-effective compared to other sources."

"Then they didn't dig deep enough," Parks said, quieter now. He was thinking of Dajwal Chellapilla, and the research he had done for NASA.

"Oh, but they did dig," Yoshi said. "Only, they dug in a way most damaging to Sugeto Mifune. They have your published works, Parks, where you propose capturing an asteroid in Earth's orbit for mining. The reporter used vivid graphics to show the military potential of having an asteroid in the skies above your enemies. They portray it as a conspiracy to develop 'Terror from the Skies', where ionosphere power beam and space rocks rain down from the heavens."

Parks gaped. Yoshi continued.

"Then, they discovered your patent. Not the US issue, but the Japanese patent which you were granted under the Patent Cooperation Treaty. You were portrayed as a greedy wheeler-dealer trying to re-direct billions of yen out of Japan. They left the suggestion that your whole plan is being run by the Pentagon, and that Mifune-san was your stooge."

"That's ridiculous," Parks said, astonished.

"The documentary went on to claim that 40 MITI engineers and scientists were told to beat the patent – to work around it – but they could not. They made Mifune-san seem a fool. He followed Soviet technology, spent freely on technology dismissed by the US, couldn't improve on an American patent, and started an arms race by developing weapons of mass destruction. They made Mifune-san seem both incompetent and dangerous – the most evil man of the millennium."

"Oh, my God," Parks said, and slumped back into the couch.

"The Prime Minister received threatening communiqués from the Chinese. China has been jealous of Japan's economic preeminence, and seems to enjoy nothing more than to see us humiliated as a nation. The political situation quickly became unbearable. Mifune-san was discharged immediately, and most unceremoniously," Yoshi said, with bitterness in his voice. "His wife left him. There are some who say he was very close to taking his own life."

"Yet he didn't," Parks said.

"No, he did not," Yoshi answered. "But there would have been honor if he had."

"Some honor, if the disgrace is based on lies," Elise said.

Yoshi gave her a patient look. "I can see how you feel that way," he told her, "and personally, I agree with you. But you must understand our culture. If he had taken his life, there would be closure to the whole matter. Our country could have dismissed the entire episode as the workings of a single misguided individual, one who had paid the ultimate price for his brash actions."

The four of them sat quietly for a moment, deep in thought.

"You said the satellite is completed," Parks began. "Who finished it? How was it funded? What is Mifune doing, and how does he want me and Afren to help?"

"So many questions," Yoshi said amicably. "Good questions. The most important, I think, is how much we need the two of you. VSI has an excellent relationship with the Kenyan government. All three of you are well-known and respected. This counts for a great deal in the Kenyan culture. Parks-san, we see this being your primary role: interfacing with the various ministries, and also working with the director at SMER. Your company will also now be responsible for disbursing funds within this country.

"Afren, your excellent relationship with the men working on the rectenna, the pumping stations, the distillation vats and the power grid is vital to the continued success of the entire project. Especially after the abrupt discontinuation one month ago, there are certain to be some upset workers. We may need to employ even more people to make up for that lost time. For this, Afren-san, we really need you."

Elise frowned deeply at this, but Yoshi did not notice.

"Thank you for your confidence in me, Yoshi," Afren said. "But you must also realize that I have an obligation to the education portion of the Vambolent's work. The cooperation with the government is based in large part upon the promise to improve literacy throughout Kenya. Elise and I still have a great deal of work to accomplish."

Yoshi looked at Afren for a full five seconds. He blinked a lot.

"We may be able to provide you with a faster means of dividing your time between the capitol and the coast."

"Like a corporate jet?" Elise asked.

Yoshi looked hard at her. "Yes," he said simply.

Elise's eyebrows raised in surprise.

"So, the other part of my questions," Parks said. "What exactly is Mifune-san doing. And how is he funding all this?"

"Mifune-san, I am happy to say, is now working as a consulting manager. He and I have divided the work into small pieces, and we are subcontracting the work to a number of defense and aerospace firms in Japan."

Afren cleared his throat.

"I take it that you are doing this to keep the work a secret?"

"Yes, at least for a little while longer. Mifune-san has structured the finances so that Vambolent Space Industries will be the visible player in Kenya. To the outside world, the entire project will be run by VSI. The Japanese contractors will produce build-to-print components, and sell them directly to VSI."

"Wow," was all Parks could think to say.

"But as I say, the timing is very critical. We must finish the demonstration quickly, before the arrangement is revealed. For if it is revealed, the calls for Mifune-san's life will escalate. It is hard to know what may happen next. But unless we achieve success quickly, it will not be good."

"I understand," Parks said. "How much time do we have?"

"The original rocket for this launch has been re-assigned," Yoshi said. "However, Mifune-san has secured a prototype rocket, the Dart, to carry the power satellite into orbit."

"A prototype?" Parks exclaimed.

"It is not so dangerous as you might think. Engine tests have all been completed. Structural tests are completed. Sub-orbital sounding flights have also been completed – all with success. Mifune-san has arranged for the first orbital insertion test to be performed at SMER. The payload will be the solar power satellite and deployable microwave antenna, both of which are already finished, and are undergoing vacuum, solar and free-fall testing."

"Isn't that risky?" Afren asked. "Sending your only payload on rocket which is not fully qualified?"

"The risk is indeed a factor. But time is working against us. This rocket is needed for deploying geosynchronous satellites, and already has full bookings for next year. The prototype must be fired within a narrow window of time. Mifune-san has arranged that this time coincide with our original launch schedule. However, we have missed more than four weeks of productive labor. This time must be made up. If we do not have all the pieces ready in time, then the demonstration will fail. Who is to know if it will ever be tried again? Mifune-san has expressed to me over and over that we must not allow any obstacles to interfere with our goal. He is fond of using the Gene

Krantz saying: 'Failure is not an option'." Yoshi smiled, and his eyes all but disappeared.

"So, let me get this straight," Elise began. "My husband's company is to buy all these parts from Japan, he is to hire hundreds of new workers, and you've basically offered to pay for a corporate jet for Afren. And this is all after Mifune was fired from his post, and all but exiled from the country. Where is he suddenly getting all this money? How does Parks know this isn't tainted money – from drugs or embezzlement or something?"

"Of course, all Mifune-san's funding is legitimate," Yoshi said in a raspy voice.

"Do you know that for a fact?" Elise asked. "Or, is that your loyalty speaking."

Yoshi's face turned a ruddy red. His jaw muscles clenched.

"Elise does not know Mifune-san personally," Parks said, trying to defuse the situation. "But she does ask an important question. This is all a bit irregular, Yoshi. For example, I have not heard from Mifune-san since they day he was fired. When he was head of MITI, our funding was bare-bones. Now, it sounds like we have carte blanche, yet we have no idea where the money comes from. Naturally, you can understand we are cautious. Even anxious. What if we are helping to set up Mifune-san for an even greater disgrace?"

Yoshi's posture became stiffer. His cheeks relaxed all the way, and his black eyes were fully-open for the first time this evening.

"I see," was all he said.

Afren spoke.

"Yoshi-san, you recall the questions little Mindy Vambolent asked? Your answers were most intriguing, especially about the sunken ruins. Civilizations rise and fall; cultures come and go. In the distant past, perhaps it was natural forces that brought ruin to a society. Within our written history, we see kingdoms toppled through corruption or cruelty, revolution or war. And now, in our modern world, we possess weapons of mass destruction, the very things Mifune-san was accused of developing. These weapons enlarge the scale and scope by which a nation, or a people, can be subjugated or destroyed. Something that has always bothered me, and I have shared this many times with Parks, is the abuse of power by a few. True, this project is for a few megawatts, yet it is only the first step towards almost unlimited energy. Energy which can be controlled or manipulated by a very few. As Lord Acton wrote: 'Power tends to corrupt, and absolute power corrupts absolutely.' So, with this Energy Project comes a great responsibility. What I believe Parks and Elise are asking is, now that MITI is no longer involved, *who* provides the money, and, perhaps more importantly, *why*."

Yoshi sat back, absorbing all this. Elise refilled tea and coffee, and Parks unloaded a fresh bottle of sake for their guest. They all drank.

"Mifune-san is very interested in philosophy now," Yoshi said, his tone and body language softer now. "In our work together, and the few times we've shared a drink after a hard day, he has told me a great many things. Things I did not know about before. I do not know if they are all true."

"What sort of things?" Elise asked.

Yoshi looked at her a moment, then spoke to Parks and Afren.

"These are the kinds of questions Mifune-san asks now: What if human civilization is not just ten thousand years old, but much, much older? Consider the possibility, and many oddities resolve themselves. Suppose there was a nation of scientist-philosophers, and they harnessed great energies drawn directly from the Earth. But in their zeal, they triggered a shift in the planet itself. Some continents were raised up, like the seaport in Central America two miles above sea level. Other continents dropped, like the walls seen under the Caribbean, the roads seen connecting islands in the Pacific, or the ruins off Yonaguni island. Maybe there was a world-wide flood, like your daughter has said, but maybe it was caused, not by natural forces, but by the hand of man?

"The civilizations in our written history can then be seen as fragments thrown off by the widespread destruction of a great civilization. The common language elements and building styles around the Atlantic ocean may have their origins in the lost society of Atlantis. Atlantis was first written of by Plato who learned of it through the Ancient Egyptians.

"Now, imagine that reincarnation is true. Then, many of those living today are the same men and women who walked and worked back then. What if some of them caused the destruction of their entire continent, and much of the world's population? And what if they are following the same course even now?"

"What do you mean?" Elise asked.

"This is what Mifune-san talks of now. A cabal of the elite who have gathered together, to learn from their earlier mistakes, and are set up to conquer the world again."

Elise leaned forward.

"And you think this is who is funding the Energy Project?"

"A private consortium has hired Mifune-san to continue the Energy Project. Where they get their funds is not something he has shared with me."

"But why?" Elise asked. "Why fund this project?"

Yoshi looked insulted. "The project has many advantages. I assume the investors are pursuing it for the same reasons that Mifune and your husband started this in the first place."

"That's not what I mean," said Elise, but looked confused. "I'm not sure what I mean."

"I think I see what you are after, Elise," said Afren. "Perhaps these people who are now funding the Energy Project are already very powerful. They may have interests in oil and nuclear power now. In addition to the humanitarian and environmental reasons for developing space solar power, they may see this as a way of extending their monopoly into the future."

Or," said Parks, "they could be setting us up to fail, to protect their existing interests."

"I do not think it is like that," said Yoshi. "From what I hear Mifune-san say, these people believe they are doing good. They want to run things because they see, like all of us do, the terrible mess we are all making of our world."

"So, they want to bring peace to the world through energy technology," said Afren. "Sort of like a benevolent dictatorship, but with no visible dictator."

Yoshi shrugged.

"What if it really is power they're after – and I don't mean electrical power," Elise said. "What if they want to use it to rule the world, and there really is some truth to the accusations leveled at Mr. Mifune?"

"Oh, my God," Parks said. "What if they want to use this technology as a weapon?" He looked at Afren, recalling the warning his friend had issued the first day they met. A shiver went up and down his spine.

"We have always planned appropriate precautions," Yoshi reminded him. "The control loop on the beam will be failsafed to prevent accidental, *or intentional*, drift of the beam."

Elise's face was deeply frowned, but she said nothing.

"You already know how *I* feel about this," Afren said to Parks.

Parks took in a breath and let it out with a heavy sigh. "What you describe has some appeal. You've asked us, as Vambolent Space Industries, to help the Energy Project continue. I have already told you of the opportunity that awaits me back home. Elise's and Afren's project is important as well, and takes up most of their time. If we gave up on those projects and worked for you and Mifune, what assurance do we have that the project won't be cancelled again? How do we know he can keep all this a secret long enough to succeed?"

Yoshi had no response.

"Yoshi," Afren said, rising, "why don't you give us some time to think over your generous offer." Yoshi rose also, as did Elise and Parks. "You realize that we are all rather surprised at this sudden turn of events."

"Of course," Yoshi said. "I hope I have not been too pushy."

Afren responded to the pun, and pushed on Yoshi's thick shoulder. The two half-grappled, but neither had the heart for it.

"I've still got the videophone," Parks said. "We'll talk. If there's any change, we'll call you right away."

"That reminds me," Yoshi said apologetically. "I need to retrieve that phone."

Parks fetched the phone, and they exchanged weak goodbyes.

When Yoshi had gone, Parks turned to his wife and his friend. He shook his head, and saw agreement mirrored in their faces. The three spontaneously hugged, a triangle of shoulders and elbows.

"I must go also," Afren said. "Good bye, my friend. I will take excellent care of your wife. Best of luck to you back home."

He left, leaving Elise and Parks standing in the foyer holding hands. They both heard a soft scurrying sound, and instantly suspected that Mindy had been eavesdropping. They looked at each other with concerned parent gazes which quickly faded to mirth. They laughed and went to their bedroom hand-in-hand.

Jomo Kenyatta Airport

Parks moved slowly in the queue to the ticket agent. The Jomo Kenyatta airport was smelly and noisy. Animals squawked from cages. Colorfully-attired families pushed elders in wheelchairs. Solitary business travelers with rumpled suits carried careworn leather valises. He pushed his own two suitcases across the gritty floor and fanned hot air across his brow with a curled magazine. In his breast pocket was a packet of tickets that promised him passage to Dulles International Airport, where he had arranged to meet with Dajwal Chellapilla. To hedge his bets, he had also set up an appointment with Winnimer Jenks the following week in Los Angeles.

Back at their rented house, the goodbyes had been teary. Elise knew it was the beginning of the end of her work in Kenya. Mindy cried silently, but so hard that Parks worried about dehydration. Her eyes were red limned, and she hugged him fiercely again and again. Eric hung back from the emotional display of the female Vambolents, giving his father a perfunctory hug just as the airport taxi pulled up in the street. Parks promised the two kids over and over that he would see them very soon, but until they were together again, he would call every day. His own voice was husky with emotion, but he acted strong to reassure them all.

At last it was his turn at the ticket counter. His bags were inspected, weighed, tagged, and placed on a large cart for loading. Parks was issued a boarding pass and directed to the security gate.

He waited another long while for the metal scanners to probe him and the X-ray machines to disgorge his carry-on. Once through, he moved on to the Customs gate, the final check before leaving the country, probably forever. While the momentum of renewed hope had carried him this far, Parks began to have doubts if he was pursuing the right course. Leaving his family in a foreign country was probably not a good idea. Elise was still working on the projects he had promised to President Mtongwe, but his part was through. He wondered briefly if he should have informed someone he was leaving, but figured the President would be too busy to bother.

The officer on duty, seated behind a glass partition, waved Parks forward without looking up. Parks strode the two giant steps from the faded yellow line to the little booth. He laid his boarding pass and opened passport on the worn countertop. The agent flipped open the passport booklet, typed Parks' number into the computer.

Only then did he look up at Parks, while he waited for the computer to confirm the information.

Looking back to his monitor, the Customs agent frowned deeply. He typed in a few more commands, then leaned to the side and reached a hand under his desktop. Parks felt a chill run up his spine, moving in counterpoint to the sweat dripping down it.

"Just a minute, please," the agent said, and continued to stare at Parks full-faced, wary and alert.

"Is there a problem?" Parks asked.

The agent gave a shrug, but did not turn his heavy-lidded gaze from Parks. Parks noticed that the agent's left hand rested on the butt of his holstered gun. The pace of his heart quickened.

A sharp clip-clop of boots on linoleum caught their attention. Parks turned to see two uniformed policemen striding briskly towards them, their hands also on their holsters. The crowd parted as the two men strode directly to the Customs booth. One stopped two arms lengths away from Parks and fixed him with a hard stare. The other bent around the glass partition and conversed quietly but urgently with the Customs agent.

Parks had read about such incidents in the travel guides. He reached casually for his wallet, with the intent to fold up some American dollars to smooth his way through Customs. But the one policeman stopped him with a vigorous shake of the head. Parks began to feel light-headed, and his knees felt weak from standing.

The other policeman was on the telephone now, leaning far into the little booth. He kept looking out at Parks as he listened. When he hung up the phone, he thanked the agent, and then addressed Parks.

"You will come with us now," he said.

"Why?" Parks asked, his voice rising in pitch. "What's the problem? What about my bags? What about my flight?"

"You are not permitted to leave the country," said the policeman.

Parks nearly collapsed from anxiety. But he did not have time to reflect.

"Let's go," said the policeman. The two of them took up positions on either side of Parks and escorted him back out of the airport, past security, past the lines for the ticket counters, past the stares of the curious, the sympathetic, and the judgmental. Perspiration ran freely from his hairline.

Outside, the policemen put him into the back of a baking-hot squad car with a rusty wire mesh separating him from the front seat. They got in, and then sped off, dodging the trucks, taxis, and bicycles on their way out to the main thoroughfare. Parks barraged them with questions, but they did not answer. When his voice got too aggressive, he received a hateful glare. He stopped talking, but looked around desperately. Parks grabbed for a door latch, considering a getaway, but there was no handle on the inside.

Further along the street, directly ahead of them, Parks could see a double fence topped by razor wire. As they pulled alongside the fence, he could see the fortress behind, and a huge carved sign over the façade: Nairobi Prison. His heart began to yammer in his chest. His muscles tensed, and his bowels simultaneously relaxed. The squad car pulled up to the gate, and they were waved in by guards with Kalashnikov AK-47 rifles.

Parks watched as his passport was handed to the man in the admitting cage. Nobody said a word as the new prisoner was processed. He stood heavily between his captors, aware of a half-dozen urgent bodily needs.

"Can I make a phone call," he blurted out. His voice was unnaturally loud, and a hush followed in the background din.

The man in the cage laughed, hard and loud, and the sound echoed through the hallways. Parks felt his vision narrowing in, and felt too large He had to sit down. Sensing him slump, the police on either side pulled him back upright.

Before his parents had been killed by the Kyoto Dragon that his father named, Parks' mother had been teaching him the power of intent. "Intention," she would say, "is the prelude to action and the pathway to the future." Although trained as an engineer herself, she told him "The laws of physics only apply in the absence of intent." As a skeptical teenager he dismissed her woo-woo ideas, but after she died he had been unable to positively dismiss them.

"Mom?" he said to himself as he was guided to the intake room. "I'm intending a positive outcome."

"Shut up," yelled a large man with a bald head. He sat Parks down in a plastic chair, and brought an oversized clipper to his head.

As his own head was shaved, Parks forced himself to remember how she told him it worked. "Fake it 'till you make it," she would say with a smile. He could even see her face in his mind's eye. All other sensory awareness dimmed. The barber pushed his head side to side, shearing off every last hair as he placed utter confidence in a positive outcome.

He was next guided to a shower room where antiseptic chemicals pierced his nose. A guard ordered him to remove his clothing. Parks moved slowly to comply, struggling to remember what else she had said. *Oh, yeah*, he remembered: "Be specific."

Parks calmly stripped naked, holding in his thoughts the image of an office at the Purdue Armstrong building, applying for and receiving NASA grants. He could not get much farther that that before he was pushed to the center of the room and sprayed with pungent insecticides.

"Believe that it is already done," he heard her voice say in his head. He was thrown a towel, and a prison uniform was placed on a shelf at the far end of the shower room. There were no shoes. Indiana felt too remote, and he could not hold his thoughts.

Toweling off, pressured from behind by the barber and a policeman, he thought of the rectenna they could be building right now on the Kwapili lands. Even that would be preferable to this! Maybe, once he got a lawyer, he could get their permission to work on the Energy Project from here?

Pulling on the too-short uniform, he felt the rough grey linen scratch along his scoured skin. The crotch was tight, and he swallowed down the bile that threatened to rise from his gullet. He saw the rectenna growing into a vast octagon. As they pressed him through the far door, and through a guarded gate into the cell floor, he saw a wide, gentle beam of microwaves cascading down from a solar power satellite. All around the rectenna life and energy blossomed.

A cell door rolled open and Parks was thrust roughly inside. He stood in the center of the room. Every space along the wall was taken, and men stared back at him with hard expressions. He turned, indignant, to the barber, and was hit in the face. He didn't even see it coming, didn't have time to react, and merely watched his body fall backwards and collapse. His tailbone hit hard on the cement floor and pain spiked up inside him. The image in his mind focused the microwaves onto an exact spot, the top of the barber's naked skull, incinerating him from the top down, until he was nothing but a pool of burbling smudge on the ground. The microwave returned to its benign spread, bringing life again, and a smile smoothed Parks' pained face. The door slammed shut and they left him.

Pulling his legs in against the agonizing protest of his hammered hip bones, Parks sat Indian-style and meditated. His cell mates hardly stirred except for their constant coughing, spitting, and moans. Flies landed on his face and walked all over it, but he commanded himself not to bother. Their too-many legs converged to his moist eyelids; their wings tickling his eyelashes. Others explored the open collar of his linen garment, blindly searching, tasting, walking, turning, leaving God-knew-what on his skin.

As his suppressed intolerance gained energy Parks visualized an L5 installation of his asteroid separator swallowing up a small asteroid. The shell of a space habitat formed in the harsh black/white radiation of the sun. Electromagnetic catapaults shipped solar panels to GEO where his payload catcher slowed each canister and steered it into a stable orbit where astronauts assembled huge solar farms. Sitting uncomfortably on the floor, aware of moisture starting to wick up through his garment, he focused hard on the station keeper apparatus which kept the energy farms turned to the greatest power source man could ever hope to tap. Solar-powered lasers blasted slag targets made by the asteroid separator, and used the reaction force to steer the panels, nudging them every half hour to get the maximum dose of solar radiation.

Then, he could take no more.

Parks jumped to his feet and began slapping at his face, brushing bugs from his neck, and hopping on one foot to favor the leg which wasn't tingled with numbness. Hands reached out to prevent him from falling on them as he went bouncing around the room, now crying out in frustrated impotence. He rushed the gate and slammed into it, gripping the bars with white-knuckled fists…

And came face to face with a tall man on the other side.

"Parks Vambolent, I presume," the man said in carefully-voiced English.

Parks stilled and calmed himself immediately, blinking as he fought to recapture his composure.

"Yes?"

"My name is Jezri LeFallon, Special Assistant to President Mtongwe."

Parks immediately brightened.

"Have you come to get me out?"

The man regarded Parks silently. A reproach. Parks sized him up. LeFallon was tall and thin-faced. Curls of ivory highlighted the mostly-dark hair cut close to his scalp. He wore a Western business suit with a pouffed silk cravat over a white shirt that protruded from sleeves held with gold cufflinks. The man's wingtip shoes reflected a gleam of light that was definitely not coming from within Parks' cell.

"You have been detained on a work permit violation," the man explained. "And if I may anticipate your next question, you must know that in Kenya, local laws are not overridden by consular assistance." His clean-trimmed eyebrows raised up, and the folds of evenly-brown skin in his forehead became perfectly parallel black lines.

Parks digested this information, disciplining himself not to fight, but to flow.

"Okay," he said. "Then why are you here?"

LeFallon narrowed his eyes briefly, then turned away.

"Wait!" Parks exclaimed. LeFallon beckoned to someone out of Parks' sight, then turned an aloof face back to the prisoner.

"Sorry," Parks said, quieting down, making himself cooperative and deferential, relaxing the body language he was giving off.

A guard came, and opened the door. His waved a thick truncheon menacingly at the remaining prisoners, letting them know he would be happy to put it to use.

Parks stepped free of the cell and reflexively drew in a deep breath. He felt small next to this man who walked tall, proud, and slowly along the cells. Parks followed behind, with the guard's black wooden baton pressed into the hollow at the base of his skull.

Through bars, gates, doors wth reinforced glass, then into an interrogation room, Parks tried hard to make sense of these events. His intention thoughts were scrambled and his emotions warred with his intellect. An overwhelming need for a water closet assaulted his body.

"May I use the facilities?" he asked.

LeFallon ignored him and sat down at the metal table, placing his palms flat on the edge. Parks sat also, grimacing at the pressure it put in his low abdomen. A loud grumble was heard in the room, and Parks realized it came from his own plumbing. He sat straighter in the heavy wooden chair. His face was tight.

"We were rather expecting to hear from you, without resorting to this." LeFallon indicated the guard with a nod of his head He watched Parks expectantly.

"I, uh… I guess you would have heard the news about Mifune and the MITI cancelling the project and assumed the deal was off," Parks said.

LeFallon watched Vambolent impassively for a moment, perhaps waiting for a better excuse.

"Well, Mr. Vambolent, President Mtongwe made a deal with you, before there was Vambolent Space Industries. It was not with MITI. Japanese decisions have no bearing on your compact. We *assumed* you must have been working on an alternative, and that you would contact our office as soon as you had one ready."

Vambolent suddenly felt ashamed of himself. Although LeFallon's voice was not scolding, he felt severely reprimanded. Reneging on a deal with the president of a country was not a matter to be easily dismissed. His mind raced with excuses, rationalizations, quickly-devised clever answers, but all of them fell short.

"So," Vambolent guessed, "it was you who put a hold on my passport, right?"

"Yes," said LeFallon. A bent smile creased his face.

"Does Mr. Mtongwe know," asked Parks in a small voice.

LeFallon shook his head slowly no, making perfect eye contact with Parks the whole time. Parks let out a shuddering sigh that made his back shivver. A small cramp appeared in a deep part of himself. He willed it into submission, but it was not going away easily. He knew it was the first of a series of increasingly urgent signals as his body reached its upper limit of having normal functions ignored. His relief over the president not knowing of his intention to flee eased the pressure, and he blinked a few to reconnect his thoughts.

"Then I still have a chance?"

LeFallon regarded Parks through narrowed eyes. "You are in a local prison. I am a federal official. I have no authority here. Perhaps you have lost your last chance."

Parks suddenly felt like he was going to let go. *I have one last chance*, he said to himself.

"What can I do?" he said. "What can I do to make it all happen, to make it all right?"

Jezri LeFallon relaxed his squared shoulders, and shifted to a more comfortable seating position moving his chair away from the table and crossing one leg over the other. He folded long fingers together over a prominent kneecap.

"My office has been in negotiations with Mr. Hanazawa and Mr. Mifune," he said.

"I see," Parks said.

"Do you? Then you know they now work for an aerospace holding firm."

"Yes, Yoshi told me about that just last night."

"And did he also tell you they have been negotiating with us, in secret, for some weeks?"

"No, he didn't mention that at all. He just said he wanted to hire me and Afren. He made it seem like it was a sudden decision."

Parks waited for further dialog from LeFallon, hoping to advance their discussion so he could be allowed to use the facilities. No answer was forthcoming and the two men stared at each other in silence for several moments. Finally, Parks figured out that he was to figure something out. What?

"Was it your idea to have them contact us?"

LeFallon smiled in return. Then he indicated Vambolent to go on, gesturing with his head.

"But why?"

"Mr. Mtongwe expressed to me a certain liking for you, Parks Vambolent," LeFallon said. But the look on his face expressed a dubious curiosity. "He also is keenly aware of the impending deadline on those initiatives upon which his presidency is contingent."

Parks rememberd how easily Yoshi had agreed to a private plane for Afren to commute between the capital and the coast. He and Mifune were obviously well-funded. He looked more closely at Jezri LeFallon. A simple bribe would have been expedient. They didn't need him at all.

"Are you friends with Afren also?" Parks guessed.

"Indeed I am," said LeFallon. "In fact, Afren and I competed on the same team at the Olympics when he won the gold. I won the bronze," he added, then sighed theatrically. "Alas, no one remembers but my own family."

"Congratulations. On the Bronze Medal, I mean," Parks said in a rushed voice. Another cramp seized up inside him, and his face grew taught.

LeFallon dismissed it with a wave of his hand, but then said, "It was a Japanese runner who took the Silver."

Vambolent's eyes showed his sudden comprehension, and a smile softened his stony facial muscles.

"So, you're giving me a second chance," Parks said.

LeFallon gave him an appraising look. "Let us simply say that the Energy Project was proceeding well when you were in charge. It was you who forged a deal with the Kwapilis. It was you who brought Afren on the job, and he who helped inspire the workers," LeFallon said. "And who do you think insisted that an *American* company lease the Kwapili lands?"

"You?"

LeFallon nodded slowly.

"So, you've been holding out on Hanazawa, waiting for me?"

"I had sincerely hoped that you would contact me first," LeFallon said, feigning a pained expression.

"I'm sorry," Parks said, and meant it.

"Your apology is accepted. So you will stay?"

"Do I have a choice?" Parks asked *sotto voce.*

LeFallon shook his head loosely side to side. Parks noticed the man's ribs convulse in a suppressed chuckle. *He's enjoying this*, he thought. He remembered Chellapilla's description of the Seven Dwarfs project. In his mind's eye, he saw Japan and the US vying to have the first Hydrogen Economy. And here was the special assistant to Kenya's president putting his own hand into play, moving the few pawns he could control. Parks felt very small and insignificant.

"Then, I will accept Yoshi's offer. Afren and I will help them complete the Energy Project." Parks pushed his chair back and implored, "Now, can I go?"

"There are a few more conditions," LeFallon said, slowly.

With a visible effort, Parks clamped down on muscles that threatened to go involuntary. "Yes," he said – it was a squeak.

"The launch facility in Malindi has experienced a drop in bookings lately. In our early meetings, Mr. Hanazawa mentioned the desire to move his launches elsewhere, since fuel costs are slightly higher when they liftoff from SMER. The launches from Kenya are important to our economy as well as to our national pride. This will be a requirement of my arrangements with Mr. Hanazawa. We would like you to visit the faclity and ensure that they are prepared to launch the Japanese rocket. Then, you will make a favorable report."

"Fine," Parks said in a tight voice, nodding rapidly. His eyes bulged slightly.

"And one more thing."

"Sure, anything!"

LeFallon unfolded his legs and leaned close to Parks. The man's genial eyes became blazing coals.

"If you try to leave again, before you are finished successfully, you will be a second offender of a work permit violation. You will be taken to Meru Prison. *Nobody* gets out of Meru Prison."

Parks sprang for the door, which the amused guard pushed open ahead of him, pointing a finger down the hallway.

CHAPTER 5

East Kenyan Coast

Parks and Yoshi stood in the harsh wind overlooking the microwave rectenna. With the footings and support posts already installed, assembly of the latticework was a simple, but labor-intensive task. Through binoculars, Parks could see individual heads of the workers all along the unfinished edge. Day by day it advanced across the land. The perimeter power rail proceeded along segmented arcs, with bends at 135 degree angles, forming a gigantic octagon on the steppes.

"It is beautiful," Yoshi said.

Parks looked at Yoshi admiring his domain. The man's eyes were hidden inside folds of skin – a sure sign of his joy. "Yes it is," Parks said. "The progress is very good. You are ahead of schedule, Yoshi. I'm quite impressed."

"Thank you very much."

Although the massive antenna would be a microscopic target from GEO, the vastness of the array was a wonder to the increasingly-frequent visitors. President Mtongwe and Jezri LeFallon came to review the progress with Vambolent, Hanazawa and Kwapili prior to bringing foreign investors. The next week they brought two carloads of Indonesian businessmen wanting to establish an operation to mine and refine lead and zinc in Kenya. They had been especially interested in watching the laborers, and had come away impressed. They were clearly excited by the Space Energy Project and hoped to have a share of its future power expansion to supply the heavy need for electricity needed by their mining operations.

Parks almost put a hand on Yoshi's shoulder, then withdrew it quickly, not wanting to get into a judo hold. Judo practice was becoming as popular as the midnight drumming sessions at the worksite. At any random moment, someone would grab you and expect you to grapple back. Only he, it seemed, had no interest in the close contact of that sport.

"Yoshi, do you want to visit the launch site tomorrow?"

Yoshi looked over at Parks, his eyes as open as Parks had ever seen them. He wondered if this meant surprise.

"That is your task. I would not want to intrude."

"Are you sure? Afren is driving me. We've got room for you…"

Yoshi smiled gently. "I will attend the launch of the satellite. That will be enough for me. I am needed here."

Parks bade Yoshi goodbye and climbed back into his company vehicle. A newly-paved access road now followed the power lines, and he sped back to Mombasa where he kept his office and *pied-a-terre*. Driving over a sturdy cement bridge, he remembered the two workers who had perished near here in the washout. Scenes of the Chalbi desert arose unbidden and hovered in his vision like heads-up displays. His back turned cold and sweaty and the image of Horse's blackened hand filled his sight.

A huge bird flew diagonally across the road directly in front of him, beating its wings and turning hard to avoid a collision. Parks jerked the wheel hard, and his right wheels went off the roadway and started skidding down a ditch embankment. He jerked the wheel back left, and hopped up onto the road too quickly. The rover went up on two wheels, careening across the road, and headed for the drop-off. Parks slammed on the brakes and rode the car off the pavement. Grating rocks and skidding rubber brought him to a shuddering stop, where the only view over the bonnet the top half of trees in the defile. His heart was hammering rapidly against his ribs and his breathing was shallow.

I almost died!, he thought.

With his foot firmly on the brake pedal, Parks shifted into reverse and activated the four wheel drive. He backed slowly up to the road surface and put the vehicle in drive. As the immediate focus provided by adrenalin lessened, and his rational mind came back, he began to wonder what would have happened if he hadn't been so lucky.

The construction would proceed apace. Yoshi had done nearly as well as Afren before him. Turnover was low, and there were many new job applicants every week. Many of the workers had re-populated Joo, which had become a vibrant community. Afren was even elected mayor. He and Yoshi gave judo demonstrations and taught open-air classes that infused the whole area with a spirit of discipline, brotherhood, and funhousing. Yoshi was loved, and his people worked hard for him. Afren was adored, and Parks mused that they would probably die for him.

But how did they perceive Parks Vambolent, their nominal boss?

His official VSI title was President, and his duties involved receiving visitors, coordinating with government officials and regulators, and negotiating terms with his Japanese subcontractors. It was the sort of work Horse was born for, but to Parks it felt hollow. He longed for his old title of just Inventor.

The project had taken on a life of its own, and his own life had changed. The work took him away from his wife, his kids, his home country, and he felt pressed into becoming something outside his nature. But maybe he had planned this – long ago. The positive visualization of his mother's teaching might be working their slow magic.

Parks set the goal of becoming a billionaire with his space inventions after reading a shelf-worth of business books. He studied the world's greatest fortunes, trying to learn the secret to self-made wealth. At the start, he expected to find that the most clever inventors became the wealthiest capitalists. Instead, he found that hyper-intelligent, opportunistic gambler-types were more often the ones to profit on a new innovation. As an inventor, history was against him. Parks was determined to be the exception to the rule.

When he, Horse, Juneau and Austin has built up the business case based on Parks' patents and patents-pending, they were all certain it would be a winner. Winnimer Jenks had informed them, regretfully he said, of the committee's decision, but encouraged them to try the idea later.

His second encounter with Jenks, over the telephone from Nairobi, had not gone well. The man had brushed him off at first, before reluctantly agreeing to a meeting, even though Elise had the strong impression of great interest. Parks now realized that Jenks may have learned about the Seven Dwarfs mission about the same time he did, and dropped Parks' prospects like a hot potato.

But Jenks did not have Horse's business plan. Parks had retrieved it from Cloy Morgan's effects using his spare key, before the authorities could secure the apartment. The packet was stored on a memory stick that Parks kept in his wallet.

He had only reviewed it briefly. The only difference between what their team created and what Horse put together was the mysterious regulus. Parks was fairly sure this was wishful thinking, and gave the matter no more thought.

Being a billionaire wasn't going to happen, he now realized. Making his patents pay off was not very likely. Making his company into anything more than a cat's paw for Sugeto Mifune was also unlikely. Getting a job in Indiana with Dajwal Chellapilla was beginning to look more attractive. If there were federal funds available, he would be willing to do almost anything to stay employed. He resolved to flee Kenya the minute LeFallon released him from his duties, and never return. Just a few more months until the launch, and then he could quit.

Mombasa

Early the next morning, Parks met Afren at the vehicle depot inside their warehouse office building. Both were well-dressed, in accordance with their duties inspecting the San Marco Equatorial Range in Malindi.

"We have a passenger to drop off along the way, Parks. A crew supervisor. I hope that is okay with you. I'm sure he won't be any trouble."

"Fine, fine," Parks said, distracted.

A tall man, perhaps of Masai descent, walked forward and gave a silent nod to Parks. He climbed into the backseat and shut himself in.

Afren drove them out of the bay doors and picked his way through the crowded streets to the bridges that led to the mainland. Parks tried several times unsuccessfully to make conversation with the crew boss, but finally decided to ignore their passenger. He wanted to talk to Afren about his plans for the future.

"I've been thinking that this project is going so well, I might want to start focusing on the next phase."

"You mean, after the work with the Japanese?"

"Yeah. Once they have this pilot running, they won't need me anymore. In fact, they don't need me already." Vambolent stopped, expecting Kwapili to object. But Kwapili just drove on, listening. Vambolent could not decide if he was just being polite, or if Kwapili agreed Vambolent was indeed becoming unnecessary.

The silence stretched longer, and Vambolent tuned in to their backseat passenger. Nothing. He could just as easily have been a statue. Parks was beginning to feel as if he no longer belonged in Kenya. Then he began wondering when he had ever really felt like he *did* belong. Then, he wondered if this was how Elise had felt when *her* project was stalled.

But now *his* project was going great. In just two months, they would launch the already-completed SPS from SMER. He was on his way to hold a planning and inspection meeting with the Italian crews who ran the launch site. The visit was more of a formality, intended to establish good relationships for the multi-cultural team that would launch, fly and operate the powersat. It felt like such a waste of time.

Maybe he was afraid of his own success, and was resisting the new role that seemed to pull on him like gravity. *Maybe I'm just going crazy*, he wondered. He took a swig of water from a bottle and looked out the window. They were driving along the coast, and he could see a long stretch of white sand behind a row of palm trees. Deep blue waters sparkled all the way to the horizon. It was beautiful. He had been in Kenya almost five months, and had never visited the beach.

"What is troubling you, my friend," Kwapili asked at last.

His caring words were like a calming breeze to Vambolent's soul. He suddenly understood why this man was so well-liked. Or rather, he didn't really know *why*, he just knew. Parks began to relax, and unload his internal worries. Until they began pouring out of his mouth, he had not realize how tightly wound he had become.

"I've been thinking a lot about what comes after the Space Energy Project," Parks said. "This whole thing is bass-akwards from my original plan. What I really want to do is capture an asteroid, and start a materials base for orbital manufacturing. From there, we can produce space stations, orbiting hotels, communications satellites, micro-g factories, and eventually permanent human habitats. But the people with money and influence in this world are too short-sighted to share that vision."

"Perhaps they have a different vision," Kwapili offered. "Or, perhaps you have not yet found the person with the same vision as you?"

"Well, nobody at NASA seemed to listen. Nobody at SAIF or ISDC seemed to listen. The few calls I made to wealthy space advocates were fruitless. Finally, I got the Japanese interested, but they are involving me only on a very small segment of what I'm really after. I've made almost zero progress towards my true goal.

"This Space Energy Project, as Mtongwe calls it, or the Hydrogen Economy, as the Japanese call it, is just a stepping stone.

Somehow I've got to get a factory in orbit. Then, I can build an infrastructure, and create a positive cash-flow industry to serve a hungry market at a time when the world is falling apart because of the alternatives. In a way, I guess it's going pretty well. It's just a hundred times slower than I ever thought it would be."

"Maybe you are afraid of your own success. Subconsciously you are pushing it away from you," Afren said. "Is that possible?"

Parks stared at him with a face torn between insult and insight. "I don't know," he said. "I don't think that's quite it. But I just can't seem to put my finger on what it is." He looked straight ahead, frowning with concentration.

"Perhaps you are afraid of losing control?"

Vambolent's head whipped to the side as he stared at his friend. His throat clenched on a quick defense. A control freak? Him? Parks knew many people who most certainly were control freaks – his wife came to mind – but he had never considered himself one. Others had called him that in the past, but he dismissed the criticism as a projection of their own sense of frustration. On the other hand, nobody is successful unless thay take some control over events, right?

But what about Kwapili? Parks' friend seemed to find success without really trying. As far as Parks could tell, Afren didn't have a care in the world – except for the cares of other people.. *Maybe that is his secret?* Parks thought. But could such a man have started this whole project: the solar power satellite, the rectenna, the hydrolysis plant for making hydrogen? Parks decided, probably not. But then again, maybe the whole idea was *still* too far ahead of its natural time, as it had been for decades. Maybe he was pushing on a rope. Maybe all his personal energy was being squandered through poor timing and unrestrained impatience. He could be spending this time with his family. Maybe his proirities were all reversed.

But at the very least, he himself had started this whole project. His arguments crystallized the idea in Mifune's mind, and now it was a going concern. He had his own business, it was making money, and they were getting something worthwhile accomplished. That should be reward enough. Most people can't even claim as much. If the work swept away from him, and somebody else became a billionaire, humanity would still win out in the long run. And that was good enough for him.

Well, maybe. Almost. At least on the outside.

The Space Energy Project now had a momentum of its own. He could die and, except for minor inconveniences to his close friends, everything would still probably happen. Even the small consolation of having started and running a self-supporting enterprise was nothing more than hubris. He was little better than useless.

As Vambolent went through this logic in his mind, there was a sudden feeling like clouds parting before the sun. He suddenly saw his role as the visionary, the initiator, the early leader. Once a project got rolling, it was time for him to move on. He had known this at some level, but because of his up-bringing, he felt an innate need to see things through to their completion. Maybe that was the key to his unsettled feelings. He needed to shed a paradigm that no longer served him.

Suddenly, he felt a whole lot better.

He breathed deeply, convulsively like when someone sitting on your chest finally gets up and leaves. Drawing in the sweet ocean air, his mood rose and he felt simply wonderful. He looked over at Kwapili with a big smile on his face. Kwapili glanced over, and Parks' joy spread to his face as well.

The tall man in the back shifted in his seat, and began humming. It was a low rumble, at times indistinguishable from the road noise, but never intrusive or demanding attention. Parks relaxed a bit more, and began to enjoy the scenery. It seemed like just moments and they were in the town of Kalifi. Driving across the bay, they could see the VSI logo on the distant inner shore, painted extra large on the gleaming intake pump. It looked beautiful. The silver edge of the solar array glinted in the sunlight. The purple-black silicon cells were pointed straight up to capture the full strength of the equatorial sun.

They exited the highway on the underpass and drove on a gravel road around the bay to the worksite. The supervisor in the back opened his door and stepped from the vehicle almost before it stopped rolling, and with long strides joined a crew of similarly tall workers standing by the office trailer. They talked a moment, and the five men turned as a unit, and approached the car, smiling, chanting and humming, and clapping their hands. Kwapili turned to Vambolent.

"They are about to give you a great honor, Parks. Just sit and enjoy. There is no need for you to do anything at all, or to thank them." Kwapili wore a beautiful smile.

Vambolent watched them fan out around the Rover, and for a brief moment, had a flashback to his deadly encounter with Shifta raiders in the Chalbi.

From outside the vehicle, they sang, and then they began to drum. Each found a different panel of the car and, with their large hands, began tapping, slapping, and banging out a complex beat to their song. For Parks, it was like being inside a drum set, with the sounds completely holding him. Each beat, and each timbre of that beat, blended together until his head swam and he felt like he was floating. Without thinking, his body moved with the music, and his rational mind just shut right off. There was not a single thought in his head, only a hyper-keen awareness of the music. It filled him so thoroughly that time itself broke away, and he was adrift in a universe of emotion, feeling, meaning and experience. Only in the climax of lovemaking had he ever experienced a sensation to compare. As they faded their song, softening their drumming, coming to a conclusion, he felt a longing pulled up from his deepest gut. It rose like a physical presence coming up from his esophagus, until it burst like a bubble from his mouth in a shuddering, quavering sigh. In his imagination, this something from inside of him broke into fragments and scattered out the window, swirling and tumbling away in the heat.

Parks sat there, stunned. The analytical, civilized part of his brain wanted to thank them. But another part of him, one that was growing stronger and more in-tune with Africa, remembered what Kwapili had told him. So he just sat there, his mouth agape, and let Afren drive him slowly back out onto the highway, and north towards Malindi.

They drove for a while in a surreal silence. Parks watched the passing scenery like an awe-struck child on his first airplane ride. Every movement of his head and hands seemed slow and thoughtful, yet he drank in the sensory input so quickly, that time passed at a different rate. In the cab of a passing truck, he watched with utter fascination as the three men inside grow closer, closer, and then zipped by. Just before they vanished from his sight, he seemed to notice an infinite number of details from the interior of their highly-decorated cab. Even this thought came on top of fresh waves of input, awareness, and connection. He rode in this style until he felt filled up. Coming out of it, he turned towards Afren, who had been patiently driving along.

"What did they do back there?" Parks asked him.

"It was a Masai ritual, Parks. I have never seen it done with a European before."

"Why does everybody keep calling me a European?" he asked next, slightly peeved.

"Europeans conquered nearly all of Africa during the Colonialization period. After World War II, they divided up the countries between themselves, and most instituted white minority rule. Nearly all whites were from Europe at that time, so the term was appropriate. Now, though, they by and large recognize that you are an American, Parks. But in East Africa, at least, the term 'European' is equivalent to 'white man'."

"Okay."

"In a similar way, people of Middle Eastern descent, even those from India, are called 'Asians'. This has started to change as more and more foreigners come, but it may take generations."

"I understand," Parks said, "but it still irks me that an American could be lumped with a European. Sometimes it seems like in Africa they recognize so very many distinct tribes and languages and religions at home, but make such broad generalizations about the rest of the world."

Afren looked at him meaningfully. "Is it not the same where you come from? Think about it," he said.

"I suppose so. New Yorkers see the whole Midwest as a wasteland you pass over on your way to California. Texans act like they are their own country. Michiganders call Canadians canooks. Yet to the rest of the world, they're all just Americans. This is probably basic to our human nature. We become so absorbed in our own immediate environment – our families, our communities, our companies – that we imagine everything beyond to contain less detail, and therefore less significance, less tangibility. These feelings predispose us to perceive our own locale as superior, somehow, to others.

"In fact, I think it applies not just to geography but also to institutions. NASA ignored my proposals because I wasn't with a university or a national lab. The Department of Energy people only listened until I said the word 'space', and immediately told me to call NASA. Every legitimate researcher at conferences was more interested in bolstering their own reputation and research, even if a few of them did listen politely. It's like we're all bouncing off of each other, with the net result being a random waste of energy where no real work gets done."

Parks fell silent, musing on the chain of events what had brought him to Kenya. The Not-Invented-Here syndrome had driven him from his own country to Japan, and thence to Kenya. Then, had come the terrible ordeal in the Chalbi, where his entire party had been wiped out, leaving only him. These blows to his ideology, to his naïve assumption that the world really would beat a path to his door for his better mousetrap, had left him sobered, more detached from outcomes. He nearly fell into the traps he had seen snare an entire older generation of space enthusiasts – disillusionment and complacency. *It's our own internal barriers*, he thought, *that are the true obstacles to our success.*

"What are you thinking about, my friend," Kwapili asked him.

"I'm thinking about what happens next. Do you mind if I just ramble, Afren; and think out loud?"

"By all means!" Kwapili answered, taking one vital hand off the wheel to make a gesture of welcome and acceptance.

Parks sat up in his seat, and organized his thoughts. A slight smile creased his face, as he felt the old thrill of developing plans and making things happen.

"Let's start from the end and work backwards," he began, adopting the pedagogic approach that so annoyed his wife. He felt in his gut that Afren would not mind, so he dropped the thought and moved on. "The goal is to develop a viable living environment in space."

Kwapili interrupted him, gently, but unexpectedly.

"Why?"

Parks was momentarily miffed, but then decided it was a very good question. He also remembered Kwapili's stance, expressed in Nbele's law office on the day they first met, that new technology could be dangerous as well as beneficial.

"Well, look at the world around us. Pakistan and India have just started the first two-way nuclear war. The bombs have destroyed habitable lands for hundreds of square miles, and the fallout has poisoned the soil across the entire region. Terrorism has already killed I don't know how many people, from the World Trade Center attack to the bombings in the Gaza Strip, the nerve gas in the subways, and the anthrax in the mail. These overt acts of sabotage to the biosphere come on top of countless sources of pollution and destruction that grow ever worse year by year. Our ozone layer continues to deplete from fluorocarbons released decades ago, and are only now rising high enough in the atmosphere to break down that fragile molecule. The release of carbon dioxide from burning of fossil fuels is accelerating past the ability of plant life to capture it, and through greenhouse warming, the global climate that has supported mankind for at least ten thousand years is getting changed. The rise to affluence of cultures around the world increases the consumption of cheap beef driving third-world cultures to clear-cut old-growth rainforest to make grazing lands that will last a few years and then turn to desert. Their appetite for fish has us clearing the oceans of everything that swims, until that vast and ancient life cycle can no longer sustain itself, and we will lose that vital food source. Phosphates and chemical fertilizer runoff, together with warmer oceans, cause algal blooms that choke off oxygenation of the water, further distressing the fish habitats which have fed coastal regions for millennia."

"Have you heard of the Gaia Theory?" Afren asked.

"A little," Parks said.

"There was a professor in England, James Lovelace, who championed a theory that the Earth's biosphere is a self-regulating organism. Like other systems that can be considered 'alive', it has the ability to adapt to change. The global environment has degraded much more slowly than many scientists predicted just twenty years ago. But the Gaia Hypothesis explains this behavior. Environmental distress in one area is moderated by a balancing response in another area."

"I see," said Parks. "Yet, if the Earth is alive, can it also be killed?"

"Yes, I suppose it can, if the stresses are severe enough, or if the means by which Gaia maintains balance are damaged."

"What I see, Afren, is that our species acts as if the Earth can accept any insult and still sustain us."

"You are correct!" Afren said. "The capitalistic goal of continued growth, and unbridled consumerism, must ultimately meet some sort of hard limit. Politicians who only think as far as the next election and corporate CEOs who think only as far as the next year's balance sheet cannot see that, as a species, we are probably headed for some major disaster. It will take us all by surprise."

"So, that's the answer to your question of why," Parks said. "Industrial pollutants, agricultural chemicals, solid waste streams, and nuclear byproducts are soiling our own nest. Global warming is affecting crop production. And on top of all this, population growth is increasing the pressure for expanding all of this. It's a self-stoking cycle. More and more people are needing more and more land to live off. But this land is being turned into deserts, it is being poisoned beyond hope, and it is being covered with waste. As the climate warms, the polar ice caps will melt – they already have started – and that will raise sea levels, pushing more and more people onto less and less land. You see, it's all tied together, Afren. And we are very rapidly approaching the breaking point."

"I feel the same way, myself. When I was in England, my friends and I started the Green Label. It was a way for companies to mark products that were made in an environmentally-friendly way. It was a popular idea, and the media actually helped us. But the exact definition of 'environmentally-friendly' became a hotly-contested topic. The government tried to step in and help, but they got nowhere. Legislation was passed, but the courts soon became bogged down in lawsuits over the name, by a whisky company, of all things. That has been years ago, and now, I think very little is happening with it."

"That's just the point," Parks exclaimed, fidgeting in his seat after they hit a particularly deep pothole. "Nobody can seem to get anything done. There is nearly universal recognition that we have a problem. Everybody has a solution. But those most able to *make* the change seem to be those most eager to keep things the same!"

"So what can one man do?" Afren asked.

"Exactly! When I look at the problems of the world, it seems there is *one* problem that causes many others, both directly and indirectly. I've always been taught that to remove the symptoms, you must first seek the root cause. Once you have identified the true source of the troubles, and can take steps to eliminate it, or at least to improve it. Then the symptoms will finally disappear."

"And that one thing is?..."

"Energy."

"Energy," Kwapili echoed pensively.

"The Industrial Revolution, which began in England, was powered by coal. The spread of the internal combustion engine was powered by oil. With the invention of the dynamo, electrical energy began to change society. For untold thousands of years, global populations grew slowly, limited naturally by the availability of regional resources. Life spans were relatively stable. Technologies were simple and developed slowly. Mankind used local materials, for the most part, and pursued sustainable practices in their homes and their trades. I am oversimplifying a bit, but the point is that suddenly, a century and a half ago, the pace of progress began to accelerate rapidly. It has all been driven, at the source, by energy.

"With the light bulb, factories were no longer limited to daylight hours, and could now produce goods around the clock. With the advent of machines, productive work was no longer limited to the reach of a man's arm, or the strength of his back. With engines – whether steam, coal, diesel, gas, or electric – transportation and agriculture were no longer limited to the speed of a horse or the power of an ox. Energy became the lever by which the world was moved. Physics teaches us that energy, the ability to do work, is equal to power expended over time. With more and more energy, more work can be done in less and less time. The growth of industrial manufacturing and consumer-driven economies has multiplied on itself many times in just the last five generations. But the root of all this growth is harnessing and using energy."

Kwapili was fascinated by this explanation, and slowed the Rover slightly in his effort of concentration. They were nearly to Malindi, and he did not want the thickening traffic to interrupt the flow of conversation.

"Energy production, at least as we define it today, starts the whole cycle, and perpetuates it at every turn. From coal mining, we have strip mines which lay waste to valuable lands, and deep mines whose tailings choke river valleys with rock. Then, from the burning of coal, we get megatons of carbon dioxide in the atmosphere, adding to the greenhouse effect. We also get tons of sulfur and nitrogen oxides, mercury and soot. The smoke from burnt coal reacts in the clouds to form acid rain, responsible for decimating forests all over the globe. From oil and gas, we have terrible ecological tragedies, from the Exxon Valdez spill to the Deepwater Horizon leak. Fisheries and coastal habitats are destroyed indiscriminately. Then, the burning of these fuels produces even more carbon dioxide, and generates ozone and particulate contaminants that create the terrible brown smog that blankets major cities and poisons the lungs of our children. From nuclear power we have acid run-off from the extraction process, and then the power plant disasters like Three Mile Island and Chernobyl and Fukushima And don't get me started about waste disposal.

"And this just addresses energy *production.* If we look at the indirect consequences, there are obvious ones like nuclear bombs. Right now, there are probably teams of terrorists looking for ways to steal the plutonium from our reactor fleets. Waste heat is another one. Each and every Carnot cycle reactor, whether coal, oil or nuclear dumps huge amounts of heat into our waterways and into the air. The enormous evaporation towers you see around power plants are needed to cool the transfer fluid to keep efficiencies high. This waste heat adds further to the warming of our planet, kills off fish and wildlife, and encourages excess algal growth. Less dramatic, but no less dangerous, oil is used to make artificial chemicals like fertilizers, pesticides and herbicides that eventually run off from farmlands, and pollute streams and rivers until the fish are no longer fit to eat, and nothing at all will grow."

They were in Malindi now, and Parks felt like he needed to wrap this up, and he hadn't even talked about the next phase in his project. But pulling together all this material was the story he would be needing soon to convince people to help him. And he had a lot of convincing to do.

"Within our grasp," he said, "are technologies to produce energy without acid rain, without nuclear bombs, without hazardous chemicals and with minimal waste heat. Using energy from the sun not only avoids all these terrible byproducts, but it is completely inexhaustible on the scale of human history. It is abundant beyond imagination. There is so much energy available that we can even use the surplus to recycle the end products of consumerism, and create a truly sustainable ecology for our home planet."

"Well," said Afren, overwhelmed, "I've never heard this all put together so well, Parks. Have you been telling this to anyone else?"

"Afren, I don't think anyone will listen, nobody important anyway. None of this is new information. And I don't want to be another zealot crying 'The End is Near'. What *I* want to do is address the problem at it's source. And I'm afraid that, rather than changing the whole system, I'm going to need to find a way around it. I'm just starting to see a way to do that.

"NASA, to their credit, has been developing some great orbital technologies. Of course there's now an orbital hotel of sorts, but only for the very rich. What matters to me are the featherdrops, the habitat development, the micro-g manufacturing, and the translunar ship designs they have and are working on. I think I see a way to tie this all together."

"We are at the San Marco Equatorial Range, Parks. See the gantries, there, through those trees?" Afren said, pointing.

"Oh, yes!"

"Should we stop the car and continue talking? I am very eager to hear more!"

"Let's save something for the trip back," Vambolent said, buoyant from the discussion.

"Very well, my good friend. But do not keep me in suspense for long. Now, here is the gate. You have our letter of invitation?"

Vambolent produced a letter from his shoulder bag and handed it to Kwapili who in turn handed it out the window. The guard caught the letter as they slowed to a stop, greeted them curtly, and looked at them closely. He went into his shack and picked up the phone. After a few minutes of waiting, the gate began to open, squeaking on rollers in need of grease, and the guard waved them inside.

San Marco Equatorial Range, Malindi

Awaiting them at the administration building, the Range Manager, Dante Riolo waved them towards an open parking spot directly in front of the main entrance. Squinting to see through the dusty windscreen against the bright glare of the sun, his bushy eyebrows suddenly leapt upwards as he recognized Vambolent. He hustled to that side of the car, and opened the door.

"*Bongiorno*, Parks. How *are* you?"

"Hello, Dante. It's nice to see you again. I'd like you to meet Afren Kwapili, our Site Director for the microwave antenna."

"*Ben arrivato*," said Riolo quickly, paying little attention to the darker man. "Can I help you with your things?" he added to Vambolent.

Vambolent looked over at Kwapili as the Italian fussed. If there was any trace of resentment there, it was well-hidden. Vambolent hoped their host did not assume Kwapili was a mere driver. He hoped there would be an opportunity for him to explain how valuable his new friend had become.

Riolo was all abuzz with excitement and chit chat as he hustled Vambolent into the building. Blubbering about how delighted they all were to be helping with the 'Energia Project', as he put it, he virtually ignored Kwapili, who followed a few steps behind.

"*Vieni*," he said, muscling open both sides of the double door.

In the foyer, an elaborate sign said 'Benvenuto, Parks Vambolent', and Riolo introduced him with the honorific 'The Great' to the receptionist secretary. With an avuncular arm around Vambolent's shoulders, he guided them through the hallway, pointing out pictures of famous rocket launches, tracking antennas and the launch facilities that make up the San Marco Equatorial Range.

"And now, we have some refreshments for you – for you both," he added, half turning, "after your long journey this morning."

With a flourish, he pushed open the swinging door into the facility canteen. Assembled there were two lines of people, flanking a table which held a large frosted cake and a bowl of dark red punch. A banner read *Buon Compleanno Parks* behind the table. As one, the people all yelled "Surprise!" and began whistling and using noisemakers.

Parks was completely taken aback. It took him a moment to realize it was a surprise birthday party. Today was his birthday, a fact he had forgotten just moments after remembering it that morning. There had been no card or call from Elise, but he *was* expecting to hear from her that evening when they returned to Mombasa.

Still somewhat dazed, Parks was introduced to the line of people, all of whom were smiling widely, and greeted him warmly. Vambolent turned to introduce Kwapili, but his friend was hanging back, letting him have his day.

He was barraged by the new names, and forgot most of them except to note that they all seemed to end in 'io' for the men, and 'a' for the women. Antonio, Quinto, Paulo, Regalio; and Maria, Silvia, Rosa, and Daniella. As he moved down the line, becoming more and more numb to the individuals, Dante Riolo brought him to a stop before the last one in line, a woman with a red checkered kerchief on her head, turned away. Something about her stance seemed familiar. She turned to him.

"Surprise, Parks!"

"Elise! Oh, my goodness!"

Dante interjected his smiling face. "And this is Elisa Vambolento, the newest honorary member of the Green Team for satellite launch and tracking."

"Oh, Elise," Parks breathed, reaching for her. She fell into his arms, and hugged him tightly. The rest of the people oohed and ahhed at the tender scene. He kissed her, leaning her back. The men whistled some more, and the women sighed and clapped their hands with joy. It was just what they had all wanted to see.

Riolo, the master of ceremonies for the occasion, demanded a small speech.

Vambolent looked again into Elise's eyes, his love for her threatening to well up into tears, and turned to the crowd.

"Thank you all. *Grazie!* I am completely taken by surprise. It is so kind of you to honor my birthday. Thank you very much."

The group cheered again, this time mostly for themselves.

"You have obviously met my beautiful wife," he said, pulling her closer to him. They cheered for her. "And I would like to introduce you to my dear friend, and an invaluable asset to our project, Mr. Afren Kwapili."

He indicated Kwapili, and the Italians who ran the San Marco Range, turned to him and gave him polite applause. They seemed genuinely warm and welcoming. One of the closer workers reached out to shake hands with him, and they started to chat.

The party grouped around the table, and broke into excited banter as the cake was cut and served with punch. Riolo hovered around Parks and Elise, and would grab his people as they wandered by, and sing their praises to the two Americans. Marcus Lafontaine, he described, would be the lead tracker for the Japanese launch, and that he was a frustrated astronaut wannabe, and would rather be inside the rocket than on the ground. Lafontaine was sheepish at the story, but in his eyes burned a fierce desire that Parks instantly recognized. As his specialty was their main reason for making the trip, the two men began to talk technical details concerning the tracking coordination, and the challenges of antenna steering for the SPS once it reached GEO. Riolo was quiet for the first time, and nodded his head sagely as the other two talked.

Elise drifted over to Kwapili, and they pressed cheeks. As they began talking about *their* own project, and their plans, more of the Italians began gravitating to that conversation. Once Elise told them that Kwapili had a gold medal from the Olympics, an air of excitement around them began to grow. One of the female range technicians had read a news report about the marathoner who ran across the country promoting a new education plan. She was amazed that this very man was here in their midst, and began chattering excitedly. More and more of the tracking team congregated around the two of them, as Elise described what they were trying to accomplish, and how Afren was such a vital part of making it all happen. Soon, the whole room of people was surrounding them, except for Parks and Lafontaine, and Dante Riolo, who was starting to look jealous of the attention.

Lafontaine described to Vambolent how the vehicles they tracked usually launched from South America. The Japanese plan to launch from SMER and come to rest at GEO almost directly overhead was not the most efficient flight plan. This would require him to coordinate tracking with other equatorial sites around the world, but that was fairly routine. They had a flight plan already computed, validated, and archived for the launch. What concerned him most was the antenna tracking routines.

"This is what confuses me, Parks," he began. "How do you hit a point just a few kilometers across from 36,000 kilometers away. That is just less than a minute of arc. With a half second round-trip delay in the control loop, how is it possible to hold the target and not waver?"

"There are multiple control inputs," Parks answered, "from inertial measurement units with micro-g accurate accelerometers and laser ring gyros with micro-radian per hour offset drift. We'll have continuous triangulation from phase-locked radio sources, so we have two independent means for dead-reckoning. And the Japanese are using an adaptive fuzzy logic control system that continuously adjusts the satellite attitude, anticipating drift even before it happens. The eighteen infinitely-variable pulsed-ion drives are positioned to give six degrees of freedom. With this system, we can rotate through the minimum energy quaternion to zero-out any appreciable error. The simulations have shown a positional drift on the order of one ten-thousandths of a degree."

The Italian tracking engineer gave a low whistle.

"That is magnificent. But still…" he looked up and to the right as he calculated in his head, "that is approximately 100 meters of wobble. Am I right?"

"Yes, Marcus. Our design constraints are set at 150 meters, and we have several safeguards built in to shut down the source if the drift starts moving close to the margins."

"And these ion drives, they are very costly, true?"

"They are. And that is one reason why this first satellite mission has just a two year life."

Riolo re-entered the conversation.

"What will happen after 2 years?"

"Well, they could re-fuel, but that takes a manned mission. I hope by that time, we are moving towards something larger and more profitable."

"Will you be giving up this location in GEO?" asked Riolo.

"Actually, I hope to expand it. With a larger array in orbit, we can beam even more energy back to the antenna in Voo."

"So, you will have many more launches planned for our fine facility?" Dante asked eagerly.

"Perhaps," Vambolent replied, being evasive, since the true answer was 'no'.

Riolo looked at him closely, and waited to hear more, but Parks just smiled and held his peace. Riolo seemed to understand and nodded with some finality, pursing his lips and furrowing his eyebrows. Lafontaine did not seem to notice his boss's crestfallen look, and launched right in with more technical questions of Vambolent.

Riolo strode over towards the other group, announcing in a loud voice that the party was over, and it was time to work again. He adopted a jovial tone, saying how there was much to do, and too many birthdays. Vambolent could not help but notice the tinge of dismay in Riolo's voice. He excused himself from Lafontaine, and raised his voice to address the whole group.

"Thank you, all of you, for such a wonderful birthday celebration. I am very glad that you will be the ones to help make the Energy Project a glowing success."

Amidst cheers and waves the party broke up and people drifted out the doors carrying half-filled cups of punch, or slices of cake on paper plates. Most of them made a point to say goodbye to Kwapili, who had become something of a hero figure among them, despite having said very little.

Vambolent, Kwapili and Riolo met for another hour to review timing, resources, backup plans and mission coordination. Elise joined them, but brought a magazine, and tuned-out the entire conversation. When they finished their business, Riolo escorted Kwapili and the two Vambolents to the front door, and they said their goodbyes. Riolo was a bit less exuberant compared to their arrival. Still, as they drove away, Parks could see him waving his whole arm in parting.

Driving south through Malindi, Elise was wedged between Parks and Afren on the front bench seat. Her head was tossed back and forth as they bounced from one pothole to the next, but she seemed ebullient, and was wearing a relaxed grin.

"So, you two had this whole thing planned. Is that it?"

"Maybe," she said coyly.

"And you, Afren, did you plan out that little ceremony at Kalifi this morning?"

"Maybe," he said, using the same sing-song tone as Elise.

"And I suppose you have something else planned in Mombasa tonight?"

"Maybe!" they said, in unison.

"I hope I don't have to worry about the two of you," he said.

Elise mock-slapped his arm, and then leaned over to kiss him. Their easy smiles infected him too, and the three of them rode for a while in a happy quiet.

As they passed south of town, on the highway proper, Afren leaned forward and spoke to Parks.

"You were getting ready to tell me about the next steps in your master project. I think you could tell us now, on the way back."

"If it's okay with Elise…"

"Certainly!"

"Okay, um, where was I, Afren?"

"You had just explained how we are soiling our nest, and energy was the root cause."

"Yes. Not the energy itself, but the way we've chosen to obtain it. Energy is a good thing, a great thing – without it, life would be impossible. So, it is not the use of energy that is bad, but the lack of concern for consequences in obtaining it that has caused so much trouble."

"H.G. Wells once said," Kwapili interjected, "that we are 'in a race between education and catastrophe'. If we could teach people of the consequences of their actions, I believe that their actions would change."

"I agree with you Afren. But I'm making it my goal to give them something to change *to*. It's like what you and Elise are doing to educate Kenyans. We had to first develop a new education system, and *then* get them to adopt it."

"I see your point," said Kwapili.

For the first time, Vambolent felt as if he had said something to make Kwapili think twice. In the months they had worked together, it seemed the erudite Kenyan already had his own thoughts figured out in advance, his philosophy secure, his viewpoint unwavering. Vambolent began to feel a glow of excitement; that if he could win this one man over to his vision, then he could convince almost anyone.

"The basic idea of beaming solar power down from space began in the late 1960s, by a consultant at Arthur D. Little corporation," Parks began. He sensed Elise tense up at his professorial tone, and decided to change his delivery style.

"Anyway, the idea never got off the ground because of launch expenses. Building the solar cells, transformers, antennas, station-keeping rockets, and all the other odds and ends is relatively easy and fairly cheap. It's the launch costs that scuttle the whole equation. Boosting all that hardware into space is horrendously expensive. After paying for freight, the cost per kilowatt-hour of electrical power you return is much more expensive than coal, oil, gas, or nuclear, and even more expensive than other clean sources like hydroelectric and wind power."

Elise asked: "But if everybody knew how much better it was, wouldn't they pay the extra for energy from space?"

"Elise, you know how corporations are," Kwapili said. "To them, profit is king. They want the cheapest power they can find, and they want it right now."

"But what about people, in their homes. Won't they pay more, if it's clean?"

"They might pay a *little* more," Kwapili answered her, "but not much. I think Parks is telling us that at first, at least, solar energy from space would be *a lot* more expensive, is that right Parks?"

"Quite a bit. Even with nuclear power costs starting to rise from the UN initiatives, and even with crude oil prices ratcheting up each year, there is still a long way to go before the *traditional* solar power satellite, or SPS concept, will be cost competitive."

"And your idea solves that problem, right, dear?"

"My invention addresses one of the two big technical hurdles: how to transform space rocks into silicon solar cells and then transport them around cislunar space. The other one is not so easy."

"That is," Kwapili offered, "how to capture the asteroid in the first place."

"Right," Parks answered, and started talking about his plans for lassoing an asteroid and returning it to Earth orbit. He first talked about the NASA approach being used for the Seven Dwarfs mission, using electric catapults to throw back bits of rock, and thereby steer the asteroid. Then explained his own proposal.

Elise had tuned him out, and he got several sentences into his monlogue before her thoughts gelled into a cogent question. She interrupted him right in the middle of his speech.

"Why not use the Moon. It's just a big asteroid, right? And it's already here!"

Parks, who disliked being interrupted, stopped dead. He was hoping to talk through the issues with Kwapili, and was resentful of what seemed an obvious question. Taking on a measured, somewhat condescending tone, he gave her the basic physics.

"Elise, dear, you're right, the Moon has plenty of silicon and aluminum for an SPS. The problem with the Moon is its gravity. While it's not nearly as strong as the Earth, the launch costs are high but for a different reason. There are no rocket fuels on the Moon. So, fuel would have to be shipped up from Earth, at exorbitant cost. You end up with the same problem."

"But what about those electric catapults, like NASA is using" she said. "Can't you just use your asteroid separator on the Moon, and then throw the stuff up into orbit?"

"Yes, Elise, that's a good point. It is possible, and that's actually been proposed by Gerard O'Neill back in the 1970s. But there's another big issue. The Moon has the same problem that the Earth has – nighttime. For two weeks out of every four, any point on the Moon is dark. So all the people and equipment you need to operate the facility would be idle, just consuming valuable resources without producing useful output. The economics are marginally feasible, but the other logistics make that a nightmare." He was aware that a nuclear power plant on the Moon could solve that problem, but the prohibition of nukes in orbit was not likely to lift anytime soon. Especially not after what had happened in Kashmir.

Parks picked up his earlier discussion where Elise had interrupted, about using explosive charges to steer asteroids. He leaned forward now, talking past Elise, and directing his voice towards Afren. Again, Elise interrupted.

"Why not just put rails on the Moon?"

"What?"

"Rails, like a train. Then you could follow the sun."

Parks, exasperated, opened his mouth to dismiss her – and then froze.

"Say that again," he said.

"You know, just lay down tracks, and put your gizmo on, like a train engine, and then you can always be in daylight."

Parks was stunned at the implications. The very nature of his device made such a proposal simple. In fact, now that she explained it, it was a perfectly logical and almost obvious solution. He felt at once humbled and exulted. This was the missing piece that tied together all his other plans.

"Honey, you're amazing!"

She just smiled back at him.

"Really, Elise. That was a stroke of genius." His head twitched and his eyes darted back and forth as he rapidly pursued the branches of logic that tumbled ahead of her insight. At dead ends he would pick up a different thread and follow it along. Elise and Afren seemed content to let him dream. He saw iron rails coming from the back of a crawler that scooped up lunar regolith. Inside, he imagined solar-powered electromagnets combing through the moondust for iron particles, and conveying them to a reaction chamber. The pig iron would be smelted into crude rails, heated by concentrated sunlight, and extruded out the back. Cross ties would need to be formed, perhaps from basaltic fibers, or microwave-sintered regolith bricks.

The first few times around the Moon, progress would be slow. Laying the rails would take many months, interrupted by the lunar night. But once those rails were laid down, then everything would speed up. The Lunar circumference is about 9,000 miles. Dividing by the number of Earth days in one Lunar day, approximately 30, gave a speed of 300 miles per day. Dividing again by 24 gave him a figure of approximately 12 miles per hour for the speed at which the sunlight traveled across the Moon. Once the Moon was girded by rails, a rolling isotope separator only needed to maintain that average speed to have continuous sunlight forever.

Except for eclipses, of course.

By piggybacking his isotope extraction device on a train, he could gather raw materials, and then deposit them at stations around the Lunar equator. The crews that serviced the separator could even travel along inside the train, and decouple one of their cars at the stations where solar-powered electric catapults could launch the materials into Earth orbit.

There were still many details to work out, such as whether to gird the equator, or to make rails closer to the poles. However, Elise's suggestion had opened his mind to the possibility of using the Moon for the first time. He began to see that the lunar rail idea was a perfect stepping stone to jump-start the whole process. Once the feasibility was shown, and the operation was generating income, then he could go chasing after an asteroid. While the Moon has many valuable minerals, the asteroids hold much greater wealth and a greater variety of raw elements. But now, that could wait a few years longer.

He looked back at Elise, and his eyes were filled with excitement. She tilted up her face and he kissed her passionately.

"I love you," he said.

"I know," she said back.

Then he remembered that Marcharelli Space, and 6 aerospace partners were already going after an asteroid, and his face fell. Only if they failed would he ever achieve his own dreams. He once again felt small and inconsequential.

Elise's arm slipped around his shoulder, and she mothered on him, sensing his mood and giving some affection. Afren's smile beamed, and they all seemed to bask in it. Their three heads bobbed in rhythm to the road.

Chapter 6

Colorado Springs, CO

New Capital Hill possessed a landscape made topographically identical to the gentle environs of the old capitol. The layout was the same, but the new buildings were stronger, and the office spaces were now underground. The new mansion for the President and the Executive Branch was a mountain fortress removed enough from the capitol building that at least one of the two would survive any single gigaton fusion explosion, yet close enough to be connected by multiple, high-speed tunnels.

Now, lawmakers were having to choose between conflicting priorities at a time when economic growth was sputtering up and down, but trending generally downward. Tax revenues were down and rates would need to be raised to support the massive relief efforts affecting millions of Americans driven from their homes and jobs. Rates of lawlessness in the northern Dixieland were escalating and federal troops were needed to keep the peace.

Alarming acceleration in cancer rates were the hottest political button now. Late Boomers, the most powerful generation the world has ever produced, were dying in droves from cancers of every type. Drugs and treatments kept improving, but cancer kept ahead. Some said if medicine had not been so strongly supported in recent decades, the world would probably be half as populated by now. Plagues were coming along faster now, driven by biological selection through improper use of antibiotics in humans and over-use of antibiotics among farm animals. Pestilence loomed everywhere, and people were afraid to eat. *More money needs to be spent*, the Late Boomers demand, and the money gets spent.

Power shifts, money shifts, and events are tugged by forces beyond the control of any one person or guiding organization. Individually and together, nations race to grab the most of finite resources. Funding for aerospace, for space, for the moon shots promised so long ago, for Mars, hoped for for so many years, would just have to wait.

Don't spend money in space. We need it here on earth, the Late Boomers said.

The spiral tightened as interconnected systems began to follow a self-stoking cycle. No amount of foresight, or good intentions, seemed able to show a way out of the circle. Money vanished into pork barrels. More and more was spent, but less and less progress was made. Stock markets and currency markets began to panic in earnest. The global economy began tipping over into a nosedive.

Taiwanese diplomats visited the Springs constantly, begging for more protection from the aggressive Chinese navy. Iranian insurgents were given more and more aid, and the US casualties mounted quickly. Another long-hidden enrichment facility had recently been discovered in a hardened bunker 300 meters below the ground.

The National Aeronautics and Space Administration suffered deep cuts in budget, and was forced to make layoff announcements. Debates raged under the manicured lawns of the capital. Across-the-board cuts came more often. More programs were dropped. The Seven Dwarfs mission was kept on, but with a lower level of support. It was left to the corporate partners to supply the deficit.

A busy Deputy to the Administrator, a now-graying Dajwal Chellapilla, made many of the public announcements of NASA press releases. The asteroid retrieval mission was NASA's first priority, he emphasized at every opportunity. Mineral resources from space had the potential to lift Earth from her troubles, he told the hungry cameras. Many believed. But those with no homes, no jobs, those with cancer, and those fighting to protect democracies and save the world from zealots took greater precedence in the overall political milieu. Waiting for a crisis to start precludes simple solutions. Emergency responses are always more costly. As Benjamin Franklin penned "A stitch in time saves nine". Foresight is a prerequisite for long-term ecomonic prosperity. Lack of vision, lack of leadership, or lack of action in time are the primary reasons for the downfall of any enterprise.

Control Room, SMER

The rocket had looked much smaller laying flat on the cargo ship. Standing under the erected launch vehicle, Parks Vambolent was awed and he was terrified. A tenacious anxiety gripped his belly muscles, and he could feel, rather than hear, the grinding of his back molars. Headphones covered his ears, and a boom microphone with noise-cancellation was tucked in at the corner of his mouth. So far, he hadn't said anything.

Dante Riolo walked shoulder to shoulder with Dr. Ruggieri, the Flight Director. Vambolent walked behind with a small group of specialists and special guests. Sugeto Mifune was flanked by two of his senior Japanese engineers. Vambolent felt out of place and in the way. Despite being thrilled to be in and among the massive structures, snaked everywhere with cables and pipelines, and draped over with the cryogenic venting of the liquefied propellants, he was nervous. All this had been started because of his dream. What if it didn't work? The enormity of this undertaking was beginning to overwhelm him.

Vambolent tried to calm himself and enjoy the pre-launch routine. With only 120 minutes until launch, this was the final on-site inspection. He had been instructed to look for ice on fittings, an occurrence which could jeopardize the launch. He had also been told, by one of the specialists, to look for any birds or animals roosting in the gantry. The specialist told him how one of the US shuttle launches had been delayed because of woodpeckers digging holes in the thick foam covering the external fuel tank. How any animal could tolerate the noise and commotion of a launch gantry seemed inconceivable to Parks. He craned his neck to look straight up the rocket fuselage. The bulge of the payload high overhead came in and out of his view through the mist of the mid-level vents.

Inside the top stage of the new Dart rocket were two deployable modules. Tucked inside were the ubiquitous silicon solar panels, novel only in their size and number, and the deployable antenna that was the heart of the experiment. So fragile was this structure, that the rocket would be doused with water just prior to ignition. The water serves as an acoustic damper to protect delicate internal structures from the launch vibrations. Once unfurled in orbit, the carbon-reinforced aluminum frame would snap together with optical-quality exactness. Only a successful deployment would allow the microwave beam to be focused tightly. Tiny actuators in the hinges of the lattice would lock the joints in precise alignment. It was a marvel of Japanese engineering, and Mifune was justifiably proud. Perhaps, Vambolent wondered, Mifune-san was as nervous as he was. Parks looked around and caught the eye of his first supporter. Mifune smiled back with a look of genuine warmth, and unshakable confidence. Vambolent took a deep breath, and felt better. Everything would go just fine, he told himself.

"Okay, we are done." Riolo's voice boomed in his headphones. "Any issues here? Lafontaine? Mastriani? Caruso? How about you, Parks. Anything?"

"Looks good!" Vambolent shouted, giving a thumbs up.

"Excellent! And our Japanese guests. Mifune-san? Any comments?"

Mifune cleared his throat, glanced quickly at one of his flight engineers, and pointed at the main engines.

"The rocket engines. They are not straight."

Riolo snapped his head around to see.

"Oh, that," he said. "Those will be gimbaled at, um… Mastriani?"

"T minus 6 minutes 30," the specialist chimed in.

"Yes. They will be gimbaled and aligned when the liquid oxygen is topped off and pressurized," Riolo explained. "It's best to keep them off-center, so the gimbals movement breaks any stickiness just before the launch and makes certain they are correctly aligned. Okay?"

Mifune's face was impassive. He seemed unconvinced.

"Okay, tell you what. Let's see if the control room can wiggle them for us, okay?"

Mifune nodded once, curtly.

With a shrug of his expressive shoulders, Riolo brought a walkie-talkie phone from his hip to his mouth, and pushed his boom microphone aside. A rubber shroud, similar to that on a combat radio, kept out much of the hissing and clanking of the rocket gear all around him. They could still hear him shouting his orders, gesticulating at the rocket as if he were standing next to his listener. After half a minute, he hipped the phone, and took a step closer to the enormous rocket nozzles. The rest of them turned with him, watching. With an audible clack, the massive metal cone nearest to them swiveled on its mount, coming straight down momentarily, hovering there, and then moving back to a 6 degree off-axis holding position.

Dante Riolo turned to them, raised his bushy eyebrows but lowered his lids. He dropped the corners of his mouth in a classic Italian look. Sugeto Mifune looked straight at him and smiled, returning with a head bow, indicating he was satisfied.

Waving an arm and speaking loud enough to be heard over the clamor, Riolo directed them back to the boat. The ferry connected the off-shore launch platform back to the mainland, and the safety of the control bunker.

In Vambolent's corner of the control room, someone had placed a hand-written placard on top of the monitor, saying "Specialisto Missione". With the user-friendly software, he could switch views from real-time videos mounted all over the SMER facility. He could view graphics of the flight path, textual descriptions of the launch sequence, manifests of payload, pyrotechnics, fuel, and navigation status, or current test results of the thousands of integrated systems that all contribute to a successful launch. The detail was overwhelming, and he had a new appreciation for the age-old honor given to rocket scientists, realizing for the first time that the honor extended also to the engineers and technicians. He felt reluctant to ask questions or interrupt anyone.

At the moment, he was also connected to the central server at the Kwapili Range. There, preparations for the rectenna were being triple-checked. At T minus 60 minutes, Riolo and Mifune came over, and they initiated a web meeting with Yoshi Hanazawa, Afren Kwapili and their chief lieutenant.

"Hello Parks, Bongiorno Signor Riolo. Konnichi-wa Mifune-san" Afren Kwapili said formally. "I have with me Yoshi Hanazawa, the site director, and our administrative manager, Mats Kafarikosi."

"Hello."

"Nice to meet you."

"How are preparations there, Afren?" Vambolent asked.

"Earlier this morning, we had a fly-over from a bush pilot who was carrying a signal generator and a microwave magnetron horn. He beamed two thousand five hundred watts down, and swept it across the antenna array. We were able to pick up over 90% of the emitted power on our meters."

"What about the other ten percent?" Riolo asked quickly.

"Yes," said Afren, after a slight delay which was the satellite link. "Some may have been consumed in Ohmic losses from cable resistance. The rest could be from calibration errors on the airplane's magnetron. It was a rather crude set-up, intended primarily to detect gross failures." He paused to see if Riolo would pursue the matter further. "I'll have Mats update you on the validation tests."

They could see Kafarikosi lean towards the camera, probably peering at their own tiny images on his computer screen. Parks raised a palm in greeting, and the man leaned back, raising a hand of his own.

"Hello," Mats said. After a pause, he began reciting the status in a clipped but sing-song voice. It was pleasant to hear his mix of British-accented English and rich East African intonation.

Vambolent felt his stomach muscles relax as the man reviewed one positive result after another. Riolo began to grow restless, obviously aware of his many other duties. When the litany was complete, Yoshi stepped forward for his report.

"Good morning. We have used diesel generators to send 750 kV through the transmission lines to Mombasa. The electrolysis station reports transmission power losses within acceptable limits. They have confirmed their step-down transformers are operational, and they can rectify AC to DC for transmission across the water to the floating platform. The tanker ship has verified hydrogen collection and all systems are go."

"Excellent," said Vambolent.

"We do have one small problem," Kwapili told them.

"Problem?" echoed Mifune, immediately alarmed. Riolo turned full face.

"Vultures," Afren said. "There are several of them soaring over the antenna array."

"Haven't you activated the screech owl noisemakers along the perimeter?" asked Parks.

"Yes, and they are effective at keeping migratory birds away. But they are ignored by the vultures. It seems we overlooked this potential impact, Parks. You see, vultures have no natural predators. They are at the top of the food chain here. They flew right over the perimeter, and are rather enjoying the giant thermal created by the antenna array. They appear to be playing."

Riolo burst out a guffaw. "Oh, for God's sake!" he said. "Why don't you shoot the ugly beasts!"

Kwapili did not answer him. Vambolent fielded the question.

"We've been very careful, Dante, to address environmental and animal rights issues with the entire project. Zealous activists can be very damaging to ventures like this. We are advocating this new technology as 'Earth-friendly', and have considered downstream consequences throughout the entire design process. The Japanese are using no hydrazine or perchlorate in their rocket engines, and we've sized the receiving antenna to keep the microwave energy density well below a lethal level for insects, birds, and small animals. Still, as an extra precaution, we installed noise-makers at the edges to frighten away migrating birds which sometimes cross over this area. But it seems we forgot about the buzzards."

"Well, who cares about buzzards, anyway?" Riolo pressed. "They are dirty and disgusting animals."

"Actually, they are highly intelligent."

Riolo let out a string of Italian curses, waving his arms upwards. "We do not have time for this silliness!" he shouted.

Mifune cleared his throat and spoke up in a gruff voice: "The satellite power will be very low when we start up. As we increase the power levels, will not these intelligent birds simply move away?"

They all considered his point for a moment, except for Riolo, who was already bobbing his head up and down vigorously. They saw Kwapili turn to his lieutenant, who, after a moment, indicated in the affirmative.

"Yes, that will probably work," he said into the web camera.

"Good. Then it is settled? We are a GO at the antenna site?" Riolo demanded.

"We are GO," Kwapili answered formally.

"Excellent," he concluded, and walked off briskly, audibly sighing with a raspy sound.

Mifune left, and Parks dismissed the men at the rectenna. Having nothing pressing to do, he stood to go get coffee. Walking out of the control room, lit mostly by the glow of computer screens, he was enthralled by the buzz of activity. At each work station, rapid-fire technical details were being discussed in Italian, in Japanese, and in English. He noticed how the scientific terms were mostly recognizable across languages. It struck him how international and diverse the whole project had become. The technology was unifying them all.

Approaching the java station, Parks breathed in the rich aromas. The Italians were intolerant of bad coffee, and the SMER facility had a gleaming espresso machine with a pleasant attendant. He ordered a latté and a biscotti, the adult version of milk and cookies. He sat down on a wire chair by himself, and collected his thoughts.

Since his birthday, the Space Energy Project had accelerated in pace, in details, and in world-wide attention. A carefully-crafted press release, given first to a sympathetic news agency, identified the primary role played by Vambolent Space Industries. With an emphasis on the international cooperation and positive environmental impact, the news had spread well, and interest was high. Malindi grew crowded with spectators and reporters present for the launch. Grandstands newly-refurbished just outside the SMER fence had been filled since early morning, and blankets covered most of the empty ground up and down the coast from the facility. Since the day before, the Kenyan Coast Guard had been busy keeping pleasure craft away from the site. The night before, after Elise and the kids had gone to sleep in their hotel room, Parks saw himself on the international news station, being interviewed for his role in the great enterprise. He had turned up the volume and tried rousing his family, but by the time their bleary eyes were open, the spot was over. The plethora of pressing problems around the world would not wait long for air time.

Sipping his drink, Parks felt mixed emotions regarding his role as a spokesman. The real benefactors of the Space Energy Project were the Japanese and the Kenyans. When this demonstration was over, neither contingent would have much need for him. At least the press coverage should give him valuable leverage back home. What he would do there was unclear. Almost certainly, Dajwal Chellapilla's offer had expired. His re-hire and subsequent rapid rise within the re-organized NASA bureaucracy was a direct continuation of what he had been doing before he returned to India – besting Parks Vambolent. The chances of Parks becoming anything more than a grunt engineer were fading. The whole idea of starting a space business with Horse now seemed ridiculously foolish and optimistic.

But then, even *this* project had seemed a long-shot. While supportive, Elise would continually remind Parks of the incredible unlikelihood that he would ever achieve all of his dreams. Her consistent advice was to sell the patents, and use the money to come up with more inventions, but meanwhile work for an established company with a generous benefits package.

Then, he had met Mifune. Without that meeting, he might still be hawking his ideas at space conferences. Without his persistence, perhaps, the Space Energy Project might not have happened here, or this soon, or be so close to success. A shivver of prickly electricity leaped up his spine. His brain experienced a rush, and he became sanquine.

Any big change in the world, like converting from fossil and nuclear fuels to space-based solar power, would never move quickly. Change had to start small. One leader, or a small group had to somehow amplify their efforts. There is no other way to overcome the innate inertia in human systems. In fact, the driving force behind human thought is the very same impetus which drove Vambolent to want to create an empire in space. The "way things are" is a balance between entropy and the creative urge. Any change in the balance took an input of energy far greater than any handful of individuals could generate. Therefore, those in economic and political control, with legions of people to command, become the ones to control the "way things are". If they are motivated by greed and power, their actions will serve to further those needs.

If instead someone were motivated by sustainability, and wanted to get a large fraction of the world to adopt a similar attitude, he or she would need to greatly leverage their efforts. They needed a "bootstrap" effect, somehow magically pulling themselves up off the ground by yanking on the straps of their own boots. For the pace of change to be accelerated, it would need to be compounded in some sort of geometric growth. For if growth is less than exponential, it will die. His task, he suddenly realized, was to find the right people and the right set of conditions, to allow his ideas to multiply on themselves.

He munched his biscuit, sipped his latté, and mused over this dynamic. Maybe it applied to more than just paradigm shifts and business start-ups. Vambolent wondered if he could apply this idea to technology as well. Maybe there was a way to make his inventions into lunar factories that were so cheap, so small, and so light, that he could find funding somewhere, to prove he could scale it up without limit. He thought of the concept of the self-replicating machine, invented by the mathematician John von Neumann. By first launching a small factory, launch costs are kept modest. Small machines, using only lunar rock and solar power, could begin fabricating their own components, so that robotic or human fabricators could arrive later with the remaining parts and make a small fleet of similar machines. Repeat the process once, then reconfigure the machines to make solar power satellites. With his isotope separator optimized for multiple output, the fabricators would build infrastructure and living quarters. Rails around the moon would allow permanent sunlight for power generation and greenhouses, which could be transmitted through the rails themselves back to the moon base, which could then be located anywhere. There could even be a network of spiraling, high-speed train lines connecting ice mines on the poles to processing facilities on the highlands, crossing the resource-rich maria, even travelling deep down inside the great rille canyons where electromagnetic catapaults launch iron canisters to points cislunar.

Parks pulled a napkin from the stack and a marker from his pocket. He began making sketches and notes, his head bent over in concentration, and his eyes failing with age. On his third napkin, he was startled from his reverie by a loudspeaker announcement.

"T minus 60 seconds, and counting."

"Oh, my!" he exclaimed. He was due back on station. Leaving the mug, but pocketing his napkins, he hurried back into the control room, sidling between the crowded technicians to get to his place.

The countdown was filled with quiet. As systems came up, they were called green, and then silence. The waiting edged everyone. Time seemed to pass at once too fast and too slow. Some fidgeted nervously. Others doodled, and a few whispered sidecracks. Many looked nervously around at each other.

Parks saw Dante Riolo walk in, and went over to Ruggieri; they conferred with bent heads. Riolo straightened and turned his blazing gaze on full power, doing a rapid 360 degree scan of every eye in the room, sending to them through body language and deliberate intimidation his strong and unstated expectations. Parks felt Riolo's eyes hitch on him, and wondered what the man was thinking. In the span of one collective heartbeat, all in the room got bent to their tasks. Riolo turned again to the FD with a great smile, slapped the man on the shoulder blade, and strode from the control room.

Vambolent wished he could see the launch directly, like his family would from the grandstands, but the control room had no window. They were too close for a window, in case the rocket should explode. But the entire front of the room, a mosaic of seamlessly connecting monitors, showed a large-as-life image of the launch pad, dotted here and there with status boxes. Plumes of gray-white vapor were already pouring out from the underside of the rocket as the countdown passed through 10 seconds remaining. At six seconds left, the main rocket engines were ignited. A flare of yellow-white intensity saturated the view, and the flames, half-hidden by smoke, burst out and spread across the launch pad. A moment later the roar of the engines passed through the control room, and they could all feel the subsonic rumble as the millions of horsepower pounded down into the earth beneath them.

The flames kept expanding, seeming a profligate waste when the rocket was still squatting on the ground, until the countdown reached the magical number of zero – the number of emptiness, the number of beginnings, the pivot point between all numbers minus and all numbers plus; "all balls" in launch lingo. At that instant, umbilicals detached, hold-downs released and gantries swung away. The rocket was allowed to rise.

The roar increased in the control room, but the image on the screen did not seem to change. Imperceptible at first, the rocket suddenly appeared to grow taller. Flames and vapor shrouded its tail, and Vambolent focused on the payload bulge where his precious hopes for the future were tightly ensconced. And then, almost without warning, the rocket's acceleration became definitely noticeable. The camera view widened, and began to follow the rocket as it lifted off the launch pad, and now they could see the flames blasting out the back of the engines, fighting against the mighty gravity that would never stop trying to pull the rocket back down.

Vambolent felt a powerful emotion rise up in his chest as the stark evidence of the rocket's flight filled him with awe. Moving more quickly now, the view of the ascending rocket threatened to recede into a distant missile, headed away. But he could still perceive the velocity rising and rising as it accelerated upward, now starting to tilt over towards an orbital insertion trajectory. It was away, taking the flames and thunder with it. The control room seemed empty now. The wavering image of the departing vehicle was becoming nothing but a condensation trail, growing slowly, and tapering to a point where it exited the atmosphere and became a spacecraft.

Without a rocket to display, the floor-to-ceiling screens now showed web-cams of the crowds outside the SMER fences. They cheered and waved at the successful launch. But inside the control room, the cheers would not come until the satellite reached GEO, unfurled itself, and began beaming power back to Earth. Their destination was a spot directly above the Kenyan equator, orbiting the planet at the same rate as the Earth turns underneath. Satellites parked at GEO appear fixed in the sky to an observer on Earth, making them very attractive for ventures such as the Space Energy Project. VSI had little trouble securing a license for this orbit from Kenya's Parliament. Mtongwe's government cited the First Meeting of Equatorial Countries, and granted VSI this prime location in space. The receiving antenna was almost directly beneath.

The launch profile was unusual, chosen partly to showcase the capabilities of the new Japanese rocket. The ascent required one full turn around the Earth to get to GEO, which took more time than a more direct flight from a point halfway around the globe. During the wait, launch technicians and flight engineers would occasionally break out for coffee. Vambolent could tell from their smiles and animated chatter that they were all happy with the launch. However, they kept their exuberance in check out of respect for the two Japanese crews overseeing deployment of the solar cells and the transmission antenna. If launch vibrations or pogoing of the rocket had damaged any of the sensitive joints, or if the Earth-based testing had failed to reveal a flaw that would only surface in zero gravity, the whole project would be a bust.

All the crews were back on station at T plus 60 minutes, when the payload and bus reached GEO. A burn to circularize the orbit brought the satellite exactly on target. The mission was complete for the Italians, who smiled broadly, but otherwise did not display their elation.

Concentrating with fierce intensity, the Japanese crews followed the control procedure established by Mifune's satellite company. Automated deployment could not be fully verified visually. Instead, integrated sensor arrays, interconnected through inductive passive-read transceiver chips, were periodically pulsed from a circuit in the main transformer unit. On the screen they all now watched complex schematics and indecipherable glyphs. Only an animated graphic illustrated the inferred progress.

The protective shielding was disengaged and folded back on the satellite bus structure. Now the solar panels deployed, stretching out accordion-style. Small attitude jets nulled the rotation, then squared the cells to the sun. They were already energized, and began charging the internal batteries. Power monitors showed available solar energy. The yellow bar rose above a blue line, and a green circle flashed to indicate full power. Now, the transmitting antenna could be unfurled.

Proper antenna operation depends on on the shape of the antenna. Small perterbations can be phased out electronically – but only small ones. The greatest challenge was the combination of size and accuracy. For three decades, Japanese and American companies had developed deployable structures. The Americans gravitated towards inflatable structures, using ultra-thin membranes. The Japanese favored machanical structures: ingenious devices with thousands of interacting elements.

An on-board camera, running on the fresh solar power, came up and was directed to the main screen. Throughout the room, many gasped in awe at the site before them. Brilliant sunlight glinted needle-like off small surfaces, and the satellite bulk seemed to appear in negative relief. The camera rotated, backed, then tilted "downward". Several people groaned, either in real or pretended nausea. At least some were faking it, as their laughter attested.

The camera zoomed smoothly onto one of the base joints on the deploy stalk. At the command of their Japanese masters in Malindi, the antenna began to unfurl in fits and starts. Vambolent closed his eyes briefly to visualize the deployment he had seen in Mifune's simulations. Each leg of the antenna was spring-loaded with an elastomer strap that was stretched to three time its nominal length. Sliding activators traveled from pivot points along the fold lines, raising each hinge past the detent so that it could snap into place. Shape-memory alloy heat-activated flex actuators at the pivot points pushed the panels open until they locked into place with thin film-coated magnetic attachment points. The magnets were created with spiral fields, so the continued but diminishing flexure of the actuators wiggled the points into precise alignment. The thin films were designed like a two-part epoxy, and would form a permanent bond within a few minutes of the intimate contact. The biggest worry was for the epoxy to set before the whole antenna had settled into place.

One of the Japanese engineers pointed to the wall screen and spoke harshly to a colleague. The whole room went deathly silent. The other man argued back, and they both leaned in close to their own monitor, silent. All eyes were on them. One man ran a hand through his hair, and leaned back. The other leaned back also, his shoulders heaving a relieved shrug. Their co-workers, waiting and watching patiently, inquired, and were rewarded with an affirmative nod. They all looked at each other, as if for permission. Then, they began standing up and shouting.

The shock startled the whole room. A collective cheer arose, and as their confidence that the deployment was really done, became a deafening roar of joy. Everyone was clapping their hands, and the control room seemed close and intimate.

Parks clapped along with them, but regretted they did not have visual confirmation of the overall assembly. He would be able to celebrate when the power connection was established, and hydrogen gas flowed into the holding tanks down in Malindi harbor.

Dante Riolo strode to the front of the room, followed by Jezri LeFallon, who had entered during the cheering. Those who had started to leave sat back down. Riolo lifted his rib cage and opened his arms as if addressing a Roman Senate.

"Friends, colleagues, citizens of the world. We are here on an historic occasion. Today we combine the talents of Italian, Japanese and American scientists to produce power and fresh water for Kenya. The Space Energy Project, first conceived by Parks Vambolent and Sugeto Mifune will soon be ready for its final test. Today with the excellent facilities and personnel at the San Marco Equatorial Range, tracking facilities and ground station, we have witnessed a flawless launch and orbital insertion. The Japanese satellite crews have successfully deployed the world's first GEO solar power satellite. And now, within the hour, we stand ready to throw the switch. When that happens, we will bring the awesome and inexhaustible power of the sun down to the Earth. Ladies and Gentlemen, I applaud you!"

Those assembled applauded themselves modestly, but many felt it was still too soon to celebrate. "And now, Dr. Ruggieri, the satellite is yours."

Technical communications chatter filled the room as teams coordinated their efforts. Marcus Lafontaine's group worked the triangulation for the satellite and were sharing their data with the first Japanese group working on attitude positioning. Their wall graphics showed the alignment error. As they cross-correlated their systems, learning the dynamics, they watched the angular error drop steadily through several orders of magnitude, and come to rest just over one ten-thousandth of a degree. They watched for several minutes as the fuzzy logic controllers maintained the alignment, demonstrating that the stability system worked as predicted, and could maintain station through the inevitable adjustments needed to keep the solar arrays facing the sun. The loose coupling between the two elements of the satellite components was an important feature, involving considerable engineering of the control algorithms and power couplings. With the antenna demonstrating stability, all that was left was to turn on the juice.

All six of the Japanese engineers looked up to await the final order. Vambolent checked in with Kwapili through his workstation video link.

"We're all set here, Parks," he announced.

Ruggieri called off the systems one by one: Guidance, Navigation & Control; Propulsion; Electrical, Environmental and Communications; Instrumentation; the Flight Dynamics Officer; and the Ground Crew at the rectenna. To each, he received a "GO!" He had green lights on every panel, and ready status from each station. There was nothing left to do but activate the magnetron and allow the microwave energy to flow from GEO to Kenya.

"Okay, Mr. Inoue, take the power to ten percent," Ruggieri said, then added: "and do it slowly – for the vultures."

Inoue entered a keystroke sequence and committed it. The signal was sent at the speed of light through their ground-based dish antenna, and reached the SPS a tenth of a second later. Inside the inverter, wired between the large solar panel array and the beaming antenna, the microwave magnetron control circuit activated. From a high-voltage rectifier, pulses of free electrons raced down the length of the Gunn diode, and from the far end, a burst of microwave energy surged out through the horn. The precisely-shaped beam bounced off the parabolic inner surface of the antenna, and was focused toward a point thirty thousand kilometers inward.

After another tenth-second delay, the microwave beam encountered the upper limits of the Earth's atmosphere, passing unimpeded through increasingly dense nitrogen and oxygen gas. With 60,000 feet to go, high-altitude cirrus clouds attenuated the beam ever so slightly, but nearly all of its power plunged downward to the desert floor. There, six feet above the sand and broken rock, a diode-studded copper grid picked up the incoming radiant energy. Copper conductors linking array sections conveyed the energy to the transformer block where a massive step-up converted it to high voltage at low frequency, suitable for power transmission. From two thick cables, protected by brown insulator segments, the electricity was carried in the space between a parallel pair of wires as an alternating electromagnetic field. It was a power transmission method first developed by Nikola Tesla for Westinghouse Electric late in the 19th century, and is still used extensively.

From the Kwapili Range to Mombasa, the steel towers held that traveling EM field over the ground, so it could travel through the gap separating the two wires. At the terminal in Kilindin harbor, an AC to DC converter turned the oscillating power to a strong current that was carried through heavily-insulated braided cable to the hydrolysis station floating on the salty waters. Sealed conduits divided the current between a number of parallel electrolysis vats. Each vat was opened below to seawater, and above each of the hundreds of anodes, a partly-submerged collection bell captured the generated hydrogen. The precious gas was drawn through flexible double-wall evacuated steel line to the massive tanker ship moored alongside. From the input port of the hydrogen feed, the gas was absorbed into enormous porous silicon-filled vessels using a gentle pushing voltage. The pure oxygen at the cathode was still a waste product, and was blown away from the electrolysis plant with an array of fans.

When the first detectable hydrogen appeared on his flow meter, the technician, who was surrounded by nearly the entire crew of the tanker, gave them the good news. They all joined in a cheer, and called Hanazawa at the Kwapili Range.

Afren Kwapili, overseeing the monitoring of each Earth-based phase of the entire project, relayed the news to the control room in Malindi. Parks piped the news into the center console on the big screen.

"Parks, we did it! It works!" Kwapili's smiling face blinked rapidly.

This was the moment they had all been waiting for, and this time, there was no suppressing the celebration. The Italians hugged each other, cried for joy, and slapped backs. The Japanese were all smiles and bobbing heads. Vambolent was swamped by well-wishers, and the noise was almost deafening. Over the din, he tried to congratulate Afren, Yoshi and Mats. Finally he gave up, sent a chat message, then shut down the video link. When their familiar faces vanished from his screen, he felt oddly lonely, and longed to leave the facility and reunite with his family.

Ruggieri ordered the magnetron shut down. The second shift skeleton crew filed in and took their positions. The five megawatt solar power satellite was put into a low-power diagnostic mode. In the morning, an exhaustive validation procedure would begin. Thorough testing of the beam alignment and failsafe systems were needed before the array could send its full rated capacity.

* * *

Riolo had staged a press conference in the cafeteria. A royal blue tapestry emblazoned with the SMER logo hung behind an oak podium at the far end of the room. Reporters crowded the door, but Riolo strode through them like tall grass, shooing them away, calling for them to "make room, make room". The press corp parted like a boat's wake, and control room staffers began filing in. When Jezri LeFallon entered, tall and handsome, local reporters swept in on him. Mifune was press-shy and had left already, but his lieutenant in charge of the two satellite deployment teams attracted a crowd of his own as he came in the doors. Parks was one of the last inside. Through the moving heads he could see Riolo, standing in the beam of the flood light illuminating the podium, mock-protesting with the reporters from Rome, but still dropping them sweet, rehearsed sound bites. He could be heard across the room.

"Parks Vambolent!" cried one reporter.

A small flock converged on Parks, both anglicans and asians, including a few who broke away from LeFallon. Microphones crowded him uncomfortably, and he could smell several flavors of foul breath. A flurry of shouted questions blurred into incomprehensibility. He was about to open his mouth, unsure of what to say, and they noticed and went silent. He hesitated.

"How do you feel?" came a female voice.

"I feel great," he said on reflex. They all fell silent, and their little spotlights were in his eyes. He squinted and blinked a few times. "Um. This is a great success for the San Marco range. The power we beamed today represents pure, clean energy inexhaustible and sustainable. It's the coolest way to run the planet."

They all paused, not wanting to stomp on the sound bite.

The woman's voice again. He had filtered on it, and the instant interruptions hardly registered.

"What's next for you?" the voice asked.

"Um, well, we have a bit more work to do here. Then, I'll probably go back to the States. I'd like to do research into lower-cost ways to put solar power satellites in orbit. I've got some plans to help build a cislunar economy, and bring nearly-free energy to the Earth."

One person called out in a quick voice "What's cislunar?" even as some others sniggered at Vambolent's point.

"Within the circle of the moon's orbit. That's cislunar," Parks said.

They paused again. It was exactly one third second, equal to two eyeblinks. Parks figured it out because several of them blinked in unison. He craned his neck back and forth to find the woman who had asked the questions. Riolo finally called them to order, and the circle parted. He saw a thin woman standing in a perfectly-cut business suit and a notepad in her cupped palm. Her face looked young and beautiful, but the whole sight of her looked older, mature and worldly. She smiled quickly and spoke to Parks in a stage whisper, closing the distance between them as people stepped aside.

"Are you aware that an Israeli university is using your ideas to map ozone-depleting chemicals in the upper atmosphere?" she said.

Parks felt his stomach rise into his throat.

"Yes, I'd heard of that," he answered. For some reason, having the question asked made him feel inadequate. He wondered what news he had missed this time.

"And did you know they plan to launch their satellite within the month, and start flying the chase plane just a few days later."

"I hadn't heard that," Parks said.

"Heidi Bayer," the woman said. She offered him a hand, held with digits tightly together, thin thumb pointed straight up. She gripped his hand hard with her tiny one, and shook it so hard her voice wavered. "Detroit Free Press."

"You're a long way from home," he said.

She smiled, showing teeth this time; and they obeyed the expressive arms of Dante Riolo waving everybody down, and sat. Bayer lowered her voice even further, projecting the small sound directly into his ear canal.

"Detroit formed a Next Energy Zone earlier in the century," she said. "We've got a Nanotechnology Zone now, and a Next Generation Transportation Zone. There are lots of projects, lots of studies, but nobody is turning it into products, into revenue, into taxes. Our paper is working with the Federal Energy Information

Adminstration to bring out stories of people generating environmentally-benign energy sources in creative ways and applying them to real-world problems." She fixed him with a fierce gaze, like that of an osprey. "Could we talk somewhere less noisy?" And inclined her head towards the back doors, where stragglers were still coming now and then.

She stood and left. Parks sat uncertainly for a few seconds, glanced at the front where Dante Riolo preened, then stood also. His passage attracted more attention than Heidi Bayer.

"Parks! Where are you going?" Riolo's voice boomed from the microphones.

All eyes swivelled to Vambolent and locked on.

"This is your celebration, too!" Riolo enormous voice called.

Parks felt trapped. He almost sat down. Then, inspiration struck.

He straightened and turned full face to Dante Riolo, who was standing before them all, sharply lit and standing out from the deep backdrop. He looked very distinguished.

"Let's hear it for the man who made everything happen today. Mr. Dante Riolo!"

Parks clapped his hands, and hammed it up, just like he'd seen Horse do in their riding days. He reached over and slapped the shoulders of a few control room workers nearby. They stood. In a whoosh, the whole room stood and cheered, waved arms, and called accolades to the SMER director. Riolo loved the attention, and so forgave Parks who discretely worked his way backwards down the aisle, and stepped out of the door held open just enough by one petite hand.

"Thanks," he breathed. "This way."

Parks took her around the corner to the espresso bar. They sat at the same table where he had worked on calculations earlier that day.

She set a small electronic voice recorder on the table, raising an eyebrow to ask permission. Parks nodded.

They talked for nearly an hour about the project, the US, and even some personal stories before Parks even thought of Elise. With a start, he interrupted her.

"You'll have to excuse me, Heidi. But I really must be going."

"When can we talk again," she said, not missing a beat.

“Tomorrow. With luck, by afternoon we’ll have reached full power. That will be the real success of this mission. That’s the real story.”

“You’re the real story,” she said.

Parks felt himself blush. A voice inside his head warned him of flattery. He stood to leave.

“You are too kind,” he said. “It has been a pleasure talking with you.”

She stood as well, offering her hand, this time more relaxed. He wondered if she was nervous.

“The pleasure has been all mine, Mr. Vambolent. I look forward to finishing our conversation tomorrow.”

“You can call me Parks,” he found himself saying. His head was curiously light, and he felt slightly disoriented. *Must be tired*, he told himself.

She gave him a charming smile and released his hand. His hand missed the contact right away and an urge within him wanted to detain her.

But she was already away, calling one more goodbye over her shoulder, and then she was at the door to the cafeteria. She slipped inside the doors without seeming to open them at all.

* * *

Thinking of Elise, Parks walked straight to the front lobby, and walked outside. A few stray reporters were smoking cigarettes and talking rapidly into headset phones. They let him pass, but Parks felt their eyes follow him.

Walking out to the VSI Range Rover, he noticed a well-dressed man leaning against a silver roadster backed into a parking slot one over from his own. The man’s eyes watched Vambolent approach. Parks thought him a reporter, or maybe an editor, but the man’s pleasant expression and non-confrontational body language made him reconsider. He appeared to be in his early sixties, with silver hair swept back from a tall, rounded brow. His eyes were close and narrow, sligthly droopy, and his nose generous with a change in direction that kept it from being too big. His lips were thin, and he was clean-shaven. The tiniest spark of recognition hit Vambolent, but more from the look than the face. He had the distinct impression this man was French.

Eyeing the man cautiously, Vambolent activated the keyless entry, and heard the satisfying *thunk* as the lock disengaged. He nodded cordially, and the man nodded back, slowly, as if inviting some comment about the launch, or the weather. Vambolent was not in the mood, and pulled open the driver door, hopping up into the seat. He reached to pull the door closed when the man spoke.

"Nobody knows, do they?"

Vambolent paused a moment, before deciding to blow off this crackpot and go celebrate with his wife and kids. He leaned out and grabbed the handle.

"That was all you, Parks Vambolent. It was all *your* idea."

Conflicting thoughts paralyzed Vambolent. His gut impression was that this man was non-threatening, so he held off the flight instinct. This was probably some sly journalist looking for a new angle, and had done enough digging to find Vambolent's influence throughout the project. He hadn't even told Heidi all of his story. He was tempted to engage in conversation, but the draw to his family was stronger at the moment. Still, the man did not look like a journalist. He had a calm yet commanding presence, much more like a corporate CEO than a stereotypical dirt digger. In his pause over how to respond, the man continued.

"While the rest of them are celebrating inside, you are already moving on to the next step. Your plans are much larger than this Energy Project. And yet, you don't know how to proceed."

Vambolent was suddenly mesmerized. The man had hooked him. But now, the flight response was threatening to activate. The man knew, or guessed, too much.

"And you are?"

"Bruno D'Argeste, a man sympathetic to your cause."

He took a step forward, extending a hand.

Vambolent warily slid off his seat and reached out with his own hand. The man was much taller than he had looked leaning back on the coupe. He towered over Vambolent like a pro basketball player. For some reason, this put Vambolent at ease. They shook hands, and the grip felt satisfying and reassuring to Vambolent, not at all like he would expect from a Frenchman. Canadian, perhaps? There was enough of an accent to suggest French as a first language, but the English seemed natural.

"Nice to meet you," he said.

"Please, the pleasure is all mine. I have been a space advocate for many, many years. And lately, I have been following your progress with amazement. How you have been able to get so much accomplished in such a short time is like a miracle!"

"I think you overestimate me, Monsieur D'Argeste. The people who did the real work are still inside, and down at the antenna site. I just came up with the ideas."

"Ah. Dismissively humble, plus, you correctly guessed me as a Frenchman! I knew I would like you." The man laughed, throwing his chin up, and looking into the pale blue sky where the satellite paced them overhead. When he looked back at Vambolent, a focused concentration filled his face. "But I have always found there is *one* mind behind all great enterprises. Without vision, the world is chaos. Where there is success, like today, there was first a dream, then a plan, and then motivation, persistence, and finally fruition. It is a formula I have used over and over myself. And now, I am so very glad to meet you, Parks Vambolent."

The man seemed eminently pleased with himself, and Vambolent began to wonder if he was not a crackpot after all.

"And what is it that you do?" he asked.

"Ah, what do I *not* do? Sometimes this is the easier question!" He laughed gently at his own remark, and then adopted a pensive look. "You have heard Archimedes' saying, 'Give me a lever long enough, and a firm place to stand, and I can move the world', yes?"

"Yes."

"Well, I have many long levers, and many firm places to stand. The important question for me then becomes, which way do we want to move the world?"

"We?"

"Yes, Parks – if I may call you that." He proceeded without waiting for an answer. "I and my associates are very interested in how the world is moved. As I said before, without vision, all is chaos. We intend to provide order. And like you, we are the ones behind the scenes, making things happen. The unsung heroes, as you Americans like to say."

He leaned forward, speaking faster now.

"Parks, you will need help for your next steps. I know quite a bit about you, but not what you plan to do next. We did not even anticipate what you did here until it was nearly done. But now, after having met you, I suspect that you have some very good ideas, and a compelling vision. If you are looking for levers, please call me. I may be able to help you."

He produced a personal card containing his name and a phone number only, and handed it to Vambolent.

"Thank you, Monsieur D'Argeste," he replied. "I will."

The taller man nodded, and reached again for Vambolent's hand. They shook briefly. D'Argeste gave a subtle heel-click, smiled and turned. In a single step he was opening the door to his roadster, and folded himself inside so quickly that Vambolent almost did a double-take. Parks climbed slowly into his own vehicle, and turned on the motor as he watched the silver car speed out towards the gatehouse.

* * *

Watching D'Argeste meet Vambolent, Jenks regretted that he did not have a parabolic microphone. From his vantage point on the distant grandstands, he could lip-read much of what Vambolent had said. But it was D'Argeste's side of the conversation which interested him more.

Jenks trained his 48x monocular scope on Vambolent's vehicle as it moved to the exit gate. The expensive piece of optics was not out-of-place at a rocket launch, although it was probably becoming noticeable that Jenks was one of the last to leave, and was not looking up. He replaced the lens caps, retracted the focus, and slid the unit into its velvet-lined case.

He sat in the fading sunlight with a contented smile on his face. Never one to rush past a success, he savored the inner joy of realizing his lifelong dream. The confirmation he had just received was hardly necessary – he already knew D'Argeste was his target – but the ritual of validating his deductions brought him a satisfying feeling of closure.

His inference web, weighted with relevance to his ultimate goal, had developed a centroid on Bruno D'Argeste just weeks before his first trip to Kenya. After the Energy Project was shut down, Jenks had returned to Los Angeles. He was satisfied that Vambolent would give up, so that Jenks' own work could resume.

When Jenks learned that the project was reactivated through private funds, he knew instinctively it would become the breakthrough he had been seeking. Discovering how D'Argeste was connected, if indeed he was, became Jenks' obsession. Using the ample research resources at his apartment, he studied the public records of the various corporations, dumping masses of information into a data mining program. Then, adding the clandestine bank data available at his office, Jenks finally found the link that broke everything open for him.

D'Argeste's name did not appear in any of his searches. However, he was able to track the account number funding the satellite program in Japan, even though he could not access actual transfer amounts. One account number in the complex chain of financial transactions differed in just one digit from a similar search of deals connected with D'Argeste and his holding company.

With that single clue, Jenks was able to form hypotheses of connection, and begin to test them. It was an exhausting effort, fraught with dead ends, but in the end it paid off. An airline ticket for B. D'Argeste, taking him to Nairobi just two days prior to the launch had been enough for Jenks. He booked his own flight immediately.

In retrospect, the trip had been a long-shot. Yet it was a calculated risk based on considerable effort, scant evidence, and gut instinct. It paid off handsomely. Jenks already knew how he would approach D'Argeste. But not here. If D'Argeste ever suspected that Jenks had discovered who he really was, there would be no second chance. With his intimate knowledge of the power structure the French businessman had erected around his still-unknown core, Jenks knew just how to get his grand proposal recognized by the top man, while revealing only pieces of it to his various underlings. It would just be a matter of time before he received an invitation to join them.

Then, his real work could begin.

SMER – Day 2

By noon, the Green Team had taken the SPS to 100% rated capacity, and characterized tranmission losses and conversion efficiencies all the way through to hydrogen production. The satellite underwent a perturbation analysis of the station keepers with the magnetron at full power. Using small, pre-programmed beam deviations, they verified the automatic shut-down devices in case the phased-array transmitting antenna drifted out of alignment. Cosine law studies were completed on the solar panel alignment, and navigation correction for solar wind pressure was confirmed. All systems tested nominal.

Dante Riolo and Dr. Ruggieri congratulated each of the teams and each worker in the control room, at the rectenna facility, and at the hydrolysis plant in Mombasa. As they had the day before, the press awaited the news. Today's crowd was slightly larger, but now included all of the larger news agencies. Brightly-painted satellite antenna trucks were parked pell-mell throughout the SMER parking lot. Reporters packed the cafeteria and were overflowing into the hallways.

Riolo flung open the control room door, and lead the charge into the fray. His booming voice was already projected ahead of him. "Ladies and gentlemen of the press, I am proud to bring you very good news today. The SPS has passed all validation tests, and is at this moment delivering five megawatts of pure sunshine power to the deserts of Kenya!" He paused to see the effects of his words, but there was no audience besides the press corps, and nobody clapped. Slightly crestfallen, but no less loudly, he called over their heads. "Now, please move into the cafeteria for a full report."

Parks had been pressured by Riolo to give some sort of speech, but Parks was hesitant to do so. Watching the SMER director from behind, he was reluctant to emerge into the bright lights. He looked around him, and found a small knot of others waiting for the doors to clear so they could exit the control room. In the far corner, Parks caught sight of someone moving away in a different direction.

Sugeto Mifune.

Excusing himself, Parks moved quickly to follow. Catching a spring-loaded door with his hand, he entered a service corridor. Mifune's back was just passing through another door at the far end. Both doors closed, and Parks held his hands out to keep from running into anything in the darkness. Reaching the far door, he grabbed for the handle, fumbled it open, and emerged into a workshop. Two technicians looked up. Vambolent excused himself, and walked across their space to the exit door as their eyes followed him silently.

He found himself coming up past the espresso bar, and decided to wait there until the press had left. Seated at the same table as yesterday was Heidi Bayer. A small white demitasse with lipstick on the rim sat on a saucer before her. She noticed Vambolent and stood quickly to greet him.

"Parks, how did it go today?" she asked.

"Nice to see you too, Heidi," he said.

"Oh, I'm sorry. I'm just excited to talk to you. From what I can hear out there, everything came off great. How do you feel about that?"

"You even talk like a reporter," Vambolent said.

"Okay, you got me. I'm turning off the reporter…now," she flicked an imaginary light switch on her head. In the same motion, she removed a clasp behind her neck and shook her hair loose, smiling up at him. Scent swirled around her. "Better?" she asked.

"Much better. Thanks. And by the way, I feel great," he replied.

"Let me get you something," she offered. "What would you like?"

"Oh, whatever you're having."

Heidi ordered quickly, then came back to the table.

"I sent a story on you to my editor yesterday. It's already on the newsstands!"

"Can I see it?"

"Sure," she said, and rummaged in her shoulder bag. She pulled out a single sheet of paper, with two narrow columns of text under the headline: "US Inventor Scores Space First".

"I wasn't first," Parks said. "Others have done this before; just not from GEO."

"But it's the whole thing you're being first at. Read on," she encouraged.

Parks started reading the article, but within just a few seconds, Heidi began talking. He tried to divide his attention between the two.

"I've got a whole series of articles planned. Each one centers on one of your inventions. I spent most of the day researching the work on the mapping of ozone-depleting chemicals."

"What did you learn?"

"Well, let's see," she said, organizing her thoughts. "At Tel Aviv University, they have the Porter School of Environmental Studies. Atmospheric researchers there received a grant from a generous entrepreneur who had a wife die of melanoma. The grant came with the stipulation that they help identify how serious the problem of skin cancer will be for their grandchildren."

"Who was this 'generous entrepreneur'?" asked Parks.

"They didn't say, and I couldn't find out. Even cross-referencing with CEOs who recently lost a spouse turned up nothing concrete. I'll need that for the story, of course, but so far – nothing. There was no mention of your name in any of the sources I looked at, Parks. I'm thinking of flying to Israel on my way back to the States and visiting those people. I wouldn't mind swapping them the true inventor of their ODC mapping method for the name of their benefactor – with your permission, of course."

"I don't mind," Parks said, and tried reading the article again.

"Anyway, they say the same thing you said in your paper. While satellites and airplanes can measure ozone concentration, they are completely blind to the amount of ozone-*depleting* chemicals in the upper atmosphere. There are theories of how those chemicals migrate there, but finally there is a way to measure it! Once the concentration and dynamics of ODCs are understood, they feel that epidemiological models can be developed to predict trends and hot spots for melanoma risk. One day, they could even have a daily forecast with this technology."

"Sounds great."

"I know this must stick in your craw, Parks. But trust me. The world will soon know your role in all this."

"I'm not sure that's what I want," he said.

Heidi looked at him closely, trying to gauge him. The girl working the espresso machine brought Parks his coffee and set it on the table.

"I guess I'm getting mixed signals from you, Parks. On the one hand, you want the world to know about your inventions and ideas. But on the other hand, you seem shy of publicity."

Vambolent just shrugged, and bent his head back over the article.

"So, anyway, they're developing a special satellite like nothing ever built or flown before. It's basically a gigantic diffraction grating, a series of very fine lines that bend light. Starting with natural sunlight, they scan a spectrum of wavelengths through the atmosphere by adjusting the width of the fine lines. A high-altitude capable aircraft, carrying a large sensor array on its back, captures the *refracted* light coming from the satellite. Since ODCs have distinctive activation energies, by measuring absorption peaks in the solar flux, they can detect the presence of ODCs at any height the plane can fly. They can even estimate concentrations. Over time, they expect to map out a 3-D image of the chemicals in Earth's atmosphere and start correlating with ozone levels and melanoma incident rates. I think it's very exciting."

"How close are they?"

"Apparently the satellite is already being built. Some small German engineering firm is fabricating the adjustable grating. The aircraft and the flight plans are being managed by the university. I tell you Parks, this has already attracted plenty of positive attention in academic communities around the world. Green groups have latched onto the idea, and even Willow Masters has thrown her support behind it. It is also being touted in the US by Jewish leaders citing this as a prime example of Israeli concern for world affairs – something notably lacking from many of the surrounding Islamic states."

Parks looked up again. A flashback to the Chalbi desert haunted him.

"This article is nice. Except for the 'balanced reporting' stuff here."

"Yeah, I knew you wouldn't like that bit," she said. "But it's all part of the game. If a reporter just focuses on the positives, pretty soon nobody buys her stories."

"But if you temper a fragile idea with a curt rejection by someone like Marcharelli Sr., lots of people won't even give it the time of day."

"How does he say it?" she asked, reaching for the sheet of paper. She read: "'He should leave this work to the established aerospace firms. Space is too complex for a one-man operation. Maybe he's just ahead of his time.' What do you think of that, Parks, 'ahead of his time'?"

"Okay, but then read what he says next."

In a flat tone, she read the last sentence of the quote. "'But by the time his ideas become economically viable, the patents will have expired.'"

"You could have left that part out," Parks said.

She scrunched her face up in a regretful grimace. "Wouldn't be balanced enough."

"Well, I've got to get going," Parks said, finishing his coffee.

"Wait a minute. What about my story. Time to turn back on the reporter!" she said, and turned her imaginary switch back on.

"I'm tired Heidi. Can you make it quick?"

"Sure," she said, and pulled her voice recorder from the shoulder bag and set it on the table. "Why don't you just tell me what happened today?"

Parks sat for a moment, but just didn't feel like talking.

"I'm sorry, Heidi. Everything went just fine, as Dante already told everyone. Not even so much as a glitch to talk about. I guess that makes for a pretty dull story."

"Then, tell me what you plan to do next. Where does the Space Energy Project go from here? What will your role be? Do you have a timetable for moving back to the States? Do you plan on heeding the alert issued by the American Consulate?"

"What alert?" he asked.

"I guess you've been too busy to watch the news, but it's growing into a big deal," she said. "The Israeli version of our CIA, called Mossad, thought they had identified a chemical warfare factory in Egypt. They prepared preemptive strike plans to take out the facility, like what they did to Iran's fledgling nuclear weapons program way back in the 1980s. However, to cover their bases, they first asked the UN to send inspectors. However, the inspectors found nothing but defensive technologies. The Mossad was told to stay out of Egypt, and to abolish all plans for an attack on the oasis facility. Oh, you should hear the politicians go on about this! Long-dormant Zionist antipathy is resurfacing."

"Oh, my God," Parks said.

"Yeah," said Heidi. "What's worse, Egypt found out about it! There's a mullah named Hamid Hababek threatening action against American interests. He wants them to stop supporting the Israeli military, stating that they threaten regional stability. He has followers not just in Egypt, but all through the Middle East, and even other parts of Africa. There have been attacks on tourists, so the consulate issued a warning that all 'non-essential personnel' should consider repatriating to the US. I wouldn't want to read about any mishaps befalling my star inventor!"

"This may not have shown up in your research, Heidi," said Parks in a subdued tone, "and I didn't tell you yesterday. But when I first came here, my survey party was attacked by Somali Shiftas. I was the only one who survived."

Heidi's eyes flew open wide. She snatched up the voice recorder to make sure the red light was on, and set it down closer to Parks. "Please, tell me more," she said, setting the unit on the table.

Parks drew in a shuddering breath, and began to recall the events of nearly a year before. It had been long enough that he no longer had nightmares, no longer felt the pressing guilt of all those deaths, and no longer saw Cloy Morgan's blackened hand. He told the story slowly. Heidi listened with rapt attention, and scribbled in a spiral notepad.

When he finished, she blinked her eyes like someone coming out of a light sleep.

"Can I publish this?" she said, breathlessly.

"I suppose," he said softly, but then his mood changed. "Look, Heidi, I'm really not comfortable with all this – all your plans to write about me. I'm no different from hundreds of other regular people, and if you try to make me look like some visionary, it's just going to be stretching the truth. You'll ruin my credibility."

"Well, okay for now," she said, "I'll keep this just between you and me. But I need something for my next article. How about going back home? There's a whole new nation to conquer."

"Been there, tried that," he said.

"Things have changed, Parks," she told him. "More and more people are waking up to the realities of modern life. Climate swings from CO_2 emissions cause droughts in Kansas and floods in Iowa. Kudzu is swallowing highways as far north as New Hampshire. Dengue fever is common now across the Southeast. Potato blight has hammered Idaho. Santa Anna winds have triggered fires all around Santa Barbara. Nobody feels safe anymore. The US finally signed the 1987 Kyoto protocol, just as a new environmental summit was announced in Ottawa. Canada is having fits. Their vast hydrocarbon resources in tar sands and frozen hydrates could fuel America's needs, but the carbon released into the atmosphere would further destabilize the climate."

"Geez, I didn't realize all that was going on," Parks said.

"There are enough pinch points, enough news coverage, that the average Joe on the street is starting to look at things a little differently. There's been a mood shift at home, Parks. People have heard enough of problems. They want solutions, innovative American solutions that will generate jobs. My editor believes this whole energy movement could really steamroll. And I'd like to get *your* ideas out there, so you can find the help you need to make it happen. Doesn't that sound exciting?"

Parks shifted in his seat. For all his attempts at humility, the thought of returning triumphant to an expecting nation was a fantasy he could not quite grow out of. A corner of his mouth moved.

Heidi caught the motion.

"So what *will* you do when you go back? What do you *want* to do?"

Parks remembered Dajwal Chellapilla. Because of the hectic schedule leading up to the launch, he had not contacted Dodge in over a month.

"I'm still deciding," he told Heidi. "I have a friend who is trying to set me up as a staff member at Purdue University, where I can work on the asteroid separator. I don't have a Ph.D., so we need to find a professor willing to work with me. They're all so busy with their own projects, that he's been having a hard time of it."

Heidi nodded. "And what about your family?" she asked.

"Well, the kids love it here. And my wife is doing some important work in the capitol. In fact, you might be interested in talking with her. Her story is actually more interesting than mine."

"What is she working on?"

"It's an education reform project. That was part of the deal President Mtongwe made with me, when he allowed this project to happen."

"You worked directly with the President?" she said.

"Yes, for a short time. Lately, I've been working with his aide, a Mr. LeFallon."

"The Olympic marathoner?" she asked.

"Yes," Parks said, laughing nervously as he remembered the interrogation at Nairobi prison.

"This is great, Parks," Heidi said. "Every time you tell me one thing, there's two more questions that I want to ask you."

"I guess that's why you're a reporter," he remarked.

"And that's why I want to report about you!" she said back. She stared at him, laugh lines scrunched up beside her eyes.

For just a moment, something stirred inside Parks. He panicked, and started to get up to leave.

"I've got to go," he said.

"Please stay," she said, and touched the back of his hand.

His skin responded with a tingle that left goose flesh on his arm. A warm sparkle moved up into his chest, and made his throat tight.

"Please?" she begged. She looked up at him, a slight pout on her lips. He could see where the lipstick had rubbed off onto her demitasse.

"Okay, but just for a little bit."

"Thank you!" she piped, watching him until he sat down again. She flipped back through her spiral notebook, her fingers animated and dexterous, scanned down one page with a red fingernail, stopped, and looked up at him. Her eyes sparkled with excitement.

"Let's start with you and Mr. Mifune," she said.

Chapter 7

West Lafayette, Indiana

A European Space Agency rocket lifted the ODC Mapper satellite the same week that the Vambolents left Kenya. The satellite was placed in an inclined circular orbit higher than low-earth orbit (LEO) but not as far out as GEO. With this orbit, the satellite tracks across much of the Earth's surface at subsonic speeds. By flying under the Mapper, the sensor aircraft rides in its refracted sunlight, collecting data. After several weeks of testing, the first flights were scheduled to start on the autumnal equinox, September 23rd.

Schools in the USA were just starting a new year. Elise enrolled Mindy in middle school and Eric in elementary school. Dajwal Chellapilla located a house for them off Northwestern Street, just a short walk from the engineering buildings of Purdue University. He had arranged for Parks to have a small, windowless cubicle on the third floor of Grissom Hall, former home of the Aeronautical and Aerospace Engineering Department. His position was called Staff Researcher, and he was assigned directly to the department head while he worked writing proposals for NASA grants to fund research on his inventions.

Dajwal's work to establish Parks at Purdue hadn't paid off until the syndicated columns by Heidi Bayer began capturing media attention. Her articles were like a spark to tinder, and had ignited a groundswell of public opinion. As Deputy Administrator, Dajwal worked in Colorado Springs with Indiana Senator Mason Barnes. Once enough of his constituents expressed interest, the Senator placed a phone call to the university Chancellor.

That same day, the Aerospace department head, Prof. L.Z. Benatz was 'asked' to cooperate.

During Dajwal's last visit to the campus, the normally gentle professor lectured him about 'cartoon engineering'.

"It's easy for amateurs to grab headlines with wild promises," he had said. "They get people all stirred up, and expecting miracles. Congress throws money at these crackpot ideas, which are nothing but hobby projects. Meanwhile, legitimate researchers, recognizing that true achievement is made by small, careful advances, watch their own budgets dwindle year by year. Instant-opinion media hype has the public expecting overnight successes, and breakthroughs 'just around the corner'. Once everyone starts expecting this, they have no patience with honest engineering, which may take decades to reach fruition. I don't want this Parks Vambolent using our facilities and our graduate students for a hobby project. And I don't want *Mr.* Vambolent to be a cartoon engineering headline-grabber who will make a laughing stock of our entire department."

Dajwal listened intensely, a pleasant smile on his face.

"Don't you have anything to say?" the professor said, exasperated.

"All of your points are very good ones," Dajwal said. "I don't doubt for a minute that your interest is for the entire department. And I understand your concerns. Frankly, I had the same concerns when I first started working with his ideas. But after closer study, I saw their potential. Certainly you are aware of the Seven Dwarfs mission? Did you know that that effort was begun *after* they learned of the work Mr. Vambolent was doing with the Japanese."

"I didn't know that."

"And did you know that Senator Barnes has nominated Parks Vambolent for the National Medal of Technology?"

"Never heard of it."

"Oh, it is the highest technical honor our government can bestow upon a citizen. The Vice President is right at this moment considering the nomination."

"Oh, God help us," said the professor.

"My counterpart at the Ames Research Center has already stated that he is expecting a grant proposal on these ideas. He has been instructed by Administrator Rinker himself that this project is to be given special consideration."

"All based on the stories by this Heidi Bayer?"

"Not based on, no sir. But her stories did raise awareness. And the need is certainly great. The people will see Purdue University as a world leader in the field of in-space resource utilization. Government officials from the United States, and many other nations, will want to see the progress that is being made. Together with the excellent work being done here already, the prestige of your department will grow considerably. And as more people recognize your leadership, it will be easier for your researchers to secure grants and contracts. I think that everyone will find some benefit in this matter."

The department head harrumphed, and frowned deeply.

“And besides,” Dajwal said, “Parks Vambolent is a very good man. He is careful, conscientious, and persistent. He was able to convince the Director-General of Japan’s mighty MITI to work with him.”

“Yeah, I read about that. And the ‘Desert Decimation’, and the ‘Space Energy Project’, and the ‘Ozone Mapper’. I saw the stories. My wife cut them out for me. Seems more like the stuff of urban legends.”

“Oh, I assure you he is the legitimate article, Professor Benatz. He is a true blue American. He is a humble guy, not an attention-seeker, not an ego-feeder. I know you will like him when he arrives at the start of the semester. Now, where did you say his office will be located?”

Purdue University

Parks Vambolent moved into his closet-sized office with a box of books and framed photos, an old laptop computer, and deep reservations about what Dajwal had promised. Everyone in the hallways recognized him. A few students had rushed up to shake his hand, eager-eyed and talkative. Pairs of people talking in the halls would pause to watch him walk by. Several of the professors he met seemed to greet him with disdain. Most were cordial, but none were openly accepting. Parks felt isolated and alone. He was not assigned any graduate students, but was paid a modest salary to write his research grants for submission to NASA, the first of which was due in three weeks.

His days passed slowly. Without an outside window for inspiration, and without a true friend on campus, Parks grew restless. In between short bursts of productive work, he would idle away reading news on his computer.

News of China had moved from the business arena to politics as their aggressive behavior towards neighboring countries continued to escalate. Their economic strength, technical prowess and space access had enabled them to develop a military strength that was second only to the US. While India was preoccupied with Pakistan, China had moved even further into lands disputed since the 1961 war.

Global warming was a perennial issue. Everyone, it seemed, had an idea to address it. Parks noticed one crazy idea of deliberately triggering volcanoes to put more dust into the atmosphere. Their theory was that the dust would cool the planet by preventing the sun's energy from reaching the surface. The number of supercomputers dedicated to modeling the Earth's climate was growing exponentially, but they still didn't have an answer. For the first time in history, the top five fastest supercomputers were all dedicated to climate research, eclipsing the giant machines used to design nuclear weapons, to model vehicle crashes, and to spy on the populace of the free world.

Heidi Bayer's series of articles had finished, and common memory of them began to fade. She had left the *Detroit Free Press* and she now had a syndicated column working out of the *Denver Post* covering the Departments of Energy, Commerce, Science and Transportation.

The Energy Project in Kenya was still providing hydrogen for Japan and distilled water for crops and animals. There were plans to expand as the hydrogen infrastructure in Japan grew more widespread. He read about Afren Kwapili's growing success with the new Kenyan Education System, and of course Elise updated him at home. The ODC mapper project was in full swing, and the reports he read showed that the concept worked nearly as well as planned. They would have the entire planet mapped in a little under two years.

An Associated Press article came in through Vambolent's newsreader program, celebrating the confluence of these two Green initiatives. His name was mentioned. They even wrote that 'Vambolent has signed on with Purdue University to develop further space technologies to improve life on Earth'. He liked that, and printed out the article, then pinned it to the wall of his office. He sat back in his creaking chair, and folded his hands behind his head as he read the rest of the article.

The article explained that a few weeks after equinox, the normal flight progression pattern would have carried the sensor aircraft too close to the microwave beam arriving four degrees south of the equator in Kenya. Instead, the flight plan filed by the Israeli university gave wide berth to the beam. The team would divert to Tel Aviv at the end of the day's run for a much-needed visit home.

The department head stuck his head into Parks' cubicle, startling him out of his reverie. Seeing Parks reading on the wall, he stopped his open mouth, and glanced at the paper. With a subtle eye roll and muted sigh, Prof. Benatz started again.

"Parks, the finance committee wants an update on your progress this Thursday. Five o'clock in the Wolf Room?"

'Sure, L.Z. I'll be there. Anything specific I need to bring?"

"Well… Results." Benatz said.

Parks slumped. He nodded, but kept his mouth shut. Benatz's head disappeared and Parks felt jittery.

He stood, headed for the bathroom again just for a distraction, but instead walked down the stairs and out into the Engineering Mall. The day was clear and bright and the sky a cerulean blue that seemed endless and deep. A cool breeze swapped places on his skin with hot sunshine. Papery leave-spiders rattled along in and among the people. Students wore the strangest clothes ever, but administrators and hot-shot professors still wore their stereotypical types of suits. The air was fresh and clean-smelling, and Parks moved slowly up-wind, leading with his nose.

He passed the aged Nuke building in the shade of a full-sized oak tree, and shuddered. His only fall-back plan, if the NASA grants didn't come through, had been be to beg his uncle for work. But when he asked, Bryce Sr. had made it abundantly clear that no one ever worked twice for Vambolent Enterprises.

Parks was nearly across campus when he realized he was already halfway home. On a whim, he continued, instead of turning around and working on his grants like he should have. Elise would be home, and the kids would be out of school soon. It was too nice a day for work, he decided. It was time to do something with his family.

With renewed resolve, he put away the nagging doubts, and walked a little faster. Moving briskly along the sidewalk, he searched the clear blue sky for that trail-less sparkle that could be the ODC mapper he originally invented. It was certainly big enough to see in daylight – if the sun hit just right.

Instead, a fiery meteor cut into the daylight sky, glowing white-hot at its core and streaking from left to right, down across his vision. Right before his eyes, the hurtling body broke apart into fragments, and each carried a new smear across the sky. They fizzled and faded, until the last one went out and all traces of the burn-up vanished. The sky was absolutely unblemished, as if nothing had happened. Parks' breath caught in his throat, and he slowed. Panic seized him. Indecision hit, and his feet came to a stop. In his head was the clear image of a meteor coming straight at him. Elise was beside him somehow. It came right for them. He blinked and opened his eyes wider at the same time.

He reeled, and felt a reality shift as the intense déjà vu competed with the fading memory of the meteor. He swayed a bit, in a sudden breeze, feeling disoriented. Parks shook his head to clear it, and took a swift, deep breath. Then, dismissing the premonition, he squared his shoulders, charged himself back up, and stepped forward again, all doubts successfully dismissed.

Over East Africa

Out in space, the satellite was a marvel of twenty-first century engineering. Within a large rectangular frame, a thin layer of low-viscosity gel was sandwiched between two high-tech thin-film membranes. Each membrane was covered with transparent piezoelectric actuators. By proper adjustment of the actuators the top and bottom surfaces could be controlled to sub-nanometer accuracy. This allowed the creation of a high precision diffraction grating, using the gel as the refractory material. Electromagnetic energy of almost any wavelength could be steered with precision to the sensors on the top fuselage of the aircraft thousands of miles below.

While monitoring the scan, David Pruitt typed in a command sequence that activated a subroutine buried deep within the assembly code of the autopilot routines. In the continuous interchange of data between the aircraft and the satellite, the hidden subroutine was transferred in three-byte increments, to hide its presence. In a few minutes, the upload would be complete, and none of the other scientists, nor their crude snooper programs, would ever detect it. The original satellite software, scrutinized by a peer review service at the insistence of the US, would now yield control to signals from a new source.

On the ground, at a secret base near the Jordan border, a satellite dish tracked the ODC Mapping satellite. At the pre-appointed time, they transmitted a test sequence, to pulse the attitude jets. A radar unit, situated miles away, detected and relayed the news that the satellite had indeed responded to their control. The slight movement was detected also by the sensor aircraft, and the course adjusted accordingly.

The scientists and engineers on board with the man called David Pruitt all noticed the small but atypical lurch of the airplane during an otherwise smooth flight. Some looked around in concern. Others passed it off as just a high-altitude gust. In the end nobody did anything about it, and went back to their data analysis and equipment monitoring. That movement was the signal to Pruitt to activate the replay of past data. He entered a new command string that would retrieve the stored scan data from the previous day and load it into RAM. A pre-programmed offset would be applied to adjust for today's flight plan, and then the modified signal would be patched into the input channels of the main processor board, instead of the on-line sensor signals. With an undetectable microsecond-long hiccup, the passengers and their monitoring computers would never know the difference.

Now, from the operations base in Israel, the mapping satellite was given a signal to modify its orbit. From below, behind, and off to the side of the beam line between the SPS and the Kwapili Range, the ODC Mapper slid upwards and vectored towards an intercept course. The maneuver would take a little over an hour. During this time, another phase of the operation began.

Kwapili Range, Kenya

Four Mossad agents hiding in the hills above the receiving antenna activated power to their robotic drones. The tiny airframes were camouflaged as vultures. They launched their six mechanical birds and retreated to the cave where they could monitor the progress of this mission.

Slowly, seemingly without purpose, the mock vultures ascended to several hundred feet, and began slowly circling the perimeter of the antenna. One by one, they drifted apart from each other, moving in two lines around the circumference. The hour passed as they made an unhurried trail to their locations above each of the fail-safe microwave detectors. Once each individual was wheeling around its sensor, it activated a microwave horn, tuned to the same frequency as the power beam streaming down from the SPS.

Intended as a backup system, in case the main control processors would go awry, these individual detectors acted as dead-man switches, safing the system against drift. Should the signal drop out from any of the detectors, that could indicate that the beam had become misaligned out of tolerance. The continuous Alignment_OK signal beamed from Kenya to GEO would be interrupted, and a circuit relay aboard the transformer would unthrow, and cut off the power.

This simple circuit was their backup assurance against any sabotage or meddling from space.

Once the drones had locked onto their targets, a signal was sent, via satellite phone, to the operations base. They were ready to begin. A return signal would be sent, timed to the beam deflection, to boost the power on the drones, so the safing circuit would never detect an interruption.

As the ODC Mapper slid into position, just beneath the SPS antenna, a reverse thrust was applied to brake the motion. This was the most critical phase of the operation, since the frame of the grating satellite would block a small fraction of the microwave beam. The drop in power might be noticed, but the expectation was that there would be no immediate cause for alarm. In a few moments, when the gel finished sloshing, the diffraction grating would be activated. The failsafes would still be giving an all clear message, even though the microwave power to the main antenna would soon drop to nearly zero. Only human intervention could shut off the power. Mossad figured that would take at least 30 seconds. Plenty of time.

At high noon, directly over Voo, Kenya, the piezoelectric actuators aboard the ODC Mapping Satellite received a new signal. Instead of forming the parallel lines of a diffraction grating, they formed instead into a Fresnel lens – concentric circles used to focus light onto a point. The gel inside the two membranes was extremely efficient at diverting the flow of microwave energy. The precise control of the lens brought a large fraction of the wide and powerful beam of the microwave antenna to a tight spot the size of a garbage can lid. From 22,000 miles up, diverting that beam a few degrees is easy.

A proportional-integral-derivative controller, with real-time feedback from camera observations by a passing Israeli spy satellite, was used to precisely aim the focused beam of microwaves. The initial spot was nearly a kilometer off target, and it took almost two full seconds for the spy satellite to mark it, and begin to zero-out the error. With each successive loop of the control routine, the aiming error dropped by a quadratic factor.

In Kenya, the drop in power set off alarms throughout the control trailer. Technicians jumped up from their routine tasks, caught entirely off-guard by the sudden break from the legato pace that their jobs had become. In the confusion, none of them thought to shut down the SPS, since the SPS seemed to be doing this on its own.

No one noticed that the failsafes were still reading a microwave beam. Confused, they called Hanazawa on their satellite phone. This delay gave the Mossad operatives the time they needed to complete their objective.

Western Desert, Egypt

Beside a small oasis, a mostly-underground facility showed up as a quadrangular compound of adobe buildings the same color as the sand. Drums of fuel were stacked in one corner of the courtyard, in the center of which stood several desert-camouflaged trucks and two black Mercedes sedans. Hidden from view outside the walls was a row of windows into offices, and several doorways. In the corner opposite the fuel drums was a large rectangular plate of sheet metal flat with the ground, hinged at two sides, and split in the middle, where an underground elevator could be raised to the surface. Four armed guards walked above the offices, giving them a view of both the compound and the land surrounding it. Across the oasis were the gardens where much of Egypt's aromatic oils were pressed from flowers. From an airborne or satellite reconnaissance view, the chemical plant would not look much different from the buildings housing the presses and bottling facilities of the perfume industry.

Whisking across the desert came a sand devil like none ever seen. The super-concentrated microwave beam heated the sand instantly, causing sparks to erupt from elemental metals in the silica matrix. The popping of the metallic components of the sand burst dust into the air, where the wind picked it up. The dust devil with fire at its base passed over a corner of the small pond which separated the perfume factories from the chemical plant. The open water vaporized instantly, creating a geyser of steam that zigzagged towards the compound at over a hundred miles an hour. The beam of energy slowed right over the compound wall, and roasted two of the unprotected guards. Their bodies were torn apart from the top of their heads downward, in a violent eruption of cooking body tissue and super-charged metal.

Into the corner of the compound, the beam came to rest for a full second on the stack of fuel drums. When the steel lids of the top row of drums were hit by the microwaves, free electrons were boiled off instantly, their kinetic energy ripping apart the iron and nickel atoms until the lid exploded in hot sparks, igniting the diesel fuel which reacted with oxygen in the air. The beam drilled down through the liquid and into the next row of drums as the superheated fuel exploded outward. The blast wave disintegrated the wall behind the drums, and blew down most of the external structure of the entire compound. The vehicles were incinerated, tumbling towards the office windows, which were blown inward from the shockwave. The exposed underbellies of the trucks detonated as their own fuel tanks were heated and burst open, adding to the conflagration and blast pressure. Beneath the fuel tanks, the underground complex was compromised as the blast tore downward through two levels of reinforced concrete floors. Burning fuel splashed through those levels, igniting everything combustible along with them in a firestorm that destroyed machinery, glassware, and plant workers indiscriminately.

After resting there momentarily, the beam walked over to the freight elevator door, causing another shower of sparks that was lost in the growing blaze. This heavier metal took a few moments before it melted. The weight of a Mercedes, which had been pushed onto one corner of the door by the first explosion, caused the softened metal to sag. The sedan tumbled down into the elevator shaft, falling past the four levels of the complex. The microwave beam continued to pour hundreds of thousands of kilowatts down into the hole, and the German automobile, which landed on its roof, now became a bomb. The fuel tank burst open. An eruption of flames ripped open the access doors, sending hunks of flaming automobile bouncing through the halls.

Chemical storage tanks, already rocked from the percussion of the fuel drums exploding above, broke loose and spilled their contents on the floor. Toxic and flammable materials combined and were set ablaze, filling the lower level with deadly fumes. Lab workers, caught completely unprepared, died of flames and fumes and asphyxiation.

When the tank of trichloroethane on the fourth level exploded, the floor to the third level broke open at the structural support and a blast of white-hot fire billowed upward, rushing through every open space, and creating new ones where the pressure was great enough and the resistance insufficient. Secondary blasts from other chemicals began burning unabated, filling the entire complex with fire and smoke. Then, just seconds after the attack began, the center of the compound gave way and collapsed downward. The weakened floors below could not withstand the extra weight, and the entire complex fell down into the bottom level, where more chemicals and gasses found sufficient oxygen and ignition sources to burn. A succession of fireballs, red, orange, yellow and white, erupted from the center of the ruined facility and rose like a bouquet of beautiful flowers unfolding over the oasis.

Microwave energy continued to pour down from space, but there was no more damage possible. The complex had been utterly and completely destroyed. Several more seconds passed by, and then the power was killed from Kenya, when they finally received the order to shut down the SPS.

PART II

Chapter 8

A wooden ship navigated the Northwest Passage for the first time since it was first anticipated by John Cabot in 1497, leaving Fundy Bay and arriving in P-K Harbor, Kamchatka. The ship arrived with great fanfare, heralded by crowds of international businessmen and tourists travelling from prime locations in now-fashionable Siberia. Climate shifts brought longer growing seasons there, and badly-needed food was being shipped everywhere from Russia's breadbasket region.

But with the arrival of corporations and resorts, many local resources became strained. Hospitals were ill-equipped to handle the dizzying array of diseases brought into the region. Water, pumped from aquifers, filled up behind old dams never built to hold so much. Local economies flourished and government coffers filled at the same time as the thin topsoil began to wash away from increased use. Eroded silt clogged streams and rivers, concentrating human and industrial waste. Displaced wildlife attacked the crops and had to be killed in huge numbers. Bloated and infected corpses of elk and deer downed by Chronic Wasting Disease became a common sight drifting past brightly-painted new hotels.

Petroleum resources in Siberia were being pumped out in great quantities, following massive investments by oil companies and the state governments. Byproducts such as fertilizers and herbicides were poured onto the land to increase crop yields even further. Runoff of the excess created algal blooms in the Arabian Sea. Fish stocks, asphixiated by the algae, experienced massive die-offs, slashing the protein food sources so needed by the region.

Droughts in central Europe brought dust-filled winds heavy with nitrates and sulphates from the long-term industrial base, denuding forests and causing respiratory problems across the continent. Incidents of cancers were rising, and the discovery of mutated insects, frogs and reptiles became increasingly common.

Atlantic storms brought dead sea creatures to the shores of the United Kingdom, starved from the consumption of the oceanic food chain by hordes of hungry krill. Dykes around Amsterdam broke through, spilling salty sea water onto 20,000 acres of farmland, homes, factories and cities. Plans to rebuild the dykes were frustrated by the rising sea levels, and many young Dutch just gave up, emigrating further inland.

Hurricanes continued to pound the south and eastern seaboards of the US, drawing more relief and aid from the American populace and their beleagured government in Colorado Springs. Northern cities became further crowded and jobless rates soared. Inflation rates rose as the industrial base of the southeast collapsed and certain goods and services became scarce.

Cries for change came from every quarter. Congress slashed budgets indiscriminately, focusing more and more on basic needs. Crumbling infrastructures went unrepaired; school systems dropped vital services; nearly all investment in science and technology was re-directed. Austerity measures were called for, and rationing began on basic commodities like grain, oil, meats, metals, coffee and orange juice. Engineers from all government agencies were being re-assigned to public works administered by the Army Corps of Engineers. National labs, foundations, and central agencies saw their people allocated in a diaspora of intellectual talent. Few agencies were so hard hit as the National Aeronautics and Space Administration, which was culled to a level never seen in the more than half-century of its existance.

West Lafayette, Indiana

Mindy Vambolent returned home from Middle School and let the screen door bump into her come-along so it wouldn't slam shut. She worked up a perky voice as she passed the den.

"Hi, Daddy."

Parks Vambolent turned part-way from the television program he was watching. His brown hair fell over sideburns that had turned gray. He had the translucent pale skin of someone who had recently put on weight, and didn't spend much time outside.

"Hi, Pumpkin. Hey, come see this on the news."

"In a minute," she called back. In the kitchen, she parked her bag and picked a pear from the fruit basket. Gathering herself up with a breath, she set her smile in place and entered the den through the swinging door.

"What is it, Daddy?" She stood, and looked at their old cabinet model TV. It was missing a few plastic pushbuttons. The screen had a grey cast.

"Remember how NASA is retrieving that asteroid?"

How could she not.

"M-hmm."

"Well, they're having some trouble with it."

The cheer in his voice was unmistakable.

"Are they going to lose it?" she asked.

Parks turned to look at his daughter. He smiled at the simple beauty of her young face. She was so full of promise. Promise for the future. But what would that future be? His face fell.

"Nothing so bad, I'm afraid. Will you watch with me?"

Mindy nodded, and sat on the ottoman. Parks raised the volume by mashing the remote button. A confident-looking spokeswoman, wearing the Seven Dwarfs mission lapel pin, was in the middle of her explanation.

"The normal wear-and-tear on the thrusters has been a little higher than we expected, operating closer to the sun as it has. This leaves the mission without enough power for its original objective. Because the asteroid has moved more slowly than planned, the Moon has swung in front of it, blocking its path. Normally, that would signal a NO GO for Earth orbit. However, NASA scientists have devised a way to salvage the mission. Actually, there are two alternatives and NASA has determined that either one would be acceptable. Our Administrator, Mr. Lewis Rinker, has allowed us to ask you, the citizens of the United States, to help us make the final decision. I'll explain the options, and then you can vote. The voting is open to anyone school age or older. You need only your Social Security number, and access to our mission website, to participate."

"What's *my* number, Daddy?"

"I don't remember, darling. We'll ask your mother when she gets home. Now, listen."

The first option filled the screen, labeled with a large letter A.

"Option A is to steer the asteroid directly at Luna, slow it down, and land it there safely for later investigation by astronauts and space researchers."

"They're not allowed to do that!" Parks exclaimed, sitting up in his chair. "That violates the Moon Treaty of 1979, Article 7, which prevents large scale destruction of the lunar environment."

"Hush, Daddy, she's explaining Option B."

"…barely missing Luna, then coming to rest in L4. We originally targeted L5, but L4 is just the mirror image, and an excellent place to park the asteroid. The only complication is that we have to use the Moon's gravity to slingshot the asteroid back to L4. This move is called a gravity assist. It's been done many times with NASA space probes that have passed by Jupiter and Saturn, and we understand the physics quite well. But the Moon is much, much lighter than Jupiter, and has much less gravity. So to perform the gravity assist, the asteroid will need to be precisely aimed, and pass within just a few miles of the lunar surface. That is option B."

"That's crazy!" Vambolent exclaimed.

Mindy looked down at the pear in her hand, and then over at her father. That's what kids at school called *him*.

The spokeswoman continued: "...between option A and option B. Of course, we also have the option to abort the mission, and keep the asteroid in an orbit around the sun. This is called a heliocentric orbit. Only this new orbit will be near enough to the Earth, that we can pick it up again in a few years when it catches up to us. That is option C."

The voting information showed on the screen. Parks grabbed a notepad and scratched the information onto an empty sheet. When he finished, he set down the notepad, picked up the remote, and pressed the mute key.

"Do you remember Africa, Mindy?"

"Yes," she said. Kenya had been great fun. But it was the return home she didn't care for. Even now, years later, a new kind of nightmare was developing for her. There weren't many families named Vambolent. By now, everyone in West Lafayette had heard of her dad, and the Oasis Massacre. Kids at school said her last name with contempt, as they must have learned from their parents. Many of the families had a mother or father who worked at the university, from which her father had been fired immediately after the Massacre. By now, there wasn't anyone in the small town who didn't know that Parks Vambolent was indirectly responsible for the event.

"Do you remember how we tried to set up an economy in space? Remember?"

Mindy had been immersed in her father's dreams so completely that even now she could not fully distinguish the real from the fantastic. While other kids were read bedtime stories by Dr. Seuss or Roald Dahl, Mindy's imagination was filled with geometric manufacturing, isotope separators, and global energy crises, all mixed in with telepathy, spiritualism, and ancient civilizations. She possessed an eidetic memory, and could still recall images of the stories her father had spun as she fell asleep in her first real bed in Nairobi. Her eyes had always been wide with wonder. She watched him now with different eyes.

"*I* was the one who got this started. Me, and Mifune-san." The name caught in his throat. Mifune had taken his own life immediately following the Oasis Massacre. No one had ever contacted Parks directly, and there was no note from Mifune. Parks learned of the suicide through the news. The guilt and grief were like a wound from a triangular blade that would not close up and heal.

He swallowed with difficulty, and spoke on. His voice was hoarse and raspy.

"They watched us. After all those years of blowing me off, they sat back and watched. It wasn't until we became successful that they decided to start a program of their own. Without me! And did they ever give me credit? No, just the opposite. When things turned sour, I was called a traitor, a bad scientist, a viewgraph engineer."

His eyes grew large.

"And now they're screwing it up. Don't you see? Don't you see how things really are? It's up to you, Mindy. You and your generation. Do you understand what I'm saying?"

"Yes, Daddy," she replied. Her lower lip quivered.

"Oh, honey, I'm scaring you. I'm sorry." He reached for her, and pulled her to him. Her pear tumbled to the floor. It rolled under the recliner.

Her voice was muffled against his chest. "It's okay, Daddy."

He held her out and looked closely into her face.

"I'm sorry, honey. It's just… It's just…" His eyes scattered this way and that, looking for something – maybe choosing from too many somethings.

"I've got some homework to do, Daddy."

"Okay, Pumpkin. You go do your homework. If you need any help, just ask, okay?"

"Okay. See ya."

He tousled her hair. She stood up and went back through the swinging door.

"Damn," he whispered under his breath, and reached for the remote. After five bad clicks, he pried at the frayed duct tape over the battery bay with shaking hands. He peeled back the tape and looked at the cheap batteries inside. He knew without checking that they had run down. They were dead.

"Damn," he said again.

He sat there a moment, then dropped the remote onto the coffee table. Rolling himself out of the permanently-reclined easy chair, he moved to the stairs and began pulling himself up the railing. His pulse pounded painfully in his chest, a long cry from the athlete's heart he never used to give a second thought.

The fourth bedroom of their home was his office cum laboratory. Nearly all of the electronic gear he had kept in storage during their stay in Kenya had since been auctioned over the internet. Two of his key patents were up for their final maintenance charges, but his household was already in debt.

The only electronics still remaining were for his Brain Wave Stimulator, an invention he had worked on after his summary dismissal from the university. Once, he had imagined it would be his cash cow, the source of funds for his more-important space inventions. But the FDA had rejected his license application on the grounds of safety. Despite many queries, he could find no venture capitalist willing to invest in a product that couldn't be licensed or sold.

All the furniture was pushed into the four corners. In the geometric center of the room sat a high-backed chair which was screwed down into the floor with four angle brackets. Mounted to the chair was an airplane-style headrest with fold-out wings.

Parks pulled down a hinged arm, fixed to the ceiling, of the type used for dental X-ray machines. At the free end of the arm was an over-sized helmet studded with small coils of wound copper wire. Dozens of insulated cables spiraled along the hinged arm, and disappeared into a hole in the ceiling. The gaps in the hole were stuffed with toilet paper.

Above him, in the attic, two parallel antenna wires mirrored another pair of wires which were hidden under the rug. These broadcast a very low frequency electric field that filled his office with energy. Elise had insisted he take down the aluminum foil he had initially used to paper the walls. Instead, he had stapled wire screen mesh from the hardware store onto the bare walls, then covered the mesh with wood paneling.

His workstation sat inside a box of bent sheet copper that had once been shiny, but was now the color of Mississippi mud, splotched with fingerprints. Stored on a DVD-ROM was a positron emission tomography map of his brain, obtained from Home hospital when he still had money to spend on his projects. The ancient machine ran a Green's Function program. By selecting a location within his skull, the Green's Function derived the configuration of magnetic fields needed to focus the energy from the long antennae onto that spot. An electromagnetic solver translated this into electric currents that were then fed into the coils on the helmet.

Parks settled into the chair, and pulled the helmet snug onto his head. A bent-metal box with several switches and dials was mounted to the armrest. Through long practice, his fingers touched lightly, seeking their rest position. To start with, he selected a pre-calculated program that filled his whole brain with electromagnetic energy at the Alpha frequency, alert yet relaxed. A pair of stereo speakers, extending down over his ears, played instrumental tones designed to harmonize the two halves of his brain. With eyes closed, he pulled in deep breaths through his nose, letting each one escape from his lips with a sigh. In less than a minute, he was in a near trance. The worry lines that were patiently carving his face into that of an old man took a short break.

Using the controls, he changed frequency in the attic generator, moving to the pace of rapid eye movement (REM) that heralds the transition between waking and sleeping. In REM, his breathing became shallower, and his body twitched here and there. Images arose spontaneously in his mind. He watched, remote and disinterested, a state he could never quite achieve in normal, unaided meditation.

His mother, in white, on the Moon. But she was Chinese. She swatted flies around her head. They were shooting stars, criss-crossing the sky behind her, growing more and more numerous by the second, until their persistent traces overlapped into a blinding brilliance of featureless nothing/everything. His mother washed into white, the white-grey of the Moon, and then she was no longer there. A clear voice spoke a quick phrase from a point inside his head: "when the new Moon is full".

It was a woman's voice, but it was not feminine. Genderless, perhaps. The feeling of a portend came with the spoken words, like that would be the sign that all was done.

"When the new Moon is full, what?" Parks thought to himself, and the thought dropped him out of REM. He was in Beta, normal waking consciousness, as he wondered, "Who was that? Why won't they say something else? What did that mean anyway?"

He hated when the voices came – it jarred him from the balance he craved and could find only here. Another fingertip command, and the machine went into Delta, the frequency of deep sleep. The helmet held his head, and in moments, he was snoring lightly.

* * *

Elise Vambolent came home with her hair in disarray from the wind. She yelled over her shoulder for her son to hurry, and came in the kitchen door. She was panting.

"Hi, honey," she said to Mindy.

"Hi, Mom. Welcome home!"

Elise kissed the top of her head, letting her nose linger briefly over the sworl where a sweet scent always floated.

"Love you."

"Love you, too, Mom."

They hugged upside down, one seated and one stooping, and made kissing noises with their lips.

Elise unloaded her burlap bag in one motion, and swung her briefcase onto a kitchen chair with the return stroke.

"How's your dad?" she asked in a stage whisper.

"He's okay, I guess. There's some NASA thing on, and he sorta wigged out."

"Hmm. Is he upstairs now?"

They both looked up. The office was right above them. Parks had of course explained about the EM fields, assuring them it was perfectly safe. Still, they had their doubts.

Eric Vambolent lumbered in. He was still small for his age, like his body was waiting for some cue to grow up. His eyes were quick and alert, but the rest of him was like a lazy cat – curious, slow, careful, prone to flopping down wherever he felt like. He slid his backpack so that it bumped into the corner, and was reaching for the refrigerator handle in the same motion. With the door fully open, he leaned one arm on it casually and stared inside.

Elise slid past him, reached over his head, grabbed a drinkable yogurt, and put it into his open hand.

"Thanks, Mom," he said, looking up at her. "I'm goin' upstairs."

"Wake your father at 6:30. Have any homework?"

"Nah." He never did.

Eric ascended the stairs, seeming to savor each step, and turned into his room. He kept his video games in standby mode. A motion sensor, installed by his dad, reactivated them. The old CRT monitor fired up, wavering unsteadily from the electric fields emitting from their attic. Eric didn't mind. It tested his edge in a new way. He sat cross-legged, logged in from the keyboard, set that down, then slipped on the glove and goggles.

An hour later, Eric saved his portion of the interactive video game and went to rouse his father. He walked in, pressed the red-colored end program button on the armrest controls, and swung the helmet up and off. Parks' head lolled sideways, and he awoke with a snort.

"C'mon, Dad. Dinner."

Parks Vambolent smacked dry lips. The neurotransmitters in his brain had only partly reset from their forced release. Then he remembered the asteroid, and bolted awake.

"Eric. Hi."

"Hey, Dad."

Murmuring to himself, Parks shut down the Brain Wave Stimulator, finally realizing it was already off. He tried to pat his son on the head, but Eric shook it away.

"You've got to watch the news with me, Eric. In your room?"

"… shure."

"Thanks."

In Eric's room, Parks sat on the bed as his son switched to the satellite feed.

"Try 137," Parks said.

Eric punched in the channel for news, and sat back to watch. It would be a few minutes before the asteroid story come on again. Parks filled him in as they waited. Even as he listened, Eric pulled up a screen-in-screen and browsed to the NASA website where they could enter their votes.

"Hold on," Parks told him. "I've got to get everyone's number."

"Don't I get to pick for myself?"

"Not this time, son."

"Not fair!"

Parks left the room and went down the stairs, blinking into the bright light of the kitchen.

"Hi," Elise said, looking over her shoulder. She was cutting onions.

"Do you have the Social Security numbers for the kids?"

She frowned.

"Why?"

In a rush, he explained it all to her. With a sigh, she rinsed her hands, and dried them on a dish towel.

"Here, take over," she said, and handed him the knife.

Moving to the secretary, she noticed Mindy's gaze, quietly watching. The two exchanged that kind of grin or grimace where you tuck in the corners of your mouth.

Parks held the onion, slicing through the root on his first cut. By the time he was chopping the long, curving segments, his eyes had filled with water. Elise handed him the blue and white cards. He left the kitchen blinking and using the shirt fabric on his shoulder to dab the tears.

Elise sighed again, and dumped the onions into the skillet, where they sizzled. Mindy bent her head back over her school books, holding a bubble gum-scented highlighter under her nose.

Upstairs, the asteroid story was being repeated. It was an exact replay of the earlier program. With Parks watching over his shoulder, Eric entered his father's number, then his mother's, then Mindy's.

"Now yours, Eric."

"But I want to vote for myself."

"Which one would you pick?"

"I wanna see them crash it into the Moon!" Eric said, a crooked smile slashing across his face.

"Give me that," Parks said, and took the keyboard from his son. For the fourth time, he selected the NO GO choice, Option C. This time, on Eric's number, he tabbed into the comment field. Typing furiously, he wrote a highly technical description of the worst-case scenario for Option B. He quoted the Moon Treaty in lambasting Option A.

"What are they thinking?" he said, mostly to himself. He pressed Enter, and absently passed the keyboard back to Eric. "They lost half their talent when India recalled the visas for their technical people. The other half is being re-assigned. NASA is drained. Rinker thinks he's got it all under control, like the other great administrators. But his entire organization is crumbling. The only objective they are working towards is continuing to have an objective to work towards."

He paused, thoughtful.

"Maybe they are just playing this for the publicity. What do you think, Sport?"

Eric had been watching him closely. He shook his head in answer.

"I didn't think so either," Parks said. "Let's go eat." He was in a good mood.

Eric held the railings and swung his feet out, dropping three steps each time. Parks followed, sniffing the aroma of Elise's cooking. They sat down at the kitchen table where steaming plates of vegetarian shephard's pie sat at each place. A mound of soy-milk mashed potatoes, ringed with microwaved corn, was topped with texturized vegetable protein grilled with garlic, onions, green peppers and spices. A meatless gravy steamed from a boat in the center of the table.

Parks sat down and started talking almost non-stop about the asteroid problem. It was the first time in a long while he had spoken about anything with such passion. After ranting for a while, he recited The Logic to them, which they all knew so well.

"One, oil production has peaked, and energy costs will continue to climb. Two, continued economic growth and geopolitical stability depend on low cost energy. Three, solar power is the only choice for sustainable, pollution-free energy. Four, the only place to collect solar power is in space. Five, the only efficient way to make space solar power is to use materials already in space. Six, the only way to process materials in space is with the asteroid separator. And, Seven…" He took a breath. Mindy and Eric finished the last sentence with him.

"*We own the patents*!"

Elise smirked and shook her head.

"Right, kids." He smiled. "And like the Chinese say…"

They played along with him, echoing in unison.

"*Dig the well before you are thirsty*!"

Elise groaned theatrically, and the children giggled. They all raised their cups in a toast, and drank. It was a family ritual they hadn't done in a long time.

Each of them loaded a mouthful of food and chewed reflectively. There was a peacefulness in the pause. Eric was watching everybody closely, chomping his too-big bite with enormous swings of his chin. Mindy said he was gross. Elise tilted her head towards him, but he just beamed her a big, squinchy smile, oozing mashed potatoes. She laughed lightly, and her gaze moved over to her husband who met her look. For a moment it flickered again – that electric shock that pricked their chests every time their eyes locked.

A piece of their happier past came alive in that moment. Their family dinners together in Nairobi had always been grand affairs that sometimes lasted all evening. Especially near the end of their stay, the Vambolent family had been on a roll. All four of them had grown so close, they became connected to each other in ways that few families ever experience.

The media orgy following the Oasis Massacre and the loss of work wore heavily on Parks. After he was asked to end his short stay in Grissom Hall, he had worked part-time from home doing technical writing. The contract company he worked through offered him a raise and a promotion, and had him teach writing skills to low-wage centers across globe. After six months conducting on-line and on-site training classes, his material was mature, and the training became rote. At the one-year review, the contractor failed to renew him. They didn't tell him, but he knew. While they had needed him to develop the course, they no longer needed him to teach it. Probably one of his own students was hired to do that at a third of the modest wages he had earned. He had been unemployed for the last five months.

Elise grabbed onto the moment while it lasted, hoping to make the most of it.

"You know, darling," she said, setting down her fork and leaning forward, "you actually *did* what you set out to do."

"How's that?"

"Remember when you were a conference rat, trying to get somebody – anybody – to listen to your ideas? Remember what you used to tell me when you came home? Remember? You said you didn't care one whit if you ever got credit for your ideas. All you wanted was for somebody to listen, for somebody to *use* them. You said it was more important *that* it gets done, than that *you* are the one to do it. Isn't that right?"

"Yes, I suppose."

"So, they're doing it, Parks. They're *doing* it. Don't you see? It's getting done. That's what you always wished for."

Her face was full of expectation.

"Yeah, Dad, you *did* it. Way to go!" Eric said.

Parks looked in surprise at the unexpected burst of words from Eric. He chuckled, and it turned into a laugh. It was a rusty laugh, and seemed to kick loose some phlegm in his lungs. The laugh turned into a coughing fit, but he was still smiling, his eyebrows pushed upward.

"You have a point," he said, and took a gulp from his wine glass. His forehead frowned deeply. "But they're doing it all wrong. Conventional mining, designed in one gravity, doesn't make sense in zero g."

"They need an asteroid separator," Elise said.

"That's right."

"And sooner or later, somebody, somewhere, will realize this and call you up. Isn't that the part Dajwal was helping with? Planting seeds? You just watch. Something will sprout. Maybe it will be like Yoshi and Afren said, after these midget space miners fail on their own, they will eventually come to find you. It's just a matter of waiting."

"Until the patents run out," Parks said under his breath.

"I miss Uncle Dodge," Mindy said suddenly. She had been beaming with happiness, and her cheeks were still a little red from smiling. Now, the blood drained back away.

"Me too," said Eric around a mouthful of food. His head dropped forward a little.

"What is Dodge doing now?" Elise asked Parks.

"We haven't spoken since last year."

Parks lapsed into dull silence, and the moment faded.

Mindy looked rapidly back and forth from her father to her mother three times, but whatever fleeting joy had come was gone. She began chatting nervously about her day at school, talking directly to her mother the whole time.

Parks finished quickly and cleared his place. He turned on the water, re-filling his wine glass as he waited for the hot. He plugged the sink, squirted soap, and grabbed the sprayer to whip it into a froth. Watching the mound of suds grow taller, he drank deeply, barely hearing the end of the meal behind him.

One by one, his family members brought their dirty plates to the soapy water. Eric pretended to bumble, and fell lightly against his dad, splashing his plate into the foamy suds. Parks grunted, and Eric playfully darted out of reach. Mindy handed hers over formally, with two hands. Elise hugged him from the side and kissed at his cheek. She seemed about to say something for a moment. Instead, she just gave him an extra squeeze around the waist, and a peck on the neck.

"I'm gonna do some e-mail," she said, and went up the back stairs to the loft over the small three-seasons room. The loft was a tiny office with a space heater, a desk, chair, and computer, surrounded by low book shelves whose contents were filled to overflowing. Pinned to the tapered walls were dozens of printouts of children's artwork or poetry. In Africa, Afren called her Aunty Lise, and under that name she was now quite famous. Some of the hundreds of e-mails she received each day were coming from Pacific rim countries too. Her automated reply included a letter of encouragement and thanks. It told them that she did read almost every email sent to her, but that she could not respond to more than a few each day.

Elise sat down and logged in. As her configurations loaded, she leaned back and worried about Parks.

*   *   *

As the days counted down to pericynthion, the asteroid's point of closest approach to the Moon, Parks grew more and more agitated. He had always talked to the television, but he no longer needed that prop to launch into a diatribe. He uncorked a wine bottle earlier in the day to dull his anger and frustration. He no longer had the patience for the Brain Wave Stimulator.

Parks could not shake off the image of this asteroid slamming into the lunar surface. What would that do? Surely the mission planners knew the fact that chunks of rock from Mars had been found in the ice cap of Antarctica. That was a 50 million mile journey. Luna is 200 times closer!

Who to tell? How to reach them?

During the brief month when he was considered for the National Medal of Technology, he had been contacted by Cabinet members, NSF staffers and NASA Public Relations people. He still had their electronic business cards, but did he dare contact them? Would his opinions ever be considered seriously again?

No, he was held up as an example of the worst type of scientist – blindly creating technologies that are easily turned into weapons. Every media network he had watched seemed to have found an expert to debunk his amateur science as dangerous, his actions as irresponsible, or his motives as un-patriotic.

Abruptly, he had found himself a pariah. The White House stopped calling, the NSF stopped e-mailing, and he was instantly labeled *persona non grata* across the entire country. Professor Benatz asked him to move his things out and lay low: "Just until things blow over", he had said to Parks.

He had never called back.

Parks considered a trip to Colorado Springs, but didn't have the money. In desperation, he finally tried calling, using the old contacts he had gathered. Indolent switchboard operators and diligent secretaries stopped him. Then, he sent e-mails. In each missive, he pleaded for the NO GO option. It was the only safe choice, he wrote. Parks sent a paper letter, express delivery, to Lewis D. Rinker, NASA Administrator, appealing for a wiser decision. Yet Parks was keenly aware that NASA needed a PR success on this mission, or face even deeper budget cuts and personnel losses in the coming fiscal year. With such a high-profile mission at stake, a mission with such great commercial potential, how could Rinker pass up the opportunity for a home run?

* * *

The night before pericynthion, Elise came into the backyard where Parks was watching the sunset with binoculars. She was wrapped in a horse blanket.

"Isn't that bad for your eyes?" she asked.

Parks started. He forced his voice to work, and to sound normal.

"It's between the sun and the Moon," he said pointing. A fingernail Moon hung over the setting sun. Clouds overhead were salmon and dusky purple. "After the sun goes behind the Earth, I should be able to see it for a few seconds, according to the Science Channel."

She squinted, her face glowing in the reddening light. Parks was sitting where the sun would set between two houses on the next hill of their neighborhood.

"I know this isn't a good time, Parks, but there's something we need to talk about."

"What?"

"Mindy wants to learn the cello."

"That's fine."

"Lessons are seventy dollars a month."

"Oh."

"Parks, we need to be pulling in more income. With just my wages at the plant, we're falling behind."

"I've tried, Elise. Nobody will hire me. After they had me teach my replacements, they failed to renew my contract. Now, I can't even find a job as a technical writer. And it's not just me," he said, turning from the sun. "Lots of other engineers are in the same boat. Jobs are hard to find everywhere. And I'm *not* going to be a Greeter at the Dollar Mart."

"Fine," she said, losing patience at last. Still, her voice was steady and strong. "Then we need to dip into our savings."

"No!" Parks shouted. "I need that money for my patents. The maintenance is already overdue. The six month grace period runs out at the end of this month."

"Then what, Parks? They already got an asteroid. All those companies are already planning to mine it. They've got it all worked out. You had a great idea, but they found another way. Don't you see? It's over, Parks. You've got to let it go."

He hid his fury from her by turning back to the west. The sun, dull red, was just sinking into a bank of clouds that hadn't been there a minute before. "Damn," he swore.

"No, Elise! I can't quit. What if they screw up?"

"They're not going to screw up, Parks." Elise's jaw tightened as she fought to keep her own anger under control. "And when they finish catching that asteroid, there are going to be plenty of people wanting a piece of it. They are going to need people like you. Even if they don't, the Vambolents are heroes in China. You need to get a job. Maybe it wouldn't be such a bad idea to look outside America. You know I'll go with you. I've done it before."

Parks slumped in his lawn chair and dropped his head into his hands.

Elise felt cold and shivered. She turned to walk inside, talking to herself; not caring if he heard. "If this doesn't end soon, I'm outta here."

* * *

The Vambolent family joined hundreds of millions of other viewers watching the telescopecast of the gravity assist maneuver on asteroid 06-401A. The spectacle itself would be less than spectacular. To evoke interest, the news channels warmed up their audiences by covering the many controversies raised by the NASA plan.

Although the Chinese Moon Base had been unmanned for years, that did not stop their government from expressing sharp concern that such a risky maneuver might threaten their habitat. A consortium of nations with space programs expounded an objection based on the Outer Space Treaty. That treaty, ratified in 1967 prevents anyone claiming a space resource as their own.

NASA had of course studied this long before the start of the Seven Dwarfs mission. They issued a policy statement jointly with UNOOSA, the United Nations Office for Outer Space Affairs. An oversight group was already in place to implement that policy, which allowed free access to the asteroid, provided it was used for peaceful purposes, and such use did not preclude others from their own work.

The consortium countered that the remote location of the asteroid was cost-prohibitive to most countries, thereby giving an unfair advantage to the wealthier nations. And so it went, round and round.

Environmental groups took issue with the mining plans. Their main objection was that processing activities at the L4 point would ruin that region of space for centuries to come. The Lagrange points 4 and 5 are stable gravity wells, they argued, and will trap much of the debris from the mining operations. That debris will accumulate over time until L4 is no longer useable for any purpose whatsoever.

Annoyed by the repetitive commentary, Parks turned off the sound. There were new batteries inside his remote. Outside their home, it was an unseasonably warm Saturday morning, an Indian summer, but he had insisted that his whole family stay inside to watch 'history in the making'. Telescopes from NOAA satellites, which normally watched weather patterns over a third of the globe, were today aimed towards the Far Side limb of Luna, where they would be the first to see the asteroid appear from behind.

"Turn the sound on, Daddy, please?" Mindy asked. The closest approach was just minutes away.

"… through the lumpy gravity of the Moon." came the commentary, growing louder. "That, plus the dramatic topology of the lunar surface make this a problem of extraordinary complexity. NASA has four powerful supercomputers constantly updating the trajectory and modifying the thrust vectors to achieve a precise gravity assist maneuver."

They all leaned closer. Shadow dominated the asteroid, leaving a wavy white crescent, which appeared to slide across the fixed stars. There was no sense of depth in such a highly-magnified view. They could just make out the thrusters, four thin bars of light poking out from the asteroid. Each thruster consisted of an electromagnetic catapult, powered by solar panels, and loaded with pellets of iron-rich mineral cored from the heart of the stony rock. The commentator counted down to the loss of visual contact, and right on cue, a wall of white swept over the asteroid in an eyeblink. The crescent of the Moon seemed to flick at the tiny asteroid. Somewhere on the Far Side, the high-speed projectile dove into gravitational turbulence, soon to re-appear on the other side of Luna.

They did not have long to wait. At extreme close-up, the expected appearance came – and then passed. Moments later, a nimbus of scintillating particles blossomed from the edge of the Moon and began spreading outward.

"Oh, my God," Parks exclaimed. "They missed!"

*   *   *

Pandemonium erupted in many places. Looting began at once in the larger cities, ringing the globe as the news spread. The UN Secretary General was barraged by charges that the US had just violated the Moon Treaty. China cited the Liability Convention of 1972, anticipating that some of the ejected rock from the crashed asteroid would fall upon their Moon base, even though it was on the complete opposite side. On Earth, governments everywhere demanded to know the trajectories and possible landfalls of any hunks of Moon rock knocked loose by the impact.

At noon, Mountain Time, NASA Administrator Lewis Rinker made a public statement. He sat at a conference table with his key staff, plus a representative from the White House, and the director of FEMA, the Federal Emergency Management Agency.

"We regret this unfortunate turn of events," he began. "On a project of this magnitude there is always a certain risk of mishap. The computations made for the lunar gravity assist have been checked, and checked again. Performance data for each of the thrusters have been verified. Topographical maps of the Moon's Far Side have been studied and re-studied. Gravity maps of the Moon, magnetic forces and solar wind have all been scrutinized to discover what may have been missed. While it is true that we were flying blind behind the Moon, our mapping and timing gave an acceptable margins of error. Our conclusion at this point is that nothing within our control that could account for the apparent impact. Nonetheless, an impact does appear to have occurred. The only explanation that we have not been able to rule out is a sudden thruster malfunction. Neither can we rule out Acts of God, such as the possible, but rare event that a micrometeorite impacted the thrusters or the solar arrays as the asteroid passed behind the Moon. Or, and I tell you people, this is how unlikely a problem could be, 06 dash 401A might have run into a mountain range that thrust up from the Moon since our latest topographical and gravimetric survey seven years prior. So, we've narrowed the option to a few, and are working full-time right now to determine the true, exact cause. He have crack teams assigned at each of our centers, and I've given them direct orders to not stop until they find it.

"However," he said in a voice that dropped in volume, pitch and tempo, "of more immediate concern is the possible landfall of any debris. I have asked the entire remainder of our staff to devote their full-time attention to this matter, and we expect to have answers as soon as humanly possible. Until then, I hope you ladies and gentlemen will please excuse me."

Rinker stood up amidst a roar of questions all shouted simultaneously. He turned his back on the assembled reporters and left by a side door, followed by the FEMA director.

*　*　*

Emergency sessions of House and Senate committees convened immediately. President Dandridge returned early from a peace brokering and terrorism reduction mission in the Middle East, and asked the Joint Chiefs of Staff to meet him upon the return of Air Force One. Most States deployed their National Guard to metropolitan areas, and it was generally expected that the President would formerly declare martial law to stem the spread of lawlessness.

The International Red Cross and the Peace Corps mobilized emergency response teams, and began deploying them towards well-spaced points around the globe, ready to provide assistance wherever it might be needed.

Doomsday fanatics, Armageddon watchers and fundamentalist Christian groups declared an end to civilization. Survivalists moved into their shelters. Opportunists scrambled to secure and hoard stockpiles of food, medicine and toilet paper. The New Age community, already disappointed at the continuation of normal existence after Y2K and the end of the Mayan Calendar both, pointed out the unlikely confluence of events and pronounced them synchronistically meaningful.

Offices emptied and churches filled. Families reunited and marriages broke apart. Online communities formed and student protesters marched. Some sat back in shock while others cried and lamented. Border disputes between nations and ethnic conflicts between races erupted in violence as people polarized. Each side hoped to be among the few survivors when millions of tons of Moon rock rained down upon their heads.

If one believed the news reports, everyone on the planet could be grouped into one of these categories. For some of those still undecided, watching the media coverage nudged them into one position or the other. Public opinion began polarizing, dividing and segmenting. Buzz grew rapidly as people sought more information. Phone lines saturated, as did internet connections. Even the larger communications networks began to grow unstable.

Parks Vambolent, perhaps alone on all the Earth, was positively gleeful.

* * *

Less than a day after the asteroid debacle, an expectant world tuned in to the analysis prepared by NASA and UNOOSA. Their first message was for people not to panic. It was already too late for that, and to make matters worse, the leading stream of high-velocity debris from the Moon began peppering the communications satellites in geosynchronous orbit some thirty thousand kilometers above the Earth. World-spanning communications links began to blink out just as Lewis D. Rinker was declaring that the problem was under control.

On the shadowed side of Earth, small bits of lunar rock began slicing into the atmosphere. Fast-moving chips of ilmenite, olivine and anorthosite the size of pin heads left streaks of white across the night skies.

Rinker announced that they were tracking some forty objects larger than a meter in diameter. This, he explained, is the approximate size of an object which might survive atmospheric entry and strike the planet surface. A dozen of these had been kicked into long, elliptical orbits that might never strike the Earth, and were of no immediate danger.

Estimated impact points had been calculated, and six, possibly seven, larger boulders were expected to hit. A map behind him showed a Mercatur projection with seven red parallelograms. Six were predominantly over ocean areas, although they overlapped continents and island nations. Central Mexico was the clear exception. It was slashed in half by a band of red, and represented the most immediate threat. Individuals in each of the seven areas were encouraged to evacuate, quickly but calmly, over the next two days.

The UNOOSA Special Director, Hasa Cladhil, spoke next.

"As Mr. Rinker has explained, the stones we are tracking are quite small by cosmic standards. The one aimed at Mexico is only slightly larger than a passenger car. As it passes through the atmosphere, the friction from the air will eat away at the stone, making it smaller still.

"We have been in constant contact with the United States on this matter, and they have been very cooperative. Together, we have developed a procedure by which large military lasers can be used to move or destroy these stones.

"By aiming a high-intensity laser beam at the stone, we expect that it will fragment into smaller pieces. Once these pieces are small enough, they will burn up harmlessly in the upper atmosphere. This is the best case scenario.

"If the laser does not break up the rock, we have another factor in our favor. The intense heat of the laser will boil minerals from the surface of the stone, and these minerals will vaporize away at high speed. This vaporization acts exactly like a rocket engine. So, if the stone does not shatter, it will be pushed away from the Earth. We can control the amount of the push by the time and power of the laser blast.

"Now listen carefully. We will not have enough power to make the stone miss the Earth entirely. But we *can* move it. We are all 100 percent confident that we can push this stone far enough to the west that it will miss Mexico completely."

He looked eminently pleased with himself.

"Therefore, we will guide the stone to a splashdown in the Pacific ocean. If this occurs, it will certainly kill a few fish. But human casualties should be exactly zero."

Cladhil beamed, showing uneven teeth. The camera swung back to Rinker.

"The Mexico fragment is due to impact in 43 hours. This gives us ample time to destroy it, or move it to a harmless impact point. Two more of the larger fragments are expected to hit in the far western Pacific, and pose a moderate danger to a number of islands there. Once the Mexico fragment is dispatched, and we have learned how to control it, we will focus our efforts on the western Pacific fragments. The remaining four or five impacts are only a risk to shipping lanes and air traffic, but you can rest assured that all such vessels will be given plenty of notice, and instructed to give the impact points wide berth."

Rinker then repeated the information in summary form, and signed off.

Meanwhile, as more and more of the fast, small detritus dove down into Earth's gravity well, the nighttime skies became criss-crossed with ionization trails. Here and there, satellites in low-Earth orbit (LEO) dropped off-line from the micrometeorite storm. NASA evacuated the renovated International Space Station, instructing the crew to return via their Soyuz capsule lifeboat. The hulls of even the safest module were not thick enough to protect them. For the first time in its long history, the ISS went vacant.

*   *   *

Two days later, the skies were nearly saturated with shooting stars. The first recorded landfall struck in Arizona, halfway across the state from Meteor Crater, near the Red Rocks area around Sedona. When it hit, the rock disintegrated, leaving a crater five meters deep and twenty five across. Quick calculations by NASA pointed out that even had it struck in mid-town Manhattan, at most only a few hundred would have perished – on a par with a single airliner disaster. Yet, given the distribution of population around the Earth, each individual had less than a one in 100 billion chance of being hit by a Moon rock. This was a far, far smaller risk than being mauled by a polar bear. It was less even than the chance of being struck by lightning twice.

Star Wars laser beams from Kirtland AFB, requiring the combined output of multiple nuclear power plants, blasted the Mexican rock as North America turned beneath it. Jetting vapors raised its orbit, but it did not break apart. When it hit the atmosphere above Baja California, the streak of its passage was clearly visible in broad daylight. It fell into the sea, giving up its kinetic energy as it drilled into deep waters. A huge cloud of super-heated steam burst into the sky in a white, roiling mushroom cloud. Concentric ring waves expanded outward, traveling at hundreds of miles per hour. When the waves reached the sparsely-populated shores nearby, they were already spread out and slowed down, surging up almost languidly past high-tide lines. The water receded, leaving small ponds of salt water behind as the only memory of the impact.

* * *

The western Pacific rocks trailed by half a revolution of the Earth. This took it outside the range of the US lasers, but it was just as well. Fine control was not possible, and the NASA-UNOOSA group quietly decided to let events take their course, while publicly stating they were doing all they could. The volcanic islands crossing the landfall zone were sparsely populated, and international airlift evacuation was already complete. More likely than not, these would also land in the ocean.

A quieter world watched and prayed as radar trajectories tracked the path. The few remaining Earth observation and spy satellites in LEO captured incredibly detailed images of the twisting rocks as they approached. Planetary scientists estimated the braking effects of the atmosphere in predicting the strike points more accurately. There was no consensus, driving a fury, and indignation, that would not be forgotten for generations.

In space, the rocks swept in, falling faster and faster, accelerated by gravity's pull. Less than an hour before the expected impact, as the world watched closely, the first western Pacific rock slammed through the defunct Japanese Solar Power Satellite, vaporizing the antenna and part of the array. It glanced off the 7-ton PMAD module of the SPS, crumpling it into space junk that went sailing down towards Earth.

In space, the collision softened the meteor's angle of attack ever so slightly.

Within seconds, radar ranging re-computed the new trajectory. It would spiral down and fall further along the planet's circumference by some 8 degrees of longitude. The new target zone was estimated to lie just east of Chongqing, deep in the heart of mainland China, in a swath that crossed the Yangtze River.

China declared war on the US.

* * *

Vambolent paced the den, talking to himself.

"You idiots!" he shouted at the television. Just which idiots he referred to was not clear.

Each news channel and each local station all watched the nightmare unfold. Satellite images of the approaching disaster were surrealistically peaceful. Stars behind the meteor moved by slowly. There was no sound, no tangible sense of danger. Not yet.

A special message from the President broke into the broadcast. Dandridge was explaining and apologizing and promising. Same President whose administration had once seriously considered, then outright rejected Parkinson Alexander Vambolent for the nation's highest technical honor. Parks turned off the sound on one pass by the television set, then turned it back on again the next.

"…to the great people of China, we commit to full compensation for…"

He turned the set off again.

"Idiots. All of you!"

Because the television was turned off, Parks missed the first of the two rocks striking the Earth. Following a parabolic trajectory, it curved downward at Mach 25, passing through the whole atmosphere in a quarter of a minute. Diving into the shallow waters just offshore of an active volcano, the meteorite pounded through Earth's thin crust there and opened up a hole to the magma and ash trapped beneath.

The mushroom cloud of steam looked little different from the ocean strike off Baja California at first. This made it easy for the media to dismiss it in their eagerness to monitor the landfall in mainland China.

Parks turned the set back on.

Watching the last large meteor fall into the atmosphere from space made great press coverage. Hundreds of eyes-in-the-skies turned to watch, catching the start of the ionization trail over Yonaguni Island. The burning rock slashed deeper into the thickening air, leaving multiple sonic booms in its wake. It was traveling far faster than the sound waves, so those in Nanking watched it pass overhead silently. They heard nothing until it was almost to their horizon. Then, the spikes of sound pressure shattered their windows and split their eardrums.

Ten thousand people cried out in terror as the rock fell upon them. Their screams were drowned by the tremendous bang of the impact. Energy equivalent to megatons of TNT was released in a millisecond as the speeding rock came to an abrupt and devastating halt 14 meters deep in the Earth.

Vambolent gasped, watching this from above, through the television in his den, with the sound off. A bright prick of light flashed as earth, water, metal and people ionized like the air surrounding a lightning bolt. A black cloud ballooned out quickly, and hung there. High prevailing winds tugged at the sphere of dust and smoke, shredding the edges. The black faded to grey-brown, and over longer moments slowly drifted away from the crater. As moments stretched to minutes, the view cleared, revealing the level devastation beneath.

It was over that fast.

But many other things had just begun.

* * *

Downriver from Chongqing was the Three Gorges Dam. This massive structure, 180 meters tall, and more than a kilometer across, supplied fully ten percent of China's electrical power, about 18,000 megawatts. When the Moon rock slammed into the nearby countryside, a ripple of seismic energy raced outward from the epicenter and flowed across the bedrock which supported 25 million cubic meters of reinforced concrete.

As the gigantic dam rose and fell with the Earth, its moorings with the walls of the gorge cracked and crumbled. At first, nothing happened. Then, hydrostatic pressure from the enormous upstream lake forced water into the tiniest of gaps. Shortly, jets of water, carrying small chips of concrete, began spewing out the sides of the dam.

More and more water tried to press through, tearing away at sharp corners. A weakened generator at the base of the dam lost integrity, and a fat gush of water surged through, roiling the river called the Chang Jiang by the local people.

The fractures opened by the meteorite impact widened. The shuddering rumble, made by the high-pressure water forcing its way through, sent stress cracks spiderwebbing throughout the matrix of the dam. Media crews had time to arrive on the scene to watch the dam get further and further compromised. Cities downriver of the Three Gorges started to evacuate, all the way to Nanking, a thousand kilometers away. They were now expecting a flood larger than any ever seen, larger even than those of ancient legends.

Chapter 9

In the days following the impact, the Vambolents watched in shock as geopolitical actions unfolded. Under orders from Beijing, the Chinese military invaded North Korea. Devastating air strikes removed key air bases and crippled missile installations.

On the opposite border, massed forces of the Peoples Liberation Army moved a parallel front further into the long-disputed border with India. China had been developing a network of road and railways over many years, especially recently. Now, they moved in to occupy the region permanently. They claimed the resources as their own in an act of eminent domain.

Declaring the NASA debacle an act of war, the Chinese President ordered all US fleet activity out of their territorial waters, as well as the entire Yellow Sea and the northern half of the South China Sea. Expatriated US citizens and all tourists were ordered to leave the country immediately. American corporate assets were seized and their factories commandeered.

Parks watched news footage of Chinese leaders demanding immediate compensation in the amount of sixty trillion renminyi for loss of life, property, and the loss of future opportunity. This staggering figure was compared to the Gross Domestic Product of the United States. It was like asking every American family to forfeit every dollar they earned for a generation. Every American company would have to forfeit every dollar of profit for decades. It was a thoroughly unreasonable request, but President Dandridge was in no position to negotiate.

A prepared statement was delivered to the world from behind the desk in the Oval Office by Eldon W. Dandridge. He was visibly shaken, bereft of his usual concerned confidence. He stumbled through his speech, clearly reading from a teleprompter, expressing deep sympathy for the many Chinese killed. He explained that diplomatic channels had been cut off, and therefore appealed publicly for a more reasonable plan of restitution. A special Presidential Commission was being formed, he said, and would study the matter in detail under full public disclosure.

From now on, it seemed, the entire exchange would be carried out under the scornful view of the world media. Such embarrassment by the US was unprecedented, despite its many blunders in the past.

Elise took vacation from work, staying home to be with Parks. He was unable to sleep for long, and showed signs of emotional collapse. After getting their kids off to school, they watched television together in the den. She wrapped her hand over his, tucking her fingers into his palm. This kept him from repeatedly clenching his fist.

Almost.

Dandridge was asking for more time. He offered humanitarian relief for survivors. He offered to pay for rebuilding of the towns and villages destroyed by the meteorite. But the sticking point was the energy lost from the crumbling dam. Even as the public negotiations fell apart, the dam gave way. Cubic miles of water burst past what had once been the largest hydroelectric project in the world. Death tolls continued to mount, damage costs continued to escalate, and whole districts of the city of Nanking were totally destroyed.

Beijing demanded that the US immediately begin to supply their nation with the equivalent electrical generating capacity lost from the Three Gorges Dam. That energy was vital to their continued economic growth. For the last few decades, as everyone knew, it was growth in the Chinese economy that had kept the entire world from slipping into a global depression. Now, all of that was in jeopardy.

Economists and energy consultants pounced on this issue with fervor. Experts of every imaginable background had opinions and recommendations. Blue ribbon panels, pulled together by news corporations, brought the issue to widespread debate. Senior Cabinet members even referred to these forums in their own public speeches. It was at once heartening and frightening as the broadcast media expanded the debate and sought out solutions, yet the US government *per se* appeared inept.

Regarding the energy issue, nuclear power was almost universally ruled out. Re-building the dam was the most popular option, but the original had taken ten years to complete. Buying oil seemed the only choice. However, oil prices were at a historic high, as low-cost sources of light sweet crude began to exhaust their reserves. The mere possibility that the US would need to dramatically increase its purchases sent the spot prices to new records, and the futures market soared. Stock markets linked with the Hang Seng, Nikkei, and the NYSE plummeted with the anticipation of an energy-induced global depression.

Almost no one took notice of the continued discharge from the broken volcano in the Pacific. Satellite photos showed a ragged band of gray dust circulating across a broad area of the central Pacific. Sunsets in Hawaii were especially gorgeous, but there were hardly any tourists there to watch them. The scientific community took notice of the dust cloud headed for North America, but policy makers were already too overloaded to give the matter much thought.

Parks and Elise had become almost numb to the images of destruction, the rhetoric of compensation, and the reports of war that raged across the east Asian continent. They watched TV in a semi-stupor.

"Hey, that guy looks familiar," Parks said, sitting upright. "Oh, my God. Elise, it's Wang Jiachi!" He fidgeted restlessly, squinting at the image. "Geez, he's gotten older."

She cut a look at him, appraising her husband. But when she heard her husband's name soken faintly behind the voice-over translation, it registered like a subliminal message. When she heard the translated statement, her eyes bugged open wide.

"*What* did he just say?" Elise asked.

The Vambolent telephone rang moments later. Parks was leaning into the TV set, soaking up every translated word of his old colleague. From their work together during Parks' stint in Beijing in his uncle's firm, Parks knew Jiachi spoke passable English. Watching him speak in Chinese with a delayed version in his own native tongue was oddly distressing. Elise got up and went to the kitchen to answer the phone. A few moments later, she brought the cordless set into the den, smiling broadly.

"It's Dajwal," she announced, handing it to Parks.

"Hey, Dodge"

'Hello, Parks. How are you doing?"

"Fine. You?" Parks said, distracted.

"Absolutely fine, thank you very much Parks. Hey, I just heard your name on the television. How does this Wang Jiachi know you?"

Parks was stupified a moment as old memories rose to mind. He and Wang had been best friends. They were young dreamers a lifetime ago, designing hot rockets over cold beers. Their interaction helped refine and solidify many of Parks' architectural concepts involving his asteroid separator. An odd feeling of disquiet, mixed with expectant anxiety came over him.

"Yeah, we heard it too," was all he could say.

"He called you a 'great innovator', Parks," Dajwal said. "He said you were a man who 'sees the future'. Isn't that great? I'll bet it sounded even more eloquent in Mandarin."

Parks made a plosive sound of dismissal. "I knew him once, long ago. Forget it, Dodge. He's just some junior engineer handed a speech that will rub America's nose in this whole mess. Surely those scriptwriters know how this whole country feels about me. They're milking this for all its worth. He's just using me," Parks said. But his words rang hollow, even to him. The Wang Jiachi he knew would never sell out.

"No, no, Parks. He's no junior engineer anymore. I'm looking him up on the internet right now as we speak. Oh, my very word, he is the Chinese Minister of Science and Technology, Parks!"

"Jiachi? Minister?"

"Oh, yes Parks. Oh, yes," Dajwal said. "He is not throwing mud. No, sir, not at all. I believe he is handing the United States a solution to their problem."

"You must be kidding." But a secret thrill snuck up his spine, even as he said it.

"Oh, boss, I assure you I am not kidding. You just watch the news for the next few hours."

"I'm not your boss anymore, Dodge."

"I have still not given up on you," Dodge said, using a serious tone for once.

"Well, thank you. But I naturally figure that's all over now," Parks said.

"But everything has changed now, Parks."

"What do you mean?"

"Don't play coy with me, Parks Vambolent. You know very well what I mean. Tell me this: what was your first reaction when the asteroid crashed into the Moon?"

Parks did not answer. But a smile did creep across his face. It was a Grinchy smile, but there was a genuine side to it as well.

"You see what I mean?" Dajwal asked, as if he could see Parks' face through the audio-only telephone. "This changes *everything*!"

Alarm bells clanged in his brain. "Dodge! What are you thinking of? Whatever it is, I beg you – don't!"

"Ah, ha," he said. "You see into my heart as well. I knew you would. It's the old Parks Vambolent now. Only crusted over a little bit, yes? Me? I am only going to stir the pot just a little. Get the juices flowing, if you know what I mean."

"Dodge! Don't do that. I'm not paying you anymore."

"Oh, that is no problem. No problem a-t'all. You see, I am now a volun*teer*!"

"No! This doesn't concern us, Dodge. It's already hard enough on my family. Please leave us alone, I'm begging you."

"Well, you know that the cat is already outside of the bag. It was not me who announced your name to the world again. It was the Technology Minister of a country with which we are at war! Who knows what will be said next along these lines. Who knows where it will lead. There are too many possibilities to predict. From where I stand, I may be able to make a well-placed word that might make what follows next more favorable to you and your family."

"How could you do that?" Parks asked.

"Maybe by doing nothing. I will watch, and wait. If there is something I can do, I will do it. My new position in NASA does not have much power or authority, despite my lofty-sounding title. But if I may say so humbly, applied carefully, my *influence* has a long reach."

"I wish I knew what that means," Parks said. "But if there is any way you can help ease the impact on my family, I would be very grateful."

"I will do what I can," Dodge said, and then asked innocently: "By the way, how *is* your family?"

Parks looked up at Elise. She had been watching him with lively, excited eyes and listening with perked ears.

"They're fine," he replied, and handed the phone back. She put it to her ear.

Elise stood, cupping the elbow of her phone hand in the palm of the other. She moved over to the window, chatting softly. Her hair fell forward to make a curtain around the mouthpiece.

Parks turned his attention back to the TV, flipping through the news channels that were still broadcasting. There was some weather news about the west coast, something about record colds under a dust cloud. It sounded bad, but he skipped through it. The next channel showed a conference room filled with press members, and a NASA logo on the back wall. He turned up the volume, and set down the remote.

A crowded conference room faced a panel holding a discussion on the problem of supplying energy to China. A moderator was fielding questions from the audience. He pointed.

A distinctive woman's voice carried easily over the crowd.

"What about Space Solar Power?" the voice said. "It seems to me that any suggestion of oil or nuclear energy is being stonewalled. Coal is too dirty and hydroelectric takes too long, so what is left? Solar is the only option; and, to be efficient, solar energy must come from space." Parks recognized that voice in the moment before she spoke his name. His heart missed that particular beat. "Didn't Parks Vambolent make that same argument years ago?" she said. "Isn't that why the Chinese Science Minister mentioned his name just twelve hours ago? What about Space Solar Power, Mr. Moderator?"

The panelists looked back and forth at each other. None of them wanted to field the question. There was a moment of dead air.

"Dr. Miller," the moderator said. "This question is aligned with some of your earlier work. Is there any potential to Space Solar Power?"

"Oh, that," Miller said. Cameras zoomed in on him. He turned aside, looking away pensively, as if too much attention might distract him. "Solar power satellites have been studied over and over since they were first proposed by Dr. Peter Glaser, who was then at Arthur D. Little, in 1967. It's a nice idea. However, every careful analysis ever performed, to my knowledge, shows that the energy costs are far, far too high to ever be economically feasible. One day, when launch costs drop down in the neighborhood of ten dollars per kilogram, *and* when solar cells have efficiencies over forty percent *and* can last 15 years, *then* SPS will be a realistic prospect."

"Just a minute!" shouted Heidi Bayer, standing up. The camera zeroed in on her face. "But with all due respect, I think you missed the point." People stopped taking voice and script notes, and now all eyes were on her. "Vambolent's idea was not to launch the solar cells, but to launch the factory that makes them. That in itself is not the new idea. But his invention, the isotope separator, is what makes the economics viable. A manned base on the Moon can produce enough electricity to replace the dam in just a few years."

Buzz in the room climbed back up until the moderator shouted for order. Of course, everyone in the room knew about Vambolent's fiasco. Compared to other possible media audiences, these particular journalists were more aware of space and energy issues, and also knew of the good ideas behind the unfortunate occurrence of the Oasis Massacre. A few more people crowded into the back of the packed room, and the noise level grew louder.

"Order, order," called the moderator. "There will be no more questions out of order. Young lady, Ms. Bayer I believe, you will restrain yourself from any further outbursts or be removed from the premises. Dr. Miller, as Moderator I advise you you are not obliged to respond to those out-of-order comments."

"It's all right," Miller said. "And yes, I've heard about these ideas. There are entire books written about it. But the fundamental problem is that the technology just doesn't exist. Someday in the future, after millions upon millions of dollars have been invested in research and development, then maybe…" he held out his hands and shrugged, "… just *maybe* it could be done. But certainly not in the time frame the Chinese are talking. It would be faster to build a new dam."

The conference room buzzed again, but the tone was down. Brief hopes seemed to crash. Parks noticed that Elise still had her ear to the phone, but was not talking. She was watching the debate, too. When he looked at her, she looked back, and her face suffused with a deep love and pride he had not seen there in a year.

"Oh, Parks, I'm so proud of you," she said, and kissed his forehead.

Parks remembered how she had been so disappointed with him in the back yard just the day before this all happened. It made him resentful of her present ebullience.

"I haven't done anything," he said. "And besides, this Doctor Miller just debunked my ideas again. As if I needed any more of that!"

She hugged his head to her breast, but said into the phone: "Sure, I'll tell him. Bye, Dodge."

"Tell me what?"

"To expect some company."

"What kind of company? What are you and Dajwal cooking up."

"Oh, he didn't tell me any details."

From her voice, Parks knew she was lying. He huffed and turned his attention back to the TV. The debate had moved on. His second fifteen minutes of fame was already over.

Malibu Canyon State Park

Jenks watched the deep red sunset above the hazy horizon from the top of a crumbly peak. The world beneath him was divided between what was above water, and what was below. He could not see below water from the mountain outcrop, but he could imagine it. Catalina Island was obscured by the smog and humidity. But he could see it in his mind's eye.

Jenks pictured the world curving out away from him in all directions. West, it was covered by a blanket of water until almost a third of the way around the world. There lay China, now the sworn enemy of the United States – the first such foe since World War II. Did that mean this was World War III?

Jenks didn't think so. He envisioned this as the tipping point he had been waiting for. This walk was to clear his head, so that he could begin the next phase of his work.

Bruno D'Argeste had invited Jenks to submit a business proposal which combined the collective merits of the various submissions Jenks had made throughout the divisions of his holding company. The invitation had been delivered by telephone. Jenks replayed the scene in his head.

"I have noticed your name from many places," D'Argeste had told him.

"How do you know they are all the same me?" Jenks quipped.

D'Argeste had laughed at once, allowing his basso rumble to linger over the phone line.

"You must be a clever man, Mr. Jenks. I assume you have been trying to reach me with some intent in mind, yes?"

"I have an important proposition. One that I would make only to you," Jenks said, coming right to the point.

"I see. Well, I would want to see some further qualifications – some further synthesis," D'Argeste had replied. "Why don't you submit your proposition to me in the form of a business plan. Will that suit you?"

"Of course that will be fine for a start," Jenks said, speaking faster than usual in his eagerness to make an impression, "so that when we meet in person, we can discuss the finer points."

"Ah, you do make a compelling case, Master Jenks. I am always on the lookout for bright young men like yourself. Send your plan to me directly," he said, and recited to Jenks a postal address. "Then, we will see about a meeting in person. If I like what I see, I may invite you to a very select conference to be held within the month. Please submit your work quickly."

"*D'Accord*," Jenks said in French. "*Merci beaucoup, monsieur.*"

That was just six days ago.

Now, the work was finished. He sent it overnight airmail, then drove up into the hills and hiked to the edge of the canyons to soar his remote-controlled Rogallo glider. The fresh air and sunshine felt good on his skin and in his hair. Today he had taken a big step forward. The plan he submitted captured everything he wanted to say to the plutocrats, and most of the solutions. Intentionally, Jenks left a few minor issues vague. Best not to appear too perfect. It might arouse suspicions.

While waiting for D'Argeste to call back with a personal invitation, Jenks had some defense to do. Vambolent's name was in the news again. The best time to apply leverage was early on, he knew. It was the only way a lone agent could make a big difference later. He knew, or guessed, the Chinese gambit in mentioning the space inventor. But he also knew the most likely US response, based on his studies of history and politics. What the Chinese probably intended as an insult would unwittingly appeal to the innate American longing for a hero-savior.

If his prognostications were inaccurate, he would simply stick with Plan A, waiting for the Vambolent patents to expire. But if they really did try to make space solar power from lunar materials, he could exploit the one crippling weakness of the entire architecture: the idea of ballistic transport, and all the fears that implies. It gave him considerable influence. Influence he would work and work, patiently and subtly, until it was time to act. Then, he would move decisively, and take the upper hand.

He breathed in deeply, and lifted his wing into the steady breeze. It rose, the balance tricky in the fragmented wind coming over the cliff's edge. Jenks worked the pitch and roll with a thumb-sized trackball. The silvery tented wing held a triangular control bar. On the control bar was a two-axis servo which activated sliding counterweights. The trick was not to overextend your balance, maneuver to maneuver, and Jenks captained competently. His craft rose and rose until it was a glint in the sky.

A Chateau outside Orléans

Bruno D'Argeste snapped open this morning's *Le Monde.* Resting the international section against a small plate of fruit and a plain croissant, he plucked the demitasse of strong coffee to his lips and sipped.

A whistling sigh of heated breath issued from his pursed lips as he scanned down the columns, skimming headlines. The coffee would sharpen his concentration soon and he could move faster, but on this page an article caught his eye.

Tensions between Israel and the Egyptians were high, and he tired of it. Egypt continued to petition the UN to make a counterstrike to retaliate for the destruction of their "legitimate" chemical processing facility two years earlier. They had rocket technology now, taught by scientists from, and bought from, the government of Ukraine. Cryogenic rockets, which could reach geosynchronous orbit or any point on Earth, were now being built outside of Cairo, according to the article.

Hamid Hababek, the noisy cleric that kept the fundamentalist Muslim Egyptians in a continual state of unrest, had announced the coming end of the world. "The Moon has split and the hour has drawn closer," he quoted. "'We then opened the gates of the sky. And we caused springs to gush out of the Earth,' saith the Quran, a body of wisdom written 1400 years ago, verses 54:1,:11, and :12."

"The end times are surely upon us", Hababek was quoted as saying, "These holy verses match precisely the sequence of events from the asteroid cleaving the Moon, the debris which opened the gates of the sky, and the springs which have gushed forward".

He referred to the Three Gorges Dam, which had finally collapsed the day before. The middle third gave way, spilling a great cascade into the ancient river valley. Deaths downriver were minimal, but property damage was enormous and still mounting. Anti-American sentiment was very intense.

"The time to act is now," Hababek was telling his followers.

"Rubbish," D'Argeste said aloud. "It is all bluster and show business." Then, he thought to himself what might be behind these threats. Would the Egyptian government really cave in to popular pressure as Saudi Arabia had done? Could they afford to lose the access to foreign aid? D'Argeste considered the possibilities. Could the crazies capture control of a rocket? Egypt had no nuclear capabilities, so in the end, they had no teeth. Something to watch, he decided, but nothing of immediate concern. He turned the page.

News of the declared war between China and the US filled the two-page spread. North Korea was no longer its own country. The Chinese warmongers were clearing out the DMZ minefield that separated them from South Korea using sniffer technology they had earlier acquired from US firms. The second rarest bird species in the word, red-crowned cranes, were displaced from their fragile habitats as the PLA plowed through. The city of Seoul, South Korea was in a state of total panic.

Taiwan braced for invasion as well. Many believed that once the US fleets finished backing off from territorial waters, the Chinese would move in. Taipei would finally be seized, and brought back after the rift which lasted since 1949. It was an outcome the US had long sought to avoid, and now they found their power ebbing away in the Asia-Pacific.

India threatened pre-emptive strikes in disputed regions, but no one expected any action. They would be killing mostly their own civilians, and of course, the Chinese had overwhelming firepower. Where the Chinese stopped their advance, the new border was established. Maps were due for an update by the world's cartographers, but they were in no hurry, since it looked as if China was not yet done.

The US still scrambled for a resolution of the energy demands, and the rare press releases from Beijing were growing terse and impatient. An oil-based short term solution seemed inevitable. Commodity prices for benzene, propylene, styrene, cyclohexane and vinyl chloride monomer pushed to many multiples of their historic highs, and the prices of plastics surged upward. Financial forecasts the world over were re-estimated downward. Stock markets sagged under the news, and the industrialized nations began a desperate scramble to secure their share of the dwindling reserves of ready petroleum.

D'Argeste knew from his financial modeling how the oil equation would drive world markets. Once the easy stuff ran out, extraction costs would start a climb that would never stop until some other energy sources became more cost effective. Only then would energy costs stabilize. It would be a new steady state for the world economy, but the transition would be painful for those not prepared. Very few industries are not utterly reliant on petroleum, but those are the very ones which will prosper when the balance tilts. He very much intended to be at the point with the greatest leverage.

A short article drew his full attention. The name 'Vambolent' grabbed his searching gaze. It was about a press release from Wang Jiachi, China's Minister of Science and Technology. More than a year had passed since D'Argeste had given a thought to the man upon whom he had once placed so much hope. Vambolent was a washout now, too much in the public eye for the sort of work he liked. *Tant pis*. He had liked Vambolent, loved his ideas. A generous offer had been made. But in the end, it had come to nothing.

D'Argeste ate, and made a smacking sound with his lips, thinking over the implications of all this news. His thoughts went back to the set of proposals he had received from a man named Winnimer Jenks. Each one was routed through a lieutenant. The man had somehow deciphered his inner power structure, and sent a message. Subtly clever, he thought. Each idea worthy, and their wholeness worth more than their sum. Perfectly timely, in each case. Masterful analysis. Clear problem solving. Thorough risk assessments. Well-thought-out consequences and work-arounds already anticipated. Incredible. He must have been well-taught.

It make him think this Jenks was too good to be believed at face value. Such did not simply drop into one's lap. Who was backing Jenks? Who was setting him up to somehow undermine the current leader of their secret society? It was a question any such leader must ask themselves at all times. And surely, the more power at stake, the greater the risk of struggle. There was almost certainly more going on behind the face of things than he knew yet.

As if on queue, an express package was delivered to him by a manservant. He acknowledged the man, and mainpulated the package – assuring himself the contents were just documents, even though extensive testing would have already been performed before the parcel came into his study. He unzipped the seal and the report slid out into his hands. The title read: "On a Means by which Controlling Stakes in Future Power Sources may be Leveraged at the Coming Tipping Point," by Winnimer Jenks.

"Very interesting," D'Argeste said, and leaned back to begin reading.

West Layafette, Indiana

A pleasantly warm breeze flew crisp, brown leaves onto the Vambolent front porch where they gathered in a sloping pile. Deciduous trees in their neighborhood were bare, but the conifers were bright green, thriving in the sunshine and periodic rains. A troop carrier growled past, an increasingly common sight now that the anti-government protests and attendant vandalism were spreading further from the campus.

A black sedan pulled to the curbside. From one side, a tall man in a military uniform unfolded himself into an erect posture, and scanned the surroundings instictively. Springing from the other side was a much shorter man of Indian heritage, dressed in a well-fitting suit of cinder gray. His eyes danced across the Vambolent home, and he strode right up the broken concrete walk, trailed slowly by the wary soldier.

Elise Vambolent answered the knock at their door jamb. Her eyes became animated, and she pushed open the screen door and stepped onto the porch.

"Dajwal!" she called, and pulled the man into an awkward hug. Over his shoulder, she noticed the other man, and her eyes narrowed in suspicion.

"Ms. Vambolent?" he offered.

"Mrs. Vambolent," she corrected. "And you are?"

"Captain Dunsworth of the Aerospace Force."

"Aerospace Force?" Elise said, releasing Dajwal and standing apart. "Never heard of it. You have credentials?"

The man fished for an ID. Dajwal stood to the side, watching with amusement.

"Our name has changed," answered the Captain, handing her his military identification wallet.

"Since when?" Elise asked.

"Since China declared war on us," he answered in a brusque tone.

"I see," Elise said, intimidated ever so slightly. "And you, Mr. Chellapilla," she said, turning to him and smiling. "You are representing NASA, or 'NASA-lite', as they call it on TV?"

"I am indeed," he answered. "Captain Dunsworth and I have been sent here from the highest levels to see your husband on official business. Also, it is very good to see you again, Elise."

She gave him a second, perfunctory hug, then invited them both inside.

Parks was in the kitchen, sipping tea, and rose when they entered. His medium build now tipped towards overweight. His clothes were new, but not of high quality. He had the stance of a man accustomed to slumped shoulders making an effort to stand straight.

"Hello, Dajwal," he said, and extended a hand to his friend. Dajwal grasped Parks' hand in both of his and shook warmly.

"Please allow me to introduce Captain Dunsworth," he said. Parks shook hands with the man who gave his full title.

"Captain Milton Dunsworth, Special Adjutant to the Secretary of Defense, Space Energy Directorate. Pleased to make your aquaintance, sir."

"Same here," Parks said. "Please, have a seat."

Elise offered coffee to Dunsworth, who replied simply "Black." She poured hot water over a tea bag for Dajwal without asking, and stirred in a teaspoon of sugar.

"So, what brings you two here," Parks said, a slight quaver in his voice.

Dajwal laughed, tiliting his chin upward. His eyes sparkled with mirth. “Oh, Parks, you are the same as ever! Never beating the bush, just coming right to the point. I like that, I like that very much.”

But Parks was looking at the officer, unwilling to open up with his old friend in the presence of a stranger. Dajwal noticed his gaze.

“Parks, Elise, I am here representing NASA, as you might be expecting. Captain Dunsworth is here because the Chinese do not trust NASA.”

Parks frowned. “What does that have to do with the price of tea?”

Dajwal’s face was instantly serious. He folded his hands together, finger by finger, and looked earnestly into Parks’ face. “As you both know, the Chinese government has made some very strong demands of the United States. I’m speaking specifically about power generation.”

“Yes,” Parks said.

“In the past, Parks, you have made public a number of proposals which have a certain… appeal, in the current crisis.”

“Dodge, we’ve talked about this again and again. Why are you talking this way.”

“I am sorry, Parks. But certain protocols must be followed, as I have been instructed.”

“Instructed by whom?”

“I will get to that in just a moment, and I appeal to your patience, so that I may present this in the proper way. Trust me, Parks.”

Parks looked like he had never trusted Dajwal less. Elise was also showing early signs of a scowl.

“Captain Dunsworth and I have been asked to make an assessment of your power generation concepts as a possible solution to this unfortunate situation.”

“Yet, NASA has studied this over and over, as that panelist described just a few days ago. Plus, plus…” Parks trailed off, too many plusses overloading his atrophied mind.

“It is the opinion of our superiors,” Dajwal continued, “that the use of Space Solar Power presents a unique… opportunity to… resolve this rather difficult scenario.”

The way Dajwal was picking his way sent warning messages through Vambolent’s brain.

"What…" Elise began, but Parks was thinking the same thing.

"What sort of opportunity? What sort of appeal?" he asked.

"It is the opinion of our best technical experts that your proposals for isotope separation are feasible, but may require a considerable development effort."

"Yes…"

"And, in the absence of a convenient asteroid for processing, the most logic site to use this isotope separator might be on the Moon – on Luna."

"Yes. Elise saw that several years ago."

Both visitors looked over at Elise. She smiled smugly at Dunsworth. He spoke next, completing the logic.

"Well, Mr. Vambolent – may I call you Parks?" he asked, looking decidedly uncomfortable with the familiar term, but using it anyway without waiting for an answer. "Well, Parks, the United States sees a certain strategic advantage, especially with the current heightened hostilities, to establishing a manned base on the Moon."

"I will not help you build a military facility on the Moon!" said Parks, rising up part-way out of his seat.

"No, sir, that's not what I meant at all. Not a military base. More of a strategic resource, you see."

"I'm not sure that I do see," said Parks, lowering himself with both arms tight, looking directly at Captain Dunsworth, and his breast display, with narrowed eyes.

"We are here to be level with you, Parks," Dajwal said, his hands open and expressing concilliation. "We're not going to tell you that the base on Luna would *never* be used by military personnel, but for the duration of the development of your isotope separator and the production of the required solar cells, the base would be a *civilian* facility."

"But you just said that development of my ideas may take a long time."

"Of course, we will plan for the work to proceed *quickly* – to meet the demands of the Chinese government. But you know, and I know, that there is a *lot* of work to do."

Elise spoke up.

"So, you want Parks to stall for you. And in the bargain, you get a Moon base for the Aero*space* Force." She put an ugly emphasis on the word 'space'.

"A moon base, on the equator, plus a space station in geosynchronous orbit," Dajwal said significantly, his eyes white and opening wide.

"Directly over China," Dunsworth added, his voice like big gravel in a tin pail.

"Parks…" Elise appealed. The center of her eyebrows arched up in alarm. It was rare to see Elise even slightly panicked, and yet now she seemed genuinely afraid.

"It sounds to me," Parks said, "that you want to set me up for a failure."

Dajwal frowned deeply, the creases in his face turning dark, almost black. "That's not the case at all, Parks. If you are *not* successful, the Chinese will *not* be appeased, and that won't help anyone."

"I see," Parks said. "So you want me to succeed, just not very *much*, and not very *fast*."

"If you succeed quickly, then everyone wins," Dajwal said, smiling again, but with an inflection in his voice that said there was more to it than just that.

"And if I succeed slowly?" Parks asked, deliberately taking the bait.

"Then America wins even more," Dunsworth interjected, smiling for the first time.

"I'm not sure I follow your logic," Parks said, turning to face Dunsworth directly.

"China has just declared war on the United States, Mr. Vambolent. Energy we supply strengthens them. They just seized a trillion dollars in US assets. It's time for you to start thinking of your *own country*."

Parks winced visibly. It took him a moment to recover from the implied insult. "I still don't like it," he said at last.

"We thought you might have some reservations, Parks," Dajwal said, glancing quickly at Dunsworth. "So, here's what we are prepared to offer you. Hear me through, and *then* you can give us your answer. Okay?"

"Shure," Parks said, in the same skeptical tone of voice his son Eric used.

"So that you can provide 20 gigawatts of baseline power for China, we are prepared to give you six moon shots, with you personally serving as the Mission Director." At this point, Parks' mouth fell open. Dajwal noted it, and his natural enthusiasm crept back into his voice. "You will have authority over a multi-billion dollar budget, and have complete unrestricted access to the full resources of a newly re-organized NASA – what we're provisionally calling NASA-lite. You, and you alone, with have full technical responsibility for the research and development work. You will command a full staff capable of carrying out your every command as fast as humanly possible. In this role, you will report to two people: the new Administrator, and to the Vice President of the United States." Dajwal paused now, thoroughly enjoying the stunned look on his friend's face. Elise was silent, but her eyes showed that her mind was racing.

"How could I say no to all that?" Parks said, semi-sarcastically. His shock dissipated quickly at the mention of the Vice President, the same man who was briefly his champion for the National Medal of Technology.

"You hardly can, Parks," Dajwal said, rising in his seat. "The President himself, Mr. Eldon Dandridge, has personally asked you to do him this special favor." From the breast pocket of his jacket, Dajwal pulled a tri-folded letter on heavy bond paper with rough, parchment edges. He opened it on the tabletop, avoiding rings of wetness, and spread it flat with the palm of his olive-colored hand. The Presidential seal dominated the top center, and under a brief message personally addressed to Parkinson Alexander Vambolent, appeared Dandrige's bold, loopy signature.

Parks could not help but chuckle. He looked once at Elise, who was now holding a slender hand over her mouth in near disbelief.

"I would be proud to accept," Vambolent heard himself say. *Who had said that*? he wondered to himself.

"Excellent," Dunsworth announced, and stood abruptly. "We have transport waiting for you. Can you be ready in thirty minutes?"

Parks and Elise looked at the soldier like he was an alien from another universe. Dajwal quickly grasped the situation, and spoke to his colleague. "Jim," he said in a soft voice. "Give us just a minute here alone, please."

Grimacing at being dismissed by a civilian, the Captain nevertheless turned on his heel and left the kitchen. The rest of them sat quietly until they heard the screen door close behind him.

Parks took a deep, shuddering breath, a gaseous stew of anxiety and excitement. His head wagged side to side as he searched for words.

"What about my family?" he asked.

"You will be provided a residence in Cimarron Hills near the new HQ facility. When your family is ready to move, I will set up all the arrangements to move them there. I know you have plenty of questions, but I beg you, Parks. We can discuss them in detail during our flight to Peterson AFB. You have a special meeting this very afternoon with Vice President Oxley to outline the project. I will help you prepare on the plane. After the meeting, you will have dinner with the new head of NASA-lite, Dr. Kevin DeRosier. He'll be making some major changes, and Parks, your project will be the heart of all their efforts. This is so exciting I am beside myself with joy for you."

"Today? Right now?" Parks said, looking down at himself.

"Yes, Parks," Dajwal said, his eyes dancing with delight. "This very day. This is your lucky day. Your very luckiest day ever."

"Dodge, this is more than we spoke of over the phone," Elise said, battling her own sense of shock.

"Yes, I'm sorry, Elise, but things are moving very fast. After so many decades of things moving very, very slow, suddenly they are moving very, very quickly. It is only because of your husband's careful planning and fortuitious circumstances that he is alone and unique in this world to take advantage of this once-in-forever opportunity. Oh, it is so exciting, I can hardly sit here in my own skin!"

His energy infected her, but internal doubts beseiged her. She took Dajwal's restless hands in hers. "Dodge, we are so grateful. Thank you, really. But we need some time to process this."

He pulled out one hand and patted hers. "Of course, Elise. Take all the time you want, but please make a decision soon." He laughed nervously at his own contradiction. Elise let it go, and Dajwal stood. He put a hand on Parks' shoulder. "This is the time, Parks, the time we spoke of ages ago in Nairobi, when you gave me a dollar and I became your employee. Now, you will officially be my boss. I sincerely hope there will be a position for me in your organization."

"Of course, Dodge," Parks said. From the look on his friend's face, he guessed that NASA-lite would probably have far fewer deputy administrators. "Of course," he repreated, and offered a hand to Dajwal. They shook, Dajwal looked relieved, and then showed himself out the door.

Alone in their humble kitchen, Parks and Elise looked straight into each others' eyes. In the first moment their gaze locked, Parks knew two things. First, his wife was scared to death. Second, she loved him again. He scooted his chair over to her and they hugged across the corner of the table. Parks pulled her closer, and she folded into his arms and tucked against his chest like she had not done since they were young. She held him with a fierce strength and he could feel the rapid beat of her pulse through her ribs. The back of her neck was warm, soft and fragrant. Parks nuzzled her, pushing his nose deeper into her hair until he felt as if she were inside him, an inseparable part of his own being. She moaned, or sighed, or purred, he couldn't tell which – but he could feel it.

She relaxed her grip and tilted her head to look at him again. "I'm scared, Parks," she said.

"I know," he told her softly.

"I love you," she said, and he could see she meant it.

He thought to himself, *So what if it's not unconditional love? As long as it's real!*

The thought made him want desperately to succeed. In that moment, all his early dreams of wealth, power and recognition meant nothing. Earning the deep, desperate love of such a beautiful woman was a greater treasure than anything else the world had to offer.

Parks lifted them to their feet, Elise feeling light and almost insubstantial. As he stood, he felt taller, straighter than he had been in years, filled with a power that had been missing for even longer. His heart beat too rapidly, but he passed it off as the thrill of Dajwal's incredible offer.

"You know," he said. "I love you more than anything, Elise. You've been great even when I've been awful. You're doing such wonderful work with Afren, even from half a world away. You're single-handedly raising our poor kids who had the dumb luck to have me as a father. And through all this you've stuck with me."

They both knew that things hadn't been all that stable. But it was a good enough rendering of history to lay a foundation for a brighter future. Elise's eyes shimmered in the early morning sunlight that came in through the curtained windows. Her beauty when happy was angelic.

"Elise," he said, the words catching in his throat, "it's the answer to my prayers."

"What if it's too good to be true," she said.

"Well… uh, it probably is," he said, and they both laughed.

"Let's go in the living room. I gotta sit down, I gotta lose some weight," Parks said.

They changed rooms and sat down. Parks noticed a smell, wrinkled his nose, and then got down onto his knees next to his broken armchair. He sniffed some more, then laid his head on the carpet, looking underneath. He reached in, awkwardly, and pulled back a withered piece of fruit that, for a moment, looked exactly like an over-ripe mango – one found under the rear seat of a Range Rover which was soaked in blood and riddled by bullets.

His heart lurched and he clutched at his chest. Breathing became fast and shallow. Angina burned under his shoulder and his eyes flew open in panicked agony.

"Parks!" Elise cried, and rushed over to kneel beside him.

They knelt on the floor as Parks regained his breathing. Hot tears dropped from Elise's chin onto Parks' shoulder. He felt the pain begin to ebb away. They sat that way for a long while, quiet, resting, healing.

"Parks, what just happened?" She whispered at length.

"I think I just had a minor heart attack," he said, also whispering. "What rotten timing."

She searched his face, looking from eye to eye, searching more deeply than she ever had. He felt electric blue lightning race through his being, energizing, restoring. Only a ghost of pain remained, and he felt fine otherwise.

"You've got to set things up right," she told him, holding one of his hands in each of hers. "You need to ask Dodge about insurance, retirement, and a savings plan, Parks."

"I will," he promised.

"And why don't you set up a company of your own, like you did in Kenya. Only this time, have the contractors go through *you*, not through your sponsor. Dodge said the Chinese hate NASA, but they love Vambolents. So you have all sorts of leverage, Parks. Make the most of it."

He looked back at her in amazement. Slumbering pathways in his brain sloughed off the crust of disuse. Avenues of possibility stretched before him the way they used to before everything collapsed around him. With a healthier dose of skepticism, his ambitions moved to a whole new level.

Moisture brimmed at the edges of his eyes as he looked upon his wife's loving face. Without words, feelings passed between them. Subtle understanding that words could never properly capture bonded them together and they both began to smile. Joy spread until the both felt suffused with lightness and energy. They stood.

"I'll pack you a bag, Parks. You go to your office and gather your space stuff."

Parks took her hand and kissed over it the way he'd seen Afren do the first time he had met Elise. They walked upstairs hand in hand. Before going to different rooms, he pulled her close again. "Thanks, Elise."

"You're welcome," she said, beaming, her own voice thick with emotion.

"How soon will you be ready to move to Cimarron Hills with me?" he asked her.

"Yesterday," she said.

Chapter 10

The French Alps

Strong winds buffeted the helicopter carrying Bruno D'Argeste and his new protégé Winnimer Jenks. An approaching storm, flowing down against the range from the northwest, pushed ahead a huge mass of air. The sharply-angled terrain of the mountains chopped the atmosphere into rough blocks, separated by turbulent mortar. One moment, they would be tilted against one piece of wind, and the next, their craft would lurch downwards several meters, and the pilot would need to angle in a totally different direction. Jenks appeared to be holding his epiglottis closed, and was pale from lack of oxygen. D'Argeste chatted on and on, communicating using throat mikes and an encrypted ultra wide band link between their headsets. They were facing rearward, so that the pilot, while thoroughly trusted, could not inadvertently see their lips.

"Yes, Jenks, we've left the Information Age behind us. Computers and communications are as fast and as easy as they need to be. The next big thing is land." D'Argeste swept his hand before the blue-grey mountains outside his window. "Even with our means of moderating population growth, there are more mouths to feed every year. With the globalization of technology, each of these mouths grows into a consumer. As consumers they require resources and they generate waste. And now, with deforestation due to clear-cutting and acid rain, plus desertification, the amount of useful land becomes less and less every year. Between agriculture, waste dumps, and living space, a great battle will soon be waged. That is why our organization is buying land."

"What about global warming?" Jenks asked, in a tight voice.

"What? Oh, that nonsense! For decades it was Global Warming. Then, scientists warned us of an Ice Age starting in Canada after a moon rock opened that volcano in the Pacific. Now it's back to global warming. All nonsense!"

A sudden lurch snapped their heads backwards. Jenks' face began to take on a greenish hue.

"*Mon Dieu*!" D'Argeste said, and pulled his throat mike away. He turned to the pilot, shouting in French over the cabin noise. "Can't you get us out of this weather?"

The pilot's response was lost in the thunderclap that exploded from the clouds right above them. The blinding flash of light illuminated the craggy mountainside below them. The pilot was pointing at something that D'Argeste could not see. D'Argeste tapped his earphones for the pilot to call him. From the encrypted transceiver box next to him, D'Argeste unplugged the jack, and inserted it into the armrest console. The pilot's voice came in, suddenly clear.

"Pardon me, monsieur, but I cannot land in this weather. I will need to run ahead of the storm. There is a small village in the next valley. Perhaps you can take a car from there."

"But we will be late. Aren't we arrived already?"

"Yes, I can see the helipad. But the storm, monsieur. They warned us of this in Grenoble."

Without waiting for a reply, the pilot turned tail to the storm and dove for speed. With the wind behind them, they raced south, down and away from the mountains. The ride quickly became smoother. D'Argeste gave a shrug, and turned up his hands, then re-plugged into the transceiver to talk with Jenks.

"What was I saying?"

"Land," Jenks prompted.

"Ah, yes. Land. What an interesting game! When you look at the Earth from above, at night, you see the lights of mankind. What do they look like? Beads of brilliance connected by threads of gold. Over time, the threads will widen, until the beads connect and merge to form splotches of light. It is very predictable. And because of this, we know which plots of land to buy now, knowing they will become valuable later. But the real secret is found deeper still, Jenks. Do you know what that might be?"

Jenks had recovered his composure quickly, and now looked completely calm and at his ease.

"Resources."

"Yes, very good. The land itself is not as important as the resources it holds. This, of course, is obvious. However, the use of land changes over time. Farmland becomes city, forests become suburbs, grazing land becomes desert, quarries become lakes, and so on and so on. Our edge is anticipating the *next* resource of the land."

"As an example," Jenks interjected, "land which became a desert, may now become a waste dump."

"Yes, that is true, but we can go deeper still," D'Argeste replied fervently. "If you use the right kind of waste, perhaps that desert waste dump could be turned into something yet again! From organic wastes, we can create methane. With inorganics, we can fill in a strip mine and build a resort hotel on top. In regions deforested by acid rain, we plant crops that thrive in low pH, such as yams. When there are blooms of ocean plankton, we harvest them directly, and market the scum as health food. The possibilities are endless!"

"What of the oceans, Bruno? Most of the planet is covered in water."

"The oceans are nearly dead, my young friend. It is a shame. My distant ancestors fished the Med. Now, there are no more fish. It is the same in many places, the same story. We have over-fished the oceans, but what can we do?"

"We can let them grow back?" Jenks offered.

"Ah, you are still an idealist. I like that." D'Argeste gave him a patronizing pat on the arm. "But no, that will not happen. The battle for the surface oceans was fought last century. The battle for what lies beneath was lost just a decade ago. Now…" he shrugged, and gave a look of resignation.

"And what about space. Once we run out of land, won't there be a push to move into space?"

D'Argeste looked directly at Jenks.

"Yes, there will. It has to be. There is no equilibrium in complex systems; everything is either growth or decline. Space is inevitable."

"Better sooner rather than later," Jenks said.

"Tell me, Jenks, do you favor the approach just announced by the Americans and this man Parks Vambolent?"

"I think it is doomed to fail."

"Yes, your report stated that opinion. But you did not quite reveal the reasons why," D'Argeste prompted.

"People are afraid of space," Jenks said. "Space is just like nuclear power, which is perfectly safe if done properly. But the world is not a perfect place, and deep down, everybody realizes that. Who wants weapons hanging over their homes at night and over their schools during the day, needing nothing but a single command to vaporize your family, your children. The threat is double, Bruno.

First, the idea of putting microwave antennae at GEO has already demonstrated its military effectiveness during the Oasis Massacre. Such force is quite unstoppable. And then there is ballistic transport. The same technology that can launch solar cells from Luna to GEO can just as easily sail them past to impact on the Earth with pinpoint precision. Need I say 'Three Gorges'?"

"Well put, Jenks. You are a shrewd man. I am taking somewhat of a risk bringing you here so early in your apprenticeship. We are not even sure if…"

Jenks interrupted him.

"We are indeed sure, Bruno. It's why we must invest now, before everything happens."

"And what of the patent issue?"

"He was too early. Twenty year patents are ridiculous on inventions like this. We can be ready in three years. The US is sure to hamstring the Lunar Energy Facility before it can deliver power. Their only true interest is to build military bases on the Moon and in orbit above China. While Parks Vambolent is crashing in public, we prepare in secret, so that we can quietly begin our deployment just after the dust settles. We'll be established years before anyone else catches on, and we will corner the market on energy, information and material manufacture. The entire global landscape and economic power balance will shift. We can…" Jenks stopped suddenly, noticing the look of alarm on Bruno D'Argeste's face.

"You are a *most* ambitious young man, Winnimer Jenks."

"Yes, sir."

D'Argeste looked at Jenks through narrowed eyes, appraising him.

"I wonder what are your ultimate plans. Not that you would tell me," he said quickly, holding up a hand. "For certainly, I play some role in them, no? But answer me this, as my apprentice: where do you expect to be ten years from now?"

"In your job," Jenks said.

"The right answer, of course. Yet, I wonder how you plan to accomplish it, for surely you have a plan. Will you parlay these space ideas into a personal fortune? Will you seize the fortune of another? Will you somehow acquire the rights held by Mr. Vambolent?"

"Do you know him?" Jenks asked.

D'Argeste looked sidelong at Jenks. He wondered if the question was as simple and logical as it seemed. He hadn't told Jenks about his meeting with Vambolent at SMER. It was not in his nature to trust people too quickly, and compartmentalization was a key element of his survival strategy. Jenks was far too smooth and aware for someone without a pedigree or specialized training. Something in D'Argeste's breadbox spread through his mid-spine and made his groin tighten up.

"Certainly I know *of* him," he answered, looking out the window at the countryside bright with light but overhung with a dark shadow of cumulonimbus anvil.

"But have you ever met him," Jenks pressed.

"No," D'Argeste said.

Somehow, he knew that Jenks saw through his lie. It was a surety, a certainty, deep in his gut. The feeling passed in a few moments, but he knew to trust that first instinct. D'Argeste began to harbor feelings of alarm over his immediate personal safety.

Jenks looked back at him, his face wearing the perfect expression of the young sycophant eagerly waiting for words of wisdom from his mentor.

D'Argeste coughed. It covered his discomfort. But the very gesture was telling to this young man who seemed to miss nothing. *What is his weakness*? D'Argeste wondered. From their talks in setting up an apprenticeship, he knew of the offer Jenks had made to Elise Vambolent at the very moment that her husband was fighting for his life in the desert. A man *that* ruthless, *that* ambitious, could surely devise a means to get whatever it was he wanted. He also appeared to lack scruples, making him doubly dangerous.

"You know that we are to downplay this whole space thing tonight," D'Argeste reminded him.

"I remember your words, Bruno. You and Kindoh only. The ten others favor a much slower pace. Same goal, but different timing. And extreme sensitivity to anything that moves too fast or too slow. Does that about size it up?"

"Indeed," D'Argeste said. He was seriously worried now. His keen sense of paranoia was activating the flow of adrenalin. His mind rewound to warnings or veiled threats he might have overlooked before, and fast-forwarded to the possibilities his death would bring. He was the leader of their organization. And when you sit atop the mountain, you make an easy target. Certainly it had been done before. What would happen after? What could Jenks be up to?

"Still, I would hate for us to miss the opportunity to get in on the ground floor, so to speak," Jenks said conversationally. "Is there any way around the group consensus?"

"Of course, we may act on our own initiative. Yet it is expected that we align our efforts to the overarching vision."

"And if you do not?"

Is he a spy? D'Argeste thought, *an assassin?* His belly rumbled. It seized up inside of him, and he felt an urgent need to move his bowels.

"Ah, you are so eager to know everything," he said, fighting for control of his rising sense of panic. "But I will tell you this: we may *prepare*. This is *your* role, my friend." D'Argeste's voice broke slightly on the last word, and he coughed again.

The helicopter tapped down gently, first on one runner and then the other. The whine and chop of the motor fell off their pitch. Jenks and D'Argeste removed their headphones, and Jenks opened the door. He jumped out into the long green wheat grass, and offered a hand to his mentor. D'Argeste straightened up and shrugged off the hand. He felt his urge diminish as soon as his feet were on solid ground. This was his backyard, and he felt slightly safer. His mind was torn between having Jenks as a powerful ally, or the very real possibility that he was an agent of one of his colleagues. If that, then Jenks was either here to gather incriminating evidence, which was most likely, or he came as an ambassador. Who could have set up this young man? He *had* to have some kind of powerful backing. But who?

The two men left the pilot to tend his craft, and started off towards the town center to find a car. The restless breeze from the approaching storm spun the leaves on the trees and washed across the verdant fields in wide swaths like the caress of an invisible hand. The sun was already being swallowed by the high-altitude haze of the thundercloud.

The meeting tonight, in his mountain redoubt, would tell them much – if he lived to get there. *No*, he thought, *not before. Probably not right away. But eventually?*

Yes, eventually, he decided. *Jenks will kill me. He brings me the dream I have treasured for all my life: finding the Regulus. If I have him killed, then there is only Vambolent left. But he doesn't stand a chance alone, and I dare not risk exposure. I need Jenks.*

Even if he kills me to get what we both want.

Cimarron Hills, Colorado

News of the visit by their new Mission Director had flashed through the new NASA headquarters. Like all buildings of the new capital, only three floors showed above the canyon floor, while the bulk of the facilities were underground. When Parks entered the building foyer with Dajwal Chellapilla and newly-approved Administrator Dr. Kevin DeRosier, crowds of NASA scientists and functionaries stood waiting.

The confident smile Parks wore to cover his nervous expression broke as he looked around at the faces staring at him. Images of Grissom Hall overlaid with the view of skeptical, dour expressions on the seasonsed veterans running NASA-lite. A few of the younger people smiled.

"Would you like to address your new team?" DeRosier asked.

Parks whipped his head to see DeRosier's face, but could see no trace of a disingenuous expression at having set him up for an impromptu speech. Looking at Dajwal, who showed genuine concern, Parks drew in a brisk breath and accepted the offer. His voice was unused to public speaking, and as he raised it, hoarseness and breaks punctuated his words.

"Hello. Thank you all for coming to welcome me. It is a great honor to be here, and I genuinely hope we can all work together to meet the Grand Challenge issued by Vice President Oxley."

If he had expected applause, it was not forthcoming. There was little or no response at all from the assembled host. Parks felt sick to his stomach. Bile rose in his throat and threatened to spew forth. Swallowing, he continued, hoping to claw his way out rather than digging a deeper grave.

"As you probably realize, we are given a task impossible by normal standards. It is not unlike the challenge issued by President Kennedy leading to the first men on the moon."

Throughout the voluminous foyer, disparaging tsks and harrumphs rose above the background hum of air circulation. Parks grew more desperate, but decided to turn things around.

"Therefore, our first goal must be to streamline our work. Reduce bureaucracy. Cut red tape. Flatten the organization, and empower the people."

More coughs and too-loud murmurring sprang forth. They had heard it all before. The same platitudes were delivered by every alternating adminstrator, from visionary to bean counter, and the tides of public opinion waned and waxed, and especially as more and more fiscal belt-tightening became the political mantra. Parks struggled to find some way to reach them.

"So, as my first order of business, everyone on the SPS project will be relieved of their present duties."

The background hum became a roar, and DeRosier stepped forward, placing a restraining hand on Vambolent's arm.

"Now, Parks," he said, for all to hear. "We are all in favor of improving operations. But let's not get carried away. Too much change too fast causes disruptions and works against the goals we all must achieve."

Parks stubbornly continued. His voice notched up in pitch, and he spoke more loudly, making echoes bounce around the chrome and glass interior.

"Your new duties, on the SPS project, will be redefined from the ground up. The work outlines prepared by Deputy Administrator Chellapilla will be your guide. Each of you will be asked to select a position within that organization, and to justify your assignment to that role."

But now, the assembled people spoke in angry voices. Some even shouted their derision. DeRosier shot Parks a withering glance and spoke beneath the roar. "If you didn't have the Vice President's ear, I'd put you out on yours!"

To his employees, he pleaded with them to calm down. "Please, people. Please! Please! Hear me. Hear me, now." The roar subsided to a grumble, and he spoke forcefully, raising himself up so that his height over Vambolent was emphasized.

"Nobody is losing their job. Your current roles are safe, although as Mr. Vambolent expresses, some of you may choose *voluntarily* to try something new. Now, you're all dismissed. Please return to your regular duties without delay."

When the three of them were in the elevator, DeRosier ripped into Vambolent verbally.

"Where in God's name do you come off with an act like that? Your little speech there set us back all the way to the beginning of our re-organization, and made even that much worse. After those shenanigans, we are likely to experience a brain-drain worse than the loss of our Indians after the Pakistani nuclear exchange!" Dajwal bristled, but DeRosier was on a roll. "If you had *any* clue how to motivate people, you would realize that you've just sabotaged your entire effort. People told me about you, that you're a…"

The elevator door had opened on the third floor, and several secretaries behind high-walled mahogany desks all turned to the sound of DeRosier's voice. He caught himself and recovered smoothly.

"…that you're an inspired genius who needs a little guidance. Now, in the future, you'll refrain from public outbursts like that, and vet your ideas for reorganization through my office. Dajwal will be our intermediary, and will be primarily responsible for ensuring that your work on the SPS project dovetails smoothly with regular operations at NASA headquarters and at our research centers. Claro?"

"Yeah, I get it," Parks said. "Where are Ned and Franklin?"

"Their cubicles are in the lower levels. Your office is over there," he said, pointing like the Grim Reaper. "Dajwal, please escort Mr. Vambolent to his office, and explain to him how things *work* around here."

"Yes, sir," Dajwal said, and put a hand above Parks' elbow. Parks looked at him, and noticed that his normally dusky skin was pale and drawn.

When they entered the new office, Dajwal closed the door and stood with his back leaning against it. "Parks, I am very concerned that you have just committed political suicide. I don't know what will be happening next, but I have heard that Dr. DeRosier is not a man to be trifled with. He will surely be looking for a way to get rid of you – of us – as soon as absolutely possible. I would not be one tiny bit surprised if he is on the telephone to the Vice President this very minute."

"Relax, Dodge," Parks said, with a confidence he did not quite feel.

"How can I just relax? For you I have spent every penny of my political capital. All the work I have done for NASA since I returned from Kenya has been building up to this very moment. Do not think I was exaggerating. I believe you are a brilliant man, a genius, perhaps. When Minister Jiachi mentioned your name, and then the journalist Heidi Bayer reminded us all of your great work, it was me, behind the scenes, who pushed back the shadow of the Oasis Massacre and got you this job. Now, in your first hour as Mission

Director, you have angered the entire staff at headquarters and made an enemy of the Administrator. Oh, woe. What am I to do?"

Parks sat behind the desk, swiveling left and right in his new chair. Somehow, Dodge's laments made him feel queerly sure of his course. In that moment, as Dajwal beamoaned his awful fate, Parks decided he would charge ahead exactly as he saw fit. He would either succeed spectacularly, or they would escort him from the building in disgrace. There would be no middle ground.

He looked up at Dodge, ashen-faced and on the verge of tears. With a blink of understanding, he realized that Dodge, a political animal through and through, would be of no further help to him. Keeping him on the SPS team would do neither of them any good.

"Dodge, can you please ask Ned and Franklin to come see me here?"

"Sure, Parks, sure. Anything else I can do for you?" Dajwal asked in a squeaky voice.

"Yes. Please tell Dr. DeRosiers that I am very grateful for your work in putting together the work outline for SPS and LEF work. He is now free to reassign you to other duties."

"But… but, Parks? Are you firing me?"

"No, Dodge. I'm saving you from certain disaster. We both know this whole project is nothing but a scam to appease the Chinese and put military bases in space. You've been too good a friend to go through all that. You've done everything I ever expected and a thousand times more. In fact, there is no greater reward I can give you except to spare you the implosion that will certainly happen."

"Oh, Parks, are you sure?" Blood had returned to Dajwal's face, but his eyes showed great confusion and anxiety.

"I've never been more sure of anything in my life," Parks said, and finally the confidence he had been faking began to take hold. "Send up Franklin and Juneau. And send me an IT expert to help set up my computer. Then, we're done, Dodge. Thank you."

Parks got up and extended a hand to his flustered friend. Dajwal reached across and accepted his grip. In a daze, he left the room.

Plopping down in the reclining executive chair, Parks put his feet up on the polished glass desk with integral data screen. Then, lifing his feet and swinging around, he wheeled over to the window and stared at the mountains in the distance. A lifetime ago, he had won the polka-dot jersey climbing the mountains of the French Alps.

Now, in the greatest battle of his life, he would ascent to a height 30,000 times taller or die trying.

Chapter 11

To Silicon Valley

The Vambolents wended through the line for airport security, passing those in front and behind them several times. Around them all, bright lights illumined their faces for the recog cameras. An endlessly cycling infomessage played on overhead monitor screens, loud enough to drown out most conversation. Parks noticed people staring at him, many of them blatantly, some of them growing quiet as they passed his family.

"Excuse me," said a man in line, tapping him on the shoulder. "Aren't you Parks Vambolent?"

Parks turned to look, noticed the man's smile, his well-fitting suit, and relaxed slightly.

"Yes. And you are?"

"I'm Peter Lausen," the man said, extending his hand. "A big fan. If you need any volunteers for your next moon shot, be sure to call me." He handed Parks a business card, and then the line took them away again.

Eric looked up at him. "Everybody thinks you're cool, Dad."

"Not everybody," Mindy said. Her brother was almost too young to remember what it had been like at school. Plus, boys like him were less attuned to what other people think.

"Who was that?" Elise asked, leaning over the business card.

Peter Lausen, the card read, Executive Director, Space Systems, Lockheed-Martin. He was based in Colorado Springs.

"I've read about him in *Space News*," Parks said. "He's a shaker and a mover. They also lost out on a big bid for us last year. I'm surprised he was so friendly."

"He probably lives near us," Elise said. "Maybe we can invite him over some time."

"That would be a conflict of interest, I think," Parks said. An airport page, louder than the informessage, muddled his words. The line brought Lausen back past them.

"Hi, I'm Elise Vambolent," she said to him.

"Nice to meet you, ma'am. I hear great things about your Education Project."

"Thank you!" she replied, clearly pleased with the compliment. "Parks and I were just saying we should invite you and Mrs. Lausen over sometime. Do you live in the Springs? Is there a Mrs. Lausen?"

"Yes, on both counts. We would like that very much."

"C'mon, Mom," Eric said, tugging her sleeve. "You're holding up the line."

The four of them were split up by the TSA agent, and they headed towards smaller lines for the metal detectors and bomb sniffers. Eric placed his skateboard in the grey plastic tub and pushed it into the X-ray machine. He was lookng forward to his first visit to California, where he planned to ride a trolley car to the top of Telegraph Hill, and ride his skateboard down to the bay. Whether he would jump off into the water, he would wait to see how the ride went.

Mindy placed her threadbare come-along on the conveyor belt. It was weighted down by stacks of books, but had plenty of room to bring back chocolate. She had saved up money to buy a kite at the piers and get some dichromic jewelry at the outdoor art fair. On Saturday, her parents were making her go to the Earth Day Festival, which she dreaded.

Elise removed her shoes, and placed them next to her purse in the bin. She looked around, and noticed Peter Lausen coming up behind her in the line. She smiled at him, and he returned the gesture. There was a small-statured woman between them, and Elise spoke over her head.

"I'll be speaking at the Earth Day Rally on Saturday. Perhaps you could come? Parks is speaking also."

Lausen frowned. "Where is that?"

"At UC Berkeley," she said. The woman behind her moved closer, urging her forward.

"I'm off to Seattle," he said, sounding disappointed.

"Oh. Well, bye." Elise said, feeling a bit foolish. She turned to go through the metal detector, turning her thoughts to the Napa Valley tour they had planned for later in the week.

Parks got selected for "random" inspection, and stood with his stockinged feet inside the yellow outlines. His shoes were being rubbed with gauze paper by a nitrile-gloved security agent. Other passengers kept filing by as he waited to be wanded. In anticipation, his arms were already held out from his sides. They were growing tired.

He thought ahead to the GCEP meeting he was to attend at Stanford University. His participation was yet another requirement that Robert Marcharelli, Sr. had placed on him in return for funding the start-up of Space Manfacturing, LLC. On Elise's suggestion, and after lengthy debate between himself, Kevin DeRosier and the new Vice President, Parks formed a holding company for receiving money from Congress to create the Lunar Energy Facility. The arrangement suited everyone except DeRosier. It mollified the Chinese, who did not trust NASA; for the Congress, it kept the entire project separate from NASA-lite's embattled funding; and for Parks, it freed him to make his organization lean and nimble – short-circuiting the bureaucratic morass of a government agency. Instead of paying back the start-up funds, Marcharelli, Sr., the self-named "East Coast Space Baron", asked that Parks include his son, Robert, Jr., in the first manned return to the moon.

Finally, an agent came over to scan Parks. As the flat wand was run between his legs, Parks noticed an older woman sneering at him. She looked well-dressed, stood straight, but wore a look of disapproval and utter derision. His shock was painful, but short-lived after long exposure. There were still plenty of people who associated him with the Oasis Massacre and the destabilization of Middle East politics that had followed. Some even blamed the Iran invasion on the Egyptian attack, not remembering that the invasion had come first.

Others portrayed him as the symbol of America kow-towing to China. The economic concessions being given to China until the energy could be supplied from space sent the US trade deficit soaring to record levels. The economy slowed from a dearth of easy credit. Inflation drove up prices, and companies laid off workers. Over-extended consumers got even more serious about low-cost products which further eroded the US manufacturing base, further strengthening China. With higher oil prices pushing up costs across the board, life was getting very difficult for a lot of people. There was pressure to abandon the space project, and Vambolent's face was the rallying point for that cause.

Environmentalists still incorrectly linked Vambolent to the Seven Dwarfs asteroid retrieval debacle and the subsequent destabilization of global weather. A steady stream of airborne soot still issued from the half-submerged volcano created by the moon meteorite. For three years, heavy storms blanketed much of the Pacific Northwest and western Canada. The dust blocked enough of the sun's radiant energy that the permafrost line had moved south, almost to pre-Industrial Age latitudes. This winter's fresh snow advanced the line further south, and there was a talk of an encroaching Ice Age over North America. The middle latitudes, including Colorado Springs, had a wild ride of swinging temperatures the whole year.

So Vambolent's face was on the newslines from sharply different angles, making him a focal point of all of them. Bad press from one side reinforced another. News reporters looking for connections had an easy time making Vambolent a target. It made him wonder sometimes if there were well-meaning kooks out there who might decide the world would be a better place without Parks Vambolent, and put a sniper bullet into him. The world was always full of crazies, but in tough times, there seemed to be ever more of them. Not for the first time, he considered buying dark sunglasses.

As a last measure, the smug agent had Parks boot up his notebook computer to prove it was not a decoy. Past security, he could see Elise trying to keep Eric from running off. Mindy sat on the floor with a book, with a decidedly unpleasant look on her face. Knowing her, she would now be reticent during the whole flight.

"I'm sorry," he said, coming up to them with his arms full. He yanked on his shoes, standing up, smashing his finger in the process, and leaving a tab of leather tucked under to rub against his ankle. "Okay, we're ready. Concourse C."

On the way down the wide concourse, Parks steered them into a gift shop for water and bagged snacks. As they piled their choices on the counter, he plucked a pair of put-over sunglasses from the rotating display, and laid them on top.

"What are those for?" Mindy asked. Her sour expression showed that she already knew. Parks didn't answer.

"Yes, Parks, what are they for?" Elise reinforced.

Parks didn't want to say that a nasty look from an old lady made him paranoid. He just muttered something about the thinning ozone layer, yanked off the plastic tag, and put them on his face.

"I can still tell it's you," Mindy said, with a slight sneer in her voice.

They boarded the plane first without anyone else approaching them. Mindy and Eric battled over the window seat.

"You're just going to read," Eric said.

"I'm older, and I'm the lady," Mindy protested. "I should get first pick."

The flight attendant came with a tray of beverages. Mindy reached for a cola and Eric used the distraction to plunk down in the window seat, fastening his lap belt quickly. He looked up at her with a jutted chin.

"Brat," she said, under her breath. She fell into the aisle seat, allowing her heavy bag to swing forcefully against Eric's legs. She knew it must have hurt, but he pretended not to notice.

The new boarding procedures filled the plane in just minutes. The doors were sealed and the plane taxied to the runway. As they built speed, Parks and Elise kissed, knowing that this was the most dangerous part of the trip. By long tradition, they kissed through the whole take-off.

"Gross!" Eric yelled from across the aisle.

The flight over the Rockies was rough, but settled out after Grand Mesa. Elise leafed through the shopping magazines while Parks worked on e-mails, sunglasses on his tray table. They were startled by the voice of their flight attendant.

"I'm sorry, ma'am. The forward lavatories are for first class passengers only."

Murmured words answered her. They turned to see her pushing shut the black cloth partition. A hand pushed it back, and another woman edged past her, holding a drink in a clear plastic cup.

"You!" she said, stopping at the Vambolent's seat. "You bastard! The whole world is screwed up. It's all your fault."

Before the harried flight attendant could stop her, the older woman from the security line tossed her drink into Parks' face at point blank range. It splashed into his eyes, and splattered everywhere. "You deserve much worse!" she yelled over her shoulder as she was ushered back.

Instantly afterwards, a uniformed Sky Marshall unlocked the cabin door, and strode down the aisleway, wielding a heavy handgun. Parks and Elise stopped mopping themselves to watch. All of first class rose in their seats to look back.

There was shouting throughout both cabins, and the Marshall descended on the lady with a vengence. In seconds she was handcuffed. With his gun holstered, he led her to the back of the plane, and whipped the rear partition closed.

The flight attendant came back to help the Vambolents clean up and answer their questions, while the lead attendant read from a pre-written emergency announcement. He told everyone not to panic, that everything was under control, and that there was no, repeat, no danger.

It was some time before people settled down again. The couple in front of Parks and Elise turned around to express their feelings of regret for them.

"You're doing a great thing," the man said. "I know who you are, of course. It's about time we went back to the moon. And it's about time we had a good reason to go there. What you're doing is not just for China. It's for everyone. Right?"

Parks nodded. "That's right. Thanks," he said.

"Don't mention it," the man said, and sat back down.

The Lunar Energy Facility, as the base would be called, had survived nearly intact from fiscal year to fiscal year, even surviving the change in Administration. There was simply no other viable and acceptable alternative to fulfilling the promises to the Chinese government.

The government of China seemed content with the arrangement, and made no fuss so long as the schedule progressed on time. The treaty they signed at the end of the war gave Taiwan provisional freedom, depending of course, on deliveries of energy beginning in year five, and fully operational in year ten.

Parks felt he could do it in eight. He would be first to implement a new idea on a grand scale. Remington had interchangeable parts. Ford had assembly lines. Vambolent had geometric manufacturing.

Finally, the announcement came telling them straighten up, buckle up and tray table up. Parks and Elise looked over at their children, cheek to cheek. Mindy rolled her eyes, and Eric, groggy from sleep, rubbed the crust from his eyes. She reached over him to undim the window.

Fog-graced hillsides and grids of houses showed the leading edge of Silicon Valley. It was the first time to California for both kids, and now they were nearly there. Eric looked for Telegraph Hill, Mindy for Ghirardelli's. Elise looked north towards wine country. Thursday!

Parks looked up at the sky, and saw a daylight half-moon. The terminator went right through Hyginus Rille, the landing site for the first manned mission. It was sunrise at the start of a long day on the Moon. And in exactly three more of those lunar days, a rocket – his company's rocket – would land there, and start building the factory he had invented almost two decades before.

Berkeley, California

The Earth Day Rally at UC Berkeley was sponsored by the newly-reformed Green Party, and held within the larger Earth Day Festival. It was the Woodstock of a new generation, and drew revelers from across the country and around the world. Rock bands SR-71, Speeding Ticket, and The Skeletons headlined, along with enough individual performers and Green Generation protest bands so the music was virtually non-stop for two and a half days.

The hilltop campus was a chaotic mosaic of color and activity, with green and the natural buff color of hemp being the dominant hues. Long-maned young people with green hair, their faces painted green, wearing hemp sunhats, headbands, shirts, purses, belts, and shoes wandered among booths selling environmentally-friendly clothing, office supplies, backpacks, cleaning solutions, kitchenware, bedroom products, and building materials. Delicious aromas wafted along with the ocean breeze, blending the smells of hand-wrapped burritos, corn-on-the-cob, sugar-fried dough drizzles, vegetarian chili, and of course, grilled cheese sandwiches. A small Farmers Market was set up for sales of locally-grown organic produce, and to take orders for free-range chicken and grass-fed beef.

A few daring corporations were bold enough to showcase their new green products, despite the generally negative attitude displayed towards big business by the Green Generation. Car displays drew crowds three rows deep. Hybrid electric SUVs sat with their doors wide open, alongside flywheel-assisted electric runabouts, and even hydrogen fuel cell commuter vehicles from Japan. The students were invited to sit in the hemp-trimmed seats, look under the recycled aluminum hood or kick the sustainably-farmed rubber tires.

It was a warm, sunny Saturday, and many people carried fruit smoothies in yard-long paper cups, or lemon shake-ups dispensed by the ounce into mugs they brought with them, and made from organic lemons, organic beet sugar, and reverse-osmosis triple-filtered water. The atmosphere was charged with love and energy. Smiling faces and held hands were everywhere, and in every color and combination imaginable. As the four o'clock hour approached, the milling crowds began to align, and to converge on the grandstand where the Rally was about to begin.

*   *   *

On the center stage, a three-color hologram of the slowly-rotating Earth floated in the shadow of the overhead scaffolding and the deep indigo of the backdrop curtains. At points on the semi-transparent planet surface, flashing red dots would appear. Flat panel displays to either side of the hologram were synchronized to the red circles, and scrolled down with the news and background information on the latest enviro-crises. The Green Party logo, a horizontal braid of white, green and blue banners, adorned the top of the stage where Willow Masters, the leader of the Green Party, was about to speak.

At the base of the stage, several television cameras were set up, and a small press box in the center of the front row held journalists and photographers with large, heavy lenses. Unknown to the crowds, there were at least three types of paid agents mingling with them. The ubiquitous FBI had several undercover agents continually following the break-away political party, noting who attended, participated, organized, and contributed monetarily. Several security guards from a private firm had been hired by Willow Masters to forestall any aggressive actions by the Freezers, an anti-activist activist group, who believed that if we simply froze the current level of technology, then everything would be just fine. Some of the younger Freezers, usually short-haired white males with plain white T-shirts and dark blue jeans, had disrupted Green Party Rallies before.

The third group, operating in deep cover, were Mad Av consultants. Young people of the Green Generation shunned any form of media advertising, taking their entertainment directly from the web, sans commercials. This market segment was the fastest growing in the country, but the advertising agencies on Madison Avenue lacked the demographic information they needed, and the buzz they required, to somehow reach these savvy shoppers. Greenies, as they were called in corporate conference rooms, tended to buy their products mail order from mom-and-pop outfits located in remote places in the Rockies, the Cascades or the Appalachians. They paid with money orders or direct transfer, making their buying habits nearly impossible to trace. It was as un-American as any other social or economic trend in the last century, flying in the face of mass marketing, volume discounts, and wholesale distribution. They were desperate to find a way to crack it open.

Without warning or introduction, the loudspeakers projected the amplified voice of Willow Masters in advance of her entrance to the stage.

"Militant Recyclers, Effluent Avengers, Smokestack Spotters, Welcome one and all!"

The crowds, which had been sobered by the dire news scrolling by in front of them, had been just quiet enough that Masters' opening lines hit them like a call to arms. They stood as one body and cheered, yelled, whistled, and whooped, waving fists in the air and shouting "All Right!" and other affirmatives of their deep resolve.

Willow Masters walked right into the holographic Earth and stood there, absorbing and acknowledging their cries of outrage, indignation, and of solidarity. In the past year, since the Green Party had become a more powerful political entity, she had pulled together local groups from around the entire North American continent, and galvanized their collective wills into a cohesive vision of the future of American politics, economy and lifestyle. The sensation she had helped create amazed even her, and she repeated publicly that it was not her doing, just that she happened to be the right person to come along when the time was ripe. This was a movement 'of the people, by the people, and for the people', she paraphrased, and she was nothing more than a spokeswoman.

As the cheering passed its peak, the hologram shifted. When the whole Earth suddenly moved sideway a meter, the noise dropped, replaced with a gasp. To their mob-mentality emotions, seeing their beloved planet move in such an un-natural way brought a quick lump to their collective throat. The globe shrank, becoming smaller and smaller and smaller. Masters held one hand out to her side, palm up. The Earth became the size of a soccer ball, and came to rest in her hand.

"The Earth is not as big as it once was," she said, and her voice echoed clearly in the hush. "The Earth has been kicked around, spat upon, dumped into, and abused for generation after generation. And we are running out of room to hide all that."

The crowd quickly recovered its ire, and cheered her words. She waited until they grew quiet again. Even her stance, while non-threatening, seemed to control the crowd. Several of the FBI agents registered amazement at this symbiotic relationship between Willow Masters and the Greenies. In any other setting, they wondered, could she pull this off?

"And now, the world is in our hands." She brought her hands forward and the hologram followed with her movement. "Each of you is the guardian and the custodian of the entire Earth, because you are intimately connected with every molecule of air, every clod of dirt, and every drop of water. For your body and the body of our planet are one. We are in continual exchange with our planet every breath, every bite, and every sip of our lives. This is your Earth, too."

As she spoke, she formed her arms into a circle, and the one globe multiplied, filling her arms with now softball-sized Earths. With an exaggerated, dramatic movement, she bent low, and then tossed her entire armload of small Earths up and out over the crowd. She followed them with her eyes, even when their apparent motion took them beyond the reach of the stage's holographic projectors, and they could no longer be seen. But every pair of eyes in the audience followed the motion as well, and everywhere in those crowds, people lifted their hands to catch the fragile Earths, to catch them before they crashed.

She walked to the very front edge of the stage, almost 2 meters above the concrete apron. She stood there, her body language conveying a captivating mixture of confidence, concern, and expectation. Without even using words, Willow Masters was communicating volumes to the audience. They already knew the words. The people assembled here were already convinced. They did not need to hear of the ecological injustices, the corporate abuses, the individual acts of neglect happening billions of times a day throughout every country of the world. She was here to align their direction, and too many words could provide points of division. Her message was simple and direct. And it was immediate.

"Men and Women of the Earth. You are the foundation of our future. Every generation before you has raped the Earth. This will not change overnight. There is too much at stake, for the stakeholders. There is too much momentum, from the shakers and movers. There is too much ignorance, in the schools, in the homes, and in the media. But they will pass away, naturally. It is up to us to decide what will come next. We are the bridge. The bridge from the old way to the Right Way."

That brought cheers from the audience. Masters repeated the statement, and pumped up the volume of the crowd, raising their shouts to a frenzy.

"We are the Bridge to the Right Way!"

With her arms, she traced a bridge over her head. The crowds yelled and whistled in appreciation. The FBI agents began to grow anxious. Although a Green Party Rally had never become violent, it was hard to see why not. Other rallies they attended could turn bloody with a lot less agitation than this!

With her arms outstretched to the side, she rotated her palms, and turned it into a gesture of welcome, of loving anticipation.

"Our role in history is to teach the next generation. We will set an example for them to follow. We will educate our friends, our neighbors, our relatives, our countrymen, and our fellow Earthers about the Need for Green."

More cheers broke out at their slogan. The "Need for Green" was appearing on bumper stickers, T-shirts, graffiti, and verbally among themselves. Masters had linked the catchy phrase with Earth-friendly activities, and helped it become a mantra of everyday life. Within the Green Generation, it was a fad, but it was also becoming a habit.

"Our goal is to start now, with ourselves, to support Earth-friendly activities in every aspect of our lives. We eat organic produce farmed in a sustainable manner. We buy clothes made of natural fabrics with minimal environmental impact. We buy in bulk, or with the least amount of recyclable packaging necessary. We live in homes built from local, natural materials, and cleaned with ecologically-sound products that biodegrade as soon as we are finished with them. And all of our refuse, all of our waste, all of our byproducts, can be recycled, composted, re-used, or at the very least, reduced, until technology catches up with us and closes the cycle. That is the Right Way!"

The audience had mellowed somewhat, and it was appropriate. They were clapping and whistling now more than yelling and whooping. She had boiled their blood, and raised their adrenaline. Now, for the next speakers, she needed them level-headed and able to sit still to hear the important messages she had organized for this critical Rally. But first, she wanted to lavish some praise.

"My dear friends, the Green Party is a modest organization. We do not receive corporate donations, nor are we bought with political favor." She let the cheers roll over the crowd for a moment. "We are an organization of volunteers and underpaid professionals who, like you, believe in the Right Way." More cheers. She smiled.

"Today, let us take a few moments to recognize our staffers, who are all here for the Green Day Rally!" She stepped to the side, and clapped for the small parade of organizers, attorneys and lobbyists walking onto the stage from behind the indigo curtain. FBI agents were busy taking pictures with their eyeglass-mounted cameras. The audience clapped until the line of party workers faced them all on the stage. Most of them looked uncomfortable with the attention.

"We all thank you! Please, take a bow!"

The 20 or so workers bowed modestly, and then began moving back off the stage, followed by the continuing applause from the crowd.

"If you have any questions, or if you would like to volunteer, the Green Party Website is shown on the monitor here," she said pointing. "And if you smell any smokestacks, need to avenge any effluent, or notice the dire need for recycling, call the number on that monitor. We will activate our worldwide network to take action as soon as possible.

"And now, without further ado, I'd like to introduce you to our first speaker."

Two roadies wheeled a podium onto stage right as she spoke.

"Today, we have a husband and wife team who share our vision. Elise Vambolent is here to discuss her pioneering method of education which has been so successful in Kenya, and is spreading across what we used to call the Third World. Her husband, Parkinson Vambolent, is the man who began the Hydrogen Economy in Japan, and is developing space-based manufacturing to help us create a sustainable future here on Earth. Ladies and Gentlemen, I give you Parks Vambolent."

The crowd cheered and applauded for the most part, but there were a number booing. Young people, wary of the establishment and leery of the government, associated Vambolent's name with the destructive power of new technologies. Willow Masters had warned him that factions within the Green Party opposed having Parks speak because of the demolition of the oasis factory in Egypt; even though it was later shown that the facility was indeed a weapons plant and the Jewish nation was exonerated. But those who booed reached a deep part of him that still felt guilt. Nonetheless, he took the stage with his shoulders back and his face calm and confident.

"Thank you, all. Thank you very much. I am honored to be here, and delighted to share with you my vision for our future." As he took a breath, a couple of the hecklers booed again. He had practiced his speech over and over, and while he did not expect to be interrupted like this, he had pre-lived this very scenario in his mind many times. He took in a deep breath, and departed from his prepared words. He walked slowly to the center of the stage, and sat down, dangling his feet over the edge.

"I am just one man," he told them. "And I care deeply for this planet, and the people on it." A few interrupted to applaud their agreement, but he pressed on. "I have a dream, a vision, and I have made this my life's work. But I underestimated the capacity for others to misuse what I created. Although we took precautions to prevent something like this happening, others were more clever. They took advantage of me. What I was doing to save lives, they turned around, and used it to take lives.

"It would be easy for me to say their use of force was justified, but I will not take the easy way out. You see, I feel personally responsible. After it happened, I had a breakdown. I wanted to quit, to give up, to let this dream fade like so many dreams do when you awake suddenly. But my wife, Elise, wouldn't let me go down that path. This has been one of the hardest lessons of my life, and I'm here to tell you I *still* don't have all the answers. But there are a few things I *have* learned.

"We can't close the door on new technology. Like Pandora's box, once opened, it cannot be undone. The whole reason we have an Earth Day is because of the destruction caused by technology, by population growth, and because of ignorance. But we can't freeze technology, as some say – it's too damaging as things stand today. My dear friends, the only way *out*, is to go *through*. We need to develop new, better, more sustainable technologies. New ways to manufacture consumer goods, new ways to produce and distribute food, new ways to generate energy to run our society. New ways that are clean, Earth-friendly, and sustainable for generation after generation."

"Yes, I have a vision. It is a good one. But I cannot do it alone. And that is why I am here before you today. I need help. The lesson we learned in Egypt means we can no longer naïvely charge forward into the future with green ecology as our only focus. We have to be smarter than those who still play the old game. We have to be more clever than the opportunists. We have to be faster than those after a quick buck. We have to be wiser than those who still think they know what's best. We have to do such a bang-up good job at everything we do, that everyone will see, and everyone will know, that we are doing things the Right Way. Isn't that how you see it?"

The audience, which had grown hushed at this heartfelt, extemporaneous talk, suddenly erupted in emotion. Affirmations of support and agreement poured forth towards the stage. There was an energy in the air that hadn't been there before. It was a beautiful moment. Vambolent put his hand to his mouth, nearly overcome with emotion himself, and deeply grateful for their response. Nobody was booing anymore, and for Vambolent, it was a watershed. He got to his feet, but stayed in the center of the stage. All his prepared slides and videos were still waiting, but he wasn't going to use them now.

"I came here with a speech to give you, but I won't. I think I just told you what I needed to say, and maybe what you needed to hear. But I do want to emphasize that we need help. We have devices operating on the Moon which right now are making the raw materials for a lunar factory. We will soon be creating a new economy in space that can generate pollution-free energy at an affordable price, and that can produce new Earth-friendly products with near-zero environmental impact. And, we can eventually build habitats that give us another basket to put some of our eggs into – your children, or maybe even your younger brothers and sisters. But *your* generation, the Green Generation, will be the first true generation in space. There is a high frontier out there, and it is wide open. I encourage those of you who are interested to study the space sciences, and to marry these to sustainable, environmentally-friendly practices. Together, we can create new technologies like the ones Willow mentioned, the ones that finish the job, and close the cycle. The technologies that allow humankind to live on a healthy, beautiful planet, for all time to come. Thank you."

Vambolent was awestruck at the overwhelming positive response he received. He stood there, soaking it in for long moments until Willow Masters came over, also clapping, and shook his hand.

"Parks Vambolent!" she shouted over the noise. She put one arm around his shoulders. The other she held out, showing him to the crowds.

The cheering rose again. They both bowed, and she kissed his cheek. She applauded him until he was off-stage.

"Okay, our next speaker is the other half of this remarkable husband and wife team. What is really amazing about Elise is the new teaching paradigm she has developed, and put into practice in Kenya. And boy, has it been successful. In only a few years, the rate of illiteracy dropped in half. Think of that folks! It's probably some kind of world record. We all know that ignorance is the enemy of change, and that literacy is the key to enlightenment. Not only is Elise helping more Kenyans to read, but the nature of her new school is teaching the children to build sustainable economic systems using local resources. The long-term potential is tremendous. I would encourage those of you who are teachers, or are planning to be teachers, to consider her new method as you teach the children of the next generation. And now, Elise Vambolent."

* * *

Elise walked onto the stage, awash with cheers, and waved to the crowd. She wore a big smile, and looked happy to be there. "Thank you, thank you!" she said, taking her spot behind the podium.

"Thank you all very much. And thank you, Willow for such a wonderful introduction. Let me first say that I can't take credit for all this myself. There have been dozens of terrific people who have pulled this together. But I must especially mention my good friend Afren Kwapili, an Olympic Gold Medalist from Kenya, who made this all possible. I wish he could be here to talk with you. He is a marathoner, and he ran all across his native country advocating the idea of Self-Schooling."

Elise explained the basic principles of her education system, and showed short videos of children teaching themselves using METAXA-based thematic learning agents developed by her world-wide staff of educators. A few charts showed great progress in literacy and standardized tests in math, science, and social studies, but her main focus was on individual success stories. In a small village by the Tana River, at the base of Mount Kenya, an 18-year old young man named Tied Umnata had helped to develop a helical turbine that could generate modest amounts of hydroelectric power without the need for dams. He and his classmates had found the invention on the internet, and had brought together the pieces to make an affordable solution for local power. They had a backpack-sized version that they were selling to the Flying Doctors, to power more advanced medical equipment in the bush. Then, they had a larger model that could be used to run Self-Schooling facilities and satellite uplink web servers in remote villages. It was a tremendous success in Kenya, and they were already exporting these units to other countries in Africa, and in Southeast Asia as well. The crowd was mesmerized by Elise, and listened as attentively as schoolchildren. Even the FBI agents were sitting now.

"We have a great program in Kenya. I don't know if it could be done here, where there is so much inertia in our education system. But there are dozens of places, all over the Earth, where Self-Schooling can make a huge difference. Like my husband Parks, I am only one person. I too, would like your help. It would be so wonderful to see Self-Schooling systems started in under-developed nations around the world. Won't you please consider volunteering your time? We don't have a website, but the excellent people at the Green Party have offered to dedicate a page to Self-Schooling until we can put our own together. Thank you for your kind attention."

The applause started right away, and kept growing. People in the front row rose to their feet, and a wave passed from front to back as the entire crowd gave her a standing ovation. Like Parks, she was a bit overwhelmed by the response, and sought him out behind the stage.

Willow Masters walked on-stage. "Elise Vambolent, people!" and they added cheers to their applause.

Masters began introducing the next speaker, who would discuss molten carbonate oxidation, a new method to completely consume plastics and feces with wheat straw and cellulose in the presence of inorganic minerals. Waste, plus oxygen, could then be converted to carbon dioxide and water, which could be reclaimed. Inorganics were converted to ash or mineral salts, which could be re-used. It required a fair amount of electricity she said, and that tied in with the clean, low-cost energy that Parks Vambolent was working on.

But Elise was barely aware of this, as she walked backstage in a daze. The cheering was like a drug to her, and she could hardly feel her feet on the ground. She heard Masters say her husband's name just as she walked up to him. He embraced her, and it felt like the safest, most loving place in the world to be. She was so happy, her heart felt like bursting. Parks rocked her back and forth, and she knew he felt it too. Something was changing for them, and for now, at least, it felt so good.

PART III

Chapter 12

The Moon

The lunar descent vehicle separated from the return module, carrying its four crew members in a low orbit around the Moon, in preparation for their landing. On the backside, they were privy to the destruction of asteroid 06-401B. Gliding over it, they saw a nickelodeon of the event, from mountaintop to mountaintop, and then to a shotgun blast. A widening cone of devastation. The disturbed moon soil was a wound that would last forever, and its ugliness lent a greater beauty to the undisturbed regolith.

From their equatorial orbit, they could see glimpses of the meandering line of gleaming iron cable that had been deposited from the back end of the lunar gleaners over the last 10 months. Robert Marcharelli, Jr. shifted over to get a better view, and caused their craft to roll.

"Sit still, Bob," pilot Dallas Barstow said quietly. He knew Marcharelli hated to be called Bob, and because of that reason, never called him anything else. In Barstow's 19 years with the space program, the only thing he had learned to hate more than turf-guarding bureaucrats were prima donna space cowboys. He knew that a crew such as theirs would never have been picked by the old NASA, but then Vambolent didn't have much of a choice, did he. If Marcharelli could pay his way to get trained and buy a berth on this mission, then who was going to talk back to the almighty dollar? Barstow brought his focus back to the navigation computer. Everything was fully automated, but they needed a smart monkey on watch to catch those stupid programming mistakes that seemed to bring down more missions every year. With millions of lines of code, it was no wonder they got a few bugs.

He glanced over at Sellenya Laurencino, their operations commander and one-stop shop for programming. That made her more valuable to the mission than he was as pilot. God, she was beautiful, despite her shaved head and sour expression. Her high eyebrows over strong cheekbones hid deep-set, penetrating eyes, and full curvy lips made her a striking beauty even after she shaved off her long, dark hair. Since their training began, Barstow had watched Marcharelli play for her six ways from Sunday, but the icy MIT professor ignored him until he gave up and went away.

Strapped in his niche, John Lomish snored. The guy was like a big cat. He could sleep anywhere at a moment's notice, and did so every opportunity he wasn't needed or having fun. He was a trained rescue worker for the few years when NASA developed their own squad for retrieving stranded astronauts. After Congress had axed the return vehicle project from NASA's budget, three researchers had died on the ISS from carbon monoxide poisoning. The Congress, in their knee-jerk, hindsight wisdom, had immediately ordered NASA to set up a ground-based rescue unit. Lomish was a rescue worker until he lost that job when the very next Congress cut the rescue unit back out of the space program. Vambolent was lucky to have such a Jack of all trades, Barstow thought. Lomish, it was widely told, could fix just about anything mechanical without specs and without a manual. With both of those, the stories went, he could fix anything else. Good guy to have along on our first manned Moon shot since 1972.

It was their last pass over the dayside of the Moon. Their touchdown spot was now near the leading edge of the terminator, giving them two full weeks to assemble the payload launcher and start up the solar cell factory before nightfall. Barstow and Laurencino exchanged technical checklist items with their ground control at Johnson Space Center in Houston. Everything was working fine. They may as well have been packed in boxes, like the cargo chords of their rocket ship, so minimal was the need for their involvement. Barstow had seen automation come a long way, and he was glad of it. Barstow might as well be sleeping along with Lomish. As 'pilot', it made him feel a bit peeved that the computer did all the work. Even the eyeball sighting that had saved the first Moon landing by Armstrong was obsolete. That work was all done by vision-recognition programs linked to artificial intelligence navigation routines. But the pay was the same, so what the stonk.

Their descent vehicle, which would soon become their habitat, was already flying bottom first. Now the thrusters were turning their tail down towards the surface, and the main engines ignited to a low thrust. After three days without gravity, it was a welcome feeling to have a 'down' again. It felt like they were sliding down an increasingly steep slope as the ship vectored more and more to the main engines, slowing their velocity, and preparing for a vertical descent to the landing site. Tiny resonances made the craft shudder through the intermediate power ranges, which was part of the reason why none of them had eaten for 8 hours. They would have breakfast on Luna.

The craft came up to 90 percent power, and the roar of the engines just below their feet drowned out all other sound. They were vertical now, and their small view ports showed only stars all around. The Moon would be entirely below them, as they were lowered on an elevator of flame to the dusty surface. Streaming wisps of scintillating dust motes licking past the viewports told them their descent was nearly over. The dust outside became an opaque grey, and then they hit – a one third gee, spine-jarring impact that marked their touchdown like the dot under an exclamation point.

The main engines shut down at once. The whine of the turbo pumps fell from a loud squeal, dropping in pitch and volume into subsonic silence. The four crew members waited for that sound to die away before anyone spoke.

"Let me be the first to welcome you to Luna," said Marcharelli, mimicking the announcement made after commercial airline flights by the relieved lead flight attendant. "The local time is zero-zero-zero-ten on our mission clock, and the outside temperature is a balmy two hundred fifteen degrees. We wish you a pleasant stay here at the Hyginus Rille, or whatever your final destination. But remember, keep your seatbelts fastened until the pilot has sounded the all-clear, and by all means, don't go outside without your Manned Maneuvering Unit."

The mood from Houston, several seconds delayed, was much more jovial. Barstow piped their response to everyone's headset, and they listened to the distant ground controllers whooping it up over the successful landing. For some reason, maybe the time-of-flight delay, or maybe that the happy voices were only in their headphones, none of the crew, except Marcharelli, seemed especially jovial. It was a poignant reminder of how far away they were, and how utterly reliant they would be on help that was a quarter of a million miles away.

Sensing the mood, Barstow made a 'ping – ping' onomatopoeia, and spoke into his curled fist to mimic speaker static. "The pilot has sounded the all-clear. You are now free to move about the cabin. But remember," he said, playing along with Marcharelli for now, "use caution when opening the overhead bins, as items may have shifted during take-off or our *bone*-crunching landing."

They all started to unbuckle. Two seconds later, the voice of their capsule commander (CAPCOM), Henry Cox, amiably jibed, "We heard that! Hey, the fuel you saved on that less-than-tender touchdown allowed you each to bring three pounds of extra gear!"

Lomish, whose deep bass voice was never less than a loud growl, spoke for the first time in hours.

"Hey Hank, I drop three pounds each time I use this pinhole toilet. How come those pencilneck geeks in Payload wouldn't let me bring along a little chaw?"

Cox's quip came back, but the effect of any joking was quelled by the unavoidable delay in the radio signals. "John, if we let you have your chaw," he said, "then we'd have to let Barstow have his guitar, Laurencino a case of merlot, and Marcharelli a ticker-tape machine!"

Cox had spent many hours training, and many hours socializing, with the crew members before the shot. It was deeply reassuring to them, hearing his familiar voice, demonstrating such an intimate knowledge of their individual tastes, mannerisms, and quirks.

"Henry, we're ready for some chow here. What's on the menu for Day One?" Barstow asked. While he waited, he folded down the tiny table that he and Laurencino would use, and the sliver of a bench to sit on.

"Okay, our 'pencilneck geeks' as Lomish so delicately calls them, have tailored your first meal to your particular tastes. The cuisine for Day One will be surpassed only by the cuisine just before your take off. Unfortunately, everything in between will be recycled from the honey pots." He paused for effect. The honey pots were where their un-recyclable wastes were stored. "Just kidding."

"Okay, for you, Mr. Barstow, we have beef-flavored soy steaks, garlic mashed potatoes, green beans with slivered almonds, and a beer-flavored malt beverage with no carbonation and no alcohol – sorry."

Barstow grimaced slightly.

"For Ms. Doctor Laurencino; hello Sellenya, I hope those gorillas are treating you well." He paused. Aboard ship she looked around at the three men suspiciously. Definitely not selected for her personality, thought Barstow. "Well, *your* repast posed a special challenge for our Payload engineers. You are the first person to request calamari, and they had to get this special-made from our meal suppliers. They have used authentic Mediterranean imports, and included this with the sauce for your spaghetti dinner. There is a biscuit flavored like garlic bread, and from our juice cellar, you may draw from the ample supply of healthy Concord grape juice. Okay?"

"Okay. Thank you Henry," she said. It had been her apparent policy to speak as little as absolutely possible, and her voice was raspy from disuse. She cleared her throat loudly.

"Okay?" Cox repeated. "Are you okay, dear?"

Laurencino cleared her throat again. It sounded like she needed to hock up a loogie, but with her refined manners, she was much more likely to just swallow it back down.

"I'm fine, Henry. It's just that the air is a bit dry here."

They were all practiced at waiting through the two second breaks in their conversation, and nobody stomped on the silence.

"Ah, we have an adjustment for that. John, can you dial up the relative humidity? Take it to 60 percent until everyone gets re-hydrated from their meal, okay?"

"Okay, boss," Lomish rumbled.

"And for you, my large-boned friend, we have a nice meal for you today before you go on your diet!"

"What diet!" he bellowed back.

"Oh, didn't they tell you," Cox joked, "Your MMU was shrunk so it could fit in the storage locker. You'll need to lose 5 kilos just to fit into it!" He paused, and Lomish pressed his lips together, not enjoying the joke.

"Just kidding you, John. We had to make several provisions for your size here, but I think you'll be pleased with the meal. It's steak, potatoes, a vegetable and near-beer."

"That sounds like the same thing Dallas is having," he said in a low, threatening tone.

"Yes, only, his is a regular size portion. Yours is double rations for today. And don't forget the humidity setting for Sellenya."

"Okay. Thanks, Hank."

"And finally, for the *pièce de resistance*, we have a fine repast prepared for our dear Mr. Marcharelli. Robert, are you ready for your menu?"

"Ready, willing and starving!" he replied enthusiastically. His hands mimicked having a fork in one fist and a knife in the other. He licked his lips anticipating the food. He really loved this astronaut business, even the freeze-dried food paste.

"For you, Robert, we have an herb-spiced Fettuccini Alfredo with a sauce made from the cream of contented cows grazing on tender grasses above the Wisconsin Dells. Glazed carrots will be your vegetable, and for your beverage, a full-flavored fruit juice from Eastern Massachusetts, pressed in the finest stainless steel equipment money can buy. And for desert: sugar cookie with cinnamon frosting awaits your eager palate. Now, doesn't that sound like a meal fit for a king?"

"It does, indeed, when do we start?"

"Let's see. Today is Tuesday. Robert, I believe you are on cook rotation today, so as soon as you would like to begin, by all means do so."

Marcharelli looked crestfallen. But then, changing into his best Ivy League Can-Do face, he stood up to get the food packets. Forgetting that Luna's gravity was so low, he flew upwards off the floor of their module. His eyes bugged open and his arms flailed up to stop himself. He narrowly missed hitting his head on the pipes and fittings of their ceiling, before guiding himself back down. More soberly now, he retrieved the food packets from the locker.

While waiting for breakfast, Lomish began checking the cabin stowage for any damage. Laurencino pulled out a keyboard and monitor, and began running diagnostics on the ship's systems. Barstow cycled the airlock and tested the seals for their first foray later that morning. Their first order of business would be to make the ship secure, and to deploy the silicon solar cells to supplant the fuel cells and recharge the batteries. Marcharelli busied himself injecting water into the meal packets where indicated, and placing them one at a time into the induction heater. As they came out, he served them unceremoniously, until all were seated and eating quietly.

If they hadn't been so still, they might never have noticed the moonquake. There was no rumble, but their craft began to shimmy ever so slightly. Each looked up, and looked at the others, to see if someone was shaking a leg under a table or something.

"What is that?" Marcharelli demanded. It faded.

Barstow and Lomish kept chewing, slowly. They guessed, but didn't want to verbalize it. It was a hard superstition to overcome when you were trained as a spacer. Problems you had no control over were best ignored until they could no longer be ignored. But Laurencino knew.

"It was a moon quake," she said simply, her voice still gravelly.

"Is she right?" Marcharelli asked of Barstow.

"Yes, Bob."

"I thought the Moon was dead. No volcanoes, no rift valleys, and no earthquakes, right?"

"Moonquakes," Barstow corrected. "Space probes suggest the Moon still has a molten core, but it's small."

"Did anybody check to see if this is an active region? I mean, did anyone set a seismometer in this exact location to make sure it's not geologically – or lunalogically active? Because if it is, this whole mission is scrubbed, right? We abandon Hyginus and the next ship goes to the Shackleton Crater like we should have in the first place!"

Nobody answered him.

"We could be sitting on the San Andreas Fault of the Moon, and nobody bothered to find out before we landed?"

Barstow pulled his boom mike back to his mouth. "Henry, you hearing this?"

They waited for his reply.

"Dallas, I note that you all felt a slight tremor at thirty eight ten on the mission clock. This is well within mission parameters. We don't anticipate any problems whatsoever. But just to be sure, I've already sent a note to our lunar seismologists. In the meantime, we have our Payload engineers looking in your manifest for some micro-g sensors we can use to keep track of seismic activity. Honestly, Robert, the risks are slim and none."

"I hear you, Henry. But for us to feel a tremor this soon after landing is not a good sign. I've studied statistics plenty, and I want a straight answer from your experts. Have them run a Poisson distribution, using any data from the Viking and Apollo seismographs, and give us an estimate of the likelihood of any bigger quakes."

"I understand your concern, Robert. And they'll be doing just that. But like you said, no volcanoes, and no rift valleys. The chances of a significant quake are probably smaller than we can measure. And like I said, as long as the quakes are small, there is no danger to either you or to the mission. Okay, Robert? I need you level-headed on this one. You've already braved many dangers on the trip over. Don't let a little moon tremor shake your confidence, right?"

Cox knew just how to handle Marcharelli. Each one of them, actually. The astronauts were all smart, really smart, and they each knew exactly what Cox was doing. But he was good at it, and it worked.

"Right. I'm fine, Henry. But I still want that Poisson distribution," he replied.

"Roger, Wilco. We'll have it to you ASAP. In the meantime, please enjoy your meal. You each have a busy day ahead of you."

Sol

To the naked eye, the sun is a shimmering disk of yellow-white. In reality, the sun is green. But with a dark filter over a small telescope, the sun's surface looks granulated red and orange. Dark red sunspots mar the image. Telescopes aboard spacecraft blank out the center, and see the solar corona as diaphanous rays. Periodically, loops and whorls of thicker plasma curl outward. While the physical surface of the sun is under 6000 degrees, the coronosphere is made up of ionized particles which may be millions of degrees. These particles eject outward from the sun and form the solar wind. A messy mixture of photons race outward at lightspeed, outpacing the 'wind' of neutrinos and electrons, protons, and neutrons in various combinations.

Deep in the sun's core, nuclear fires generate enormous quantities of energy that must work their way slowly to the surface through the densely-packed matter of the star. Moving charges generate magnetic fields. Lines of magnetic force interact with the physical movement of star-stuff in an unstable equilibrium dance. Currents of matter, charge and thermal energy boil and bubble. Where field lines draw energy out from the sun, darker, cooler regions form which can be as large as a planet and last for a week or two.

Satellites around Earth and also in orbits around the sun monitor the sun constantly, since the impact to humankind can be so great if space weather turns bad. Sunspot activity moves through 11 year cycles, and when they are numerous, the swirling patterns of energy in the sun can bunch up dangerously. When magnetic fields around sunspots grow too strained, they can reconnect suddenly. These reconnections can release energy on a scale inconceivable to man. Within a few minutes, temperatures can flare to 20 million degrees and send 10 billion tons of electrified gas hurtling outward from the sun at speeds of 40 kilometers per second. The energy equivalent of billions of megaton atomic bombs explodes outward in a superflare called a coronal mass ejection. Prediction of CMEs is a black art, and even when the "all clear" signal has been given, conditions can change quickly.

The CME which blew out from the sun at the end of the astronaut's first day on the moon first announced itself through a high flux of X-rays. The size of this flare rivaled that of the largest ever recorded. With humans outside of the Earth's magnetopause, there was nothing to shield them from the full onslaught of the high-energy particles that would soon follow. In GEO, satellites prepared to shut down to avoid power surges. On Earth, power grids in the north braced for overloads from the pelting of electrical energy into the Earth's own, much weaker, magnetic field.

Johnson Space Center, Houston

Henry Cox, the lead CAPCOM, walked into the daily Mission Review at 10 minutes past the hour. He looked harried, but his voice was strong and confident.

"Sorry I'm late folks. We had a minor funny. Everything's under control."

He sat down along the back wall. At the head of the conference table, on the other end of the room sat Kevin DeRosier, Administrator of NASA-lite since the reorganization almost five years earlier. To have accomplished so much in so short a time was a tremendous credit to his effectiveness. Re-use of proven designs, and elaborate systems integration were hallmarks of his craft. Massively parallel efforts, cross-correlated and linked through remote co-simulation brought the launch vehicle, the habitat and the factories all together faster than anyone thought possible. He was widely praised outside the organization, but almost universally loathed within.

Next to DeRosier was his deputy Colleen Whiteside, Program Coordinator for the manned Lunar Energy Facility missions and primary contact for Vambolent's company. The Flight Director, Amy Genova, and her lieutenant, Jared Ventris, filled out the NASA side of the table.

Opposite them were three Space Manufacture representatives, including one of the gleaner and rail engineers who led the team which laid iron rails across the surface of the Moon. Two empty seats were at the front. Vambolent stood at the screen, but interrupted his material to acknowledge Cox.

DeRosier scowled at his tardy employee. Not one to let bad news go lightly, he challenged Cox.

"How minor, Henry? What exactly *was* the problem?"

"Oh, just normal stuff. Happens all the time if you live in California. Tiny moon tremor. They'd have never noticed if they weren't sitting still at breakfast."

"A moon tremor?" DeRosier said, aghast.

"Anyway, they were a little edgy," Henry continued, "I'm sure you can appreciate that, being all that way from home, and two of them never been all the way into space before. We just needed to put their fears to rest. No big deal."

"Henry," DeRosier shot back, "I have found you to be the king of understatement. Tell me now how bad this is. I've never heard of the Moon quaking. Is this an anomaly? A whifferdill? Or is this the harbinger of more danger to come? Is the mission in danger, Henry? Everything rides on your answer. So come clean."

Anger nearly flashed in Henry's face, but he breathed a defeated sigh. His shoulders sagged – he had hardly realized how tightly he was holding them. Part of his mind knew it was better not to say anything, but his mind was exhausted. The moonquake happened late last night, and he had been doing assessment ever since. The worse they dug, the worse it looked. They discovered the last bit of news just before he arrived, which made him late, which gave DeRosier an opening to probe.

Henry put on a confident face. He personally didn't believe the astronauts were in any real danger. A whole lot of hullabaloo would do everyone more harm than good. Things were tense enough as it was. But how to get DeRosier off his back?

"No, Kevin, the mission is in no danger," Cox said. There was a collective sigh from many of the people in the room. Henry continued. "We have seismic data from Surveyor and Apollo. Nearly all moonquakes are very mild, with displacements of several microns – imperceptible to humans. This one was a little bigger than average, but still well below the tolerance of our factory and habitat parameters. Really, Kevin, the biggest concern I have is for the crew's emotional state. We need them sharp and focused to complete all the work facing them in the next 13 days."

Colleen Whiteside began doodling a script number 13 on her notepad.

DeRosier did not look convinced. The Flight Director knew enough to be alarmed, and was staring focused at Cox.

Vambolent spoke up in the breath before DeRosier could dig more deeply.

"Senator Barnes is still absent, but we have a lot of material to cover and a limited time slot, so perhaps we should resume."

"Okay, I'm ready," said Cox, a bright too brightly.

DeRosier stared at Cox while Vambolent went through prepared slides with the latest analysis of their system readiness and mission profile. His report was glowing with news that the NASA teams had done a great job at completing complex tasks on time. Nature had cooperated with the launch. The solar activity they were watching carefully was still nominal. The landing had been perfect and all systems were in working states with green diagnostic indicators. His enthusiasm caught the attention of the room and the mood seemed to lift. There were questions and give-and-take concerning a broad range of details and dependencies.

Every day they assessed risk and delays due to natural disasters, human error, equipment fatigue or failure, sabotage, inaccurate specifications, code bugs and incomplete validation. Any one of these could jeopardize the mission. With so many interdependent parts, they were keenly aware of the race car drivers' old saw that it was the 25-cent parts that got you.

Probability charts showed mean time to failure and infant mortality rates, which all looked great. Radiation tags with dosimeters were recorded. Heart, lungs, pulse and brain activity were recorded on each of the four people on the Moon: Barstow, Lomish, Marcharelli, and Laurencino. Vambolent finished up all the gory details of reliability engineering. There were few questions. It was time for a vote to determine if the review was accepted, and the mission could continue to the next phase of lowering the habitat and assembling the factory.

"I'll call the roll. We're still missing Senator Barnes. He wanted to be here," Parks said, glancing involuntarily at his watch. "So, call out your votes…"

Right on cue, Mason Barnes threw open the door.

"I vote Yes!" he announced. An aide swept in behind him, winded. The Senator plopped down into the seat reserved for him, and waved a hand towards DeRosier. They shook hands, and Barnes said: "What's your vote, Kevin?"

Silence, for a minute. DeRosier appeared to be deep in concentration.

"Kevin, how do you vote?" Vambolent asked.

"I'm thinking, Parks – thinking. This moonquakes thing has me reconsidering the risks. I gather from what Henry over there told us is that we didn't really expect something that large. So, it makes me wonder what else we might not have considered. I mean, don't get me wrong, Parks, nor you Colleen," he said turning, "but despite your detailed study, we just don't have any firsthand knowledge of the Moon that isn't from the Bronze Age of space travel. I mean, astronauts playing golf, and driving a buggy across the Moon is nothing compared to what we are undertaking here. Am I right, Senator?"

Barnes gave a mysterious look, which could have meant anything.

DeRosier was just getting warmed up. "I mean, this whole business of supplying energy to China is a bit over the top. No one has objected yet, but when we start using ballistic transport to move payloads around, people are going to notice. Personally, I think this whole program is an exercise in futility. We're putting lives at risk for political gain. It doesn't seem to me that we're following the kind of careful, exacting science that is demanded by the harsh and unforgiving environment of outer space."

Cox slumped in his chair, looking straight ahead through half-closed eyes. Vambolent sat down slowly, as it appeared DeRosier was about to launch into a speech. Barnes had folded his hands across his belly, and appeared to be studying them.

"Let's take a for-instance," DeRosier continued. "Henry, tell us more about moonquakes. I'm somehow certain there is more you know than you've shared with us so far. What is it? What have you found out?"

Cox answered in a quiet monotone. He was like a prisoner in a kangaroo court, where nothing he could possibly say would exonerate him.

"Moonquakes come and go in cycles," he said. "There are as many as five cycles, mostly harmonics of the lunar sidereal month.

By extrapolating these cycles from the last hard data, we suspect that three of these cycles are near their peaks."

"You mean that right now, today, is the peak of seismic activity?" DeRosier pressed.

"Thursday, actually, the thirteenth."

Whiteside looked up and the whites of her eyes showed large through her glasses. Her doodles had become very elaborate.

"So, they already felt a 'perceptible' tremor, a large one by lunar standards, and it's not yet the peak of the cycle?"

Cox nodded wanly.

"Why did you not know this before?"

"The N-body problem has not been solved, Kevin. We should have dropped a tri-ax on Hyginus a year ago when the gleaners landed."

"This is just grand," DeRosier said, shaking his head. What else, Dr. Cox? What else have you found?"

Cox didn't respond for a few seconds. DeRosier simply waited, frozen like a statue that would never fade away.

"Friday is the Full Moon. Plus, it's a syzygy."

"Oh, crap!" Whiteside exclaimed, and whipped a hand up to cover her mouth. She had made the '3' into the Earth and the Sun lined up, and the '1' into the habitat. As her hand dropped past her heart, the Sign of the Cross was barely a waver.

DeRosier seemed actually to be enjoying this. He sat forward on his chair.

"Wonderful! Now, we have an alignment of the Sun and the Earth coinciding with a peak in the seismic cycle. And what will that do to your risk factors, Dr. Cox?"

A slow fire which had been smoldering unnoticed in Cox suddenly rose into his eyes, and he looked straight at his boss.

"Sure, there's more, DeRosier." He spoke forcefully. He was fed up with the bullshit and wanted to get on with the job. Rather than play cat and mouse, where he was the mouse, he just gave it up. Maybe an easy win would shut the man up.

"It's the rille system. We picked the location halfway between Apollo 11 and 15 where there is a steep cliff wall. Our plan is to use that wall to anchor the electric catapult for the ballistic transport to GEO, right? It's a good spot, close to the equator. But as we've been really researching moonquakes in the last 19 hours, we've uncovered some alternative theories.

"There is a small maria, or lunar sea, called Vaporum just north of our landing site. It sits between two mountain ranges. At the convergence of the ranges is Mount Hadley, nearly 14,000 feet high. Some volcanologists have proposed that this maria may be riddled with lava tubes."

Nobody interrupted, but several muttered 'lava tubes' under their breath, and Colleen Whiteside wrote the words on her yellow legal pad.

"Flowing magma," he explained, "pushed ahead of the emerging mountains, may have formed a skin or shell when exposed to the cold of space, especially if this happened during lunar night. That's why they think the Near Side of the Moon is so different from the Far Side. Then, either the lava just drained away into a lower elevation, or perhaps it formed a highly-porous pumice that later collapsed, leaving an underground tunnel."

He waited for anyone to jump in with questions, but they didn't. He gave a shrug with his eyebrows and continued.

"These same scientist also think that pumice would make a perfect hydrogen sponge, better even than porous silicon. It could be an import from the Moon."

He chuckled, but it echoed, and he continued in a more professional tone.

"Hyginus Rille is just south of Mare Vaporum, as is the Ariadaeus Rille. What makes this new theory plausible is the shape of Hyginus Rille." He stood and began sketching on a whiteboard at the back of the room.

"The rille is in the shape of a boomerang, with one end pointing towards Mt. Hadley, and the other end paralleling Ariadaeus Rille. Right in the bend is a crater: Hyginus Crater. You see, the crater is smack-dab in the middle of the Rille. This suggests, to some, that the same impact which caused Hyginus Crater also *formed* or *created* Hyginus Rille itself." He paused, seeming uncertain of how to proceed. The fire in his eyes had faded, and only concern for the astronauts showed on his tired face.

Vambolent asked quickly: "Aren't those features millions of years old, Henry?"

"Yes, Parks. Probably hundreds of millions. And there's a very good chance that everything is stable."

"So, why do you seem down?" Vambolent asked.

Cox looked over at DeRosier. The man looked shell-shocked. The fire of his witch hunt had died. *Good*, thought Cox, and he continued in response.

"To the southwest of Hyginus crater, there are several features; lines, shallow depressions in the lunar surface that are only visible when the terminator is nearby. They are little more than slightly concave regions. You could be standing in one, and not even know it. What matters to us though, is that these seem to step from Crater Chladni about 75 miles away."

Vambolent shuddered. The sound of 'Chladni' struck a chord with 'Chalbi' and the association evoked a visceral response. He tried to shake off the feeling, and prodded Cox a little further.

"But those Apollo 17 images, I saw them," said Parks. "Those features end well shy of our mining plat."

"Yes, Parks, they do. But according to this new theory, that's only because they haven't been disturbed yet."

Barnes spoke up, but without his usual bluster. "So, what you're saying is that we could be sitting right on top of one of those tunnels, am I correct, Mr. Cox?"

"Yes, Senator."

"Can we move to a back-up site?"

"Not enough fuel to move the whole ship," said the Flight Director, Amy Genova.

"And aren't our operations far, far softer than a meteorite impact, barely a feather drop?" Barnes asked.

"Yes, sir," Cox replied automatically.

"Then, what's the problem?" Barnes asked, not unkindly.

DeRosier came alive again. He leaned onto the table and his elbow jarred the legal pad. Whiteside shot him a nasty look.

"I'll tell you what's wrong, Senator. We've got ourselves a genuine worst-case stack-up. Any one of these factors alone might be small enough to ignore. In fact, we did ignore to some extent during our too-fast mission planning." His eyes flicked briefly in Whiteside's direction, implying her guilt. "But we didn't consider the unlikely event they would all happen together". DeRosier held up his fingers, and began ticking off points.

"First, we made a hard landing – to save fuel, right? Well, who knows, we might have landed right on top of one of those hollow magma channels, just like the Senator said. That brings us to two, the tremor felt by the astros. It could have been triggered by the landing.

Hell, it could have been the last in a series of tremors. It could be an early warning of a really big shaking on Thursday. Because, three, we have positive interference from three seismic cycles, making moonquakes even more likely. Four, we have the Sun and Earth both pulling the Moon in the same direction, putting it into tension, I presume. Just the thing you don't want when planets start shifting. And in the middle of all this, we have our people slated to start driving metal rods into the side wall of the rille. Isn't that right, Henry?" He had all five fingers played wide counting the dangers, aiming his hand accusingly at Cox from across the room.

The look of smoldering coals was back in Cox's eyes. He didn't answer. This time, DeRosier seemed to have caught the tiniest trace of intimidation. He dropped his hand and sat back, then turned to look at the senator.

"Maybe it's just as well," Barnes said in a quiet voice.

"What do you mean?" Vambolent asked.

"It's why I was late. The Appropriations subcommittee has asked us to delay missions 3 and 4 until next fiscal year. I was late getting here trying to filibuster their decision." He looked at Parks, an apology in his face. "I couldn't do it. We're losing the support we've been counting on. The Green Party has a strong following and there's pressure across Congress to fund carbon reduction and pollution prevention work. Plus there's tremendous domestic focus with the economy in shambles and the jobless rate soaring. People just don't seem to realize that spending money on space is spending money inside the United States. They can't seem to remember that we're not doing this just for China, but eventually for our own people as well. Our supporters realize that our work will ultimately help solve the same problems they're trying to patch short term. But they just can't ignore the voices of the voters. I did the best I could."

The room fell quiet. Vambolent was in shock, even though he had been expecting this all along. In all of their contingency plans, alternate scenarios and mission adaptability, this was one outcome they did not dwell on.

Vambolent pushed away the malaise threatening to engulf him, and rose to his feet, a little unsteady. He stood next to the projector for a moment, and then returned to the vote.

"DeRosier?"

"Yes."

"Whiteside?"

"I vote yes, Parks. Don't let this slow you down," she encouraged. "It's only a delay, right Senator?"

Barnes nodded slowly, but was no longer making eye contact with Vambolent, who finished the roll.

"Well, it's unanimous. The review passes," he said, without any enthusiasm. "The mission proceeds." Vambolent sat down again heavily, and began gathering his things.

There was the sense that something more needed to be said, but nobody was speaking. Neither was anyone else preparing to leave. Several times, Barnes parted his lips as if to speak, but did not. The scratching of Whiteside's pen against the paper became the only sound in the room. She stopped self-consciously.

A small rap at the door galvanized everyone's attention. Opening it a crack, a young man put his head into the room. He looked around and quickly located Cox.

"Mr. Cox, you're needed in the control room, please."

"I'm off-duty there, Jeremy. Broderick has the link."

"I think you'd better come anyway, sir."

"Is it a moonquake?" DeRosier asked in a loud voice.

"Oh, Mr. DeRosier, sir. I'm sorry sir, a moonquake?"

"Yes, man, a moonquake. Is that what you're here about?"

"No, sir."

"Hmm."

"What is it, Jeremy. You can speak freely here," Cox told him.

The young man looked around the room cautiously. When he saw Senator Barnes, his eyes stopped and opened wide. "Are you sure?"

"Speak!" Cox blurted out, frustration beginning to tell.

"Those sunspots we've been watching. Well, they blew. It's, um… it's a CME, sir. Twenty plus. You should come right away."

He left then, closing the door extra softly behind him.

Colleen Whiteside put a hand to her forehead. DeRosier rubbed his face with his hands. Barnes voiced the question begging to be asked.

"Mr. Cox. What is a CME, and how does it affect our people up there?"

"Well, Senator, CME stands for Coronal Mass Ejection. A solar flare, if you will. It's a massive emission of material, mostly hydrogen and helium nuclei, from the outer surface of the sun.

CME's are unexpected, of course. We had hoped not to have one right now, before the habitat is covered." Cox was beginning to look utterly defeated. His voice was breaking and growing rough. He cleared his throat loudly.

"He said 'twenty plus'," Barnes said. "What does that mean?"

"That means it's a big one, Senator. A killer."

Whiteside spoke up, addressing the Senator.

"What it means sir, is that we need to re-adjust the astronaut's duties. We already made provisions for this. It just means we have to engage a different work schedule. Protocol A has them burying their habitat last, just before they take off on the Ascent Spindle. That gives us the most time to make the mission a success. But anticipating a solar flare or CME, we have a fully-developed protocol for them to bury at least part of the habitat, and take shelter until the storm has passed."

"I have a question," Barnes said. "I thought light from the sun only took about 8 minutes to reach Earth. Won't those people get fried before they can get the job done? Hell, they'll be outside in their suits, won't they?"

"The CME travels slower than light speed, but it is still pretty fast," Whiteside explained. "It may strike in two or possibly three days. The radiation will be deadly for several hours. But I think we can handle this, Senator. Isn't that right, Flight?"

Flight Director Genova outlined the new mission profile briefly and clearly. It was bum luck, but not entirely unsuspected. Seismic readings of the sun's surface had indicated the potential for flares, so they were ready. However, the new profile put more emphasis on safety than on productivity, something that made Vambolent even more anxious.

Genova asked that the meeting be adjourned and reconvened after two hours. During that time, the new profile would be justified to the current time line, and the astronauts briefed on the change of procedure. They all needed to get busy, which was better than dwelling on the raft of bad news.

"C'mon, Henry, I'll walk with you to the control room," Whiteside offered, walking over to him. It was a fair guess that they were going to gripe about DeRosier as soon as they reached the hallway. Cox stood, they left, and the room began to empty out.

DeRosier stayed and turned to the Senator. "Senator Barnes, it's been a pleasure to see you again. I can't tell you how happy our whole staff is here at JSC for your help in getting this project started. It's given our organization purpose!"

"It's this man right here," Barnes pointed at Vambolent, "who got this started. If you want more funding, you make him successful. Right now, he's your customer. Make him happy."

DeRosier looked at Vambolent and his nostrils flared slightly, then turned back to the Senator.

"If that will make *you* happy, Senator, then that's what we shall do."

"And by all means, make sure those astronauts are safe. If they fall into a sink hole, or die from radiation sickness, I'm coming back here to kick some ass and take some names. I'm sure you understand my meaning, Kevin."

"Yes, sir. Quite clearly," DeRosier said, clearly dismissed. He exited.

Vambolent walked over to Barnes, and they shook hands warmly.

"How's Elise doing, Parks?"

"Fine. And you? How's the family?"

"Well, my son keeps coming home with his Greenie friends. I understand their point, but I can't talk to them. It's like they speak a different language. You know what I mean?"

"Isn't that what teenagers are all about, Mason? They're carving out their own niche, and establishing their own independence."

"I hear you spoke at the Earth Day Rally in Berkeley," Barnes said, apparently changing the subject. "With the Green Party," he added.

Vambolent was suddenly alert to danger.

"Is that a problem?

"Maybe yes, and maybe no. But that point got brought up at the Appropriations meeting. It puts me in an awkward position, Parks."

"I see."

"I doubt if you do, Parks. There's a lot more going on here than I'm able to share with you. There are wheels within wheels, my friend."

Parks blinked once before responding. "Well, could you at least give me a heads-up when one of them is about to run me over?"

Barnes chuckled, and his belly relaxed.

“Yes, Parks. That, I can do. That much, I *can* do.” He clapped Vambolent on the shoulder with one hand, shook hands with the other, and then left, trailed by his aide.

Time was running out for him.

And for his program.

He realized, with a start, that he didn’t know what to do next. There was plenty to do, but it was all in the background somewhere. A part of him longed for the oblivion of the broken down recliner and the duct-taped remote control. His mind blanked out for a short while. Then he was suddenly aware that a gap in time had just passed.

Had it started already?

He abruptly swept up his materials and fled the room.

Hyginus Rille

Near the northeast margin of Central Bay, a tall white spacecraft perched above the sparkling gray-brown dust of the lunar surface. Four space-suited figures hopped around the landing site, arranging solar cells, and dismantling the cargo chords of the cylindrical craft. Their ship, now called Habitat A, was a rectangular prism with four crescent-shaped cargo bays fixed to each face. The habitat inside was designed to lay over, so it could be covered with lunar regolith. Regolith is a layer of pulverized rock that settled back to the Moon after impacts by asteroids and comets. Smaller meteorites simply burrow into it. This moon dust covers the top 50 centimeters of the surface in most places. It is an exotic blend of silicates and basalts, including tiny spheres of glass and small nuggets of free iron. The top layer is fine and loose, an easily-disturbed grit that gets everywhere. Deeper down, the regolith is a firmly packed aggregate of rocks and dust lying on the lunar bedrock. A layer of regolith two meters deep is sufficient to protect humans from anything the sun can throw at them, and most of the galatic cosmic rays arriving from supernovae halfway across the universe. Now, with an intense burst of sub-atomic particles flying towards them at near-relativistic speeds, their over-riding priority was to get some regolith over their heads.

Under the direction of Colleen Whiteside, the mission managers had quickly re-structured the mission to Protocol B. Not only would the habitat need to be prepared first, but they would have to do it in record time to avoid the invisible but lethal damage the sun had already thrown at them. The cargo chords were detached, and left lying on the ground. Later, if there was time, the chords would be covered by regolith too, to protect the electronics until they were fully installed with their own shielding.

John Lomish finished assembling the rover-dozer, a one-man multi-purpose vehicle for preparing their site. From behind, it looked like a Roman chariot. Lomish stood in the rear, and operated the device with a set of handlebars. Oversized control switches allowed him to switch forward to reverse, and to gear up or down. Under high-torque mode, and using his own generous mass to anchor the craft to the ground, he could push a dozer blade deep into the regolith. The woven piano-wire tires allowed the vehicle to travel fairly rapidly over the uneven lunar surface, but for today's work, he had attached the "chains". The chains consisted of a cable mesh belt with thick protruding wires, several centimeters long, which would give the wheels more purchase. An efficient electric motor powered by fuel cells and supplemented by a battery with integral solar charger, powered the mighty rover-dozer.

Dallas Barstow and Robert Marcharelli worked in tandem to remove the rocket motor and Ascent Spindle from the habitat. The same rocket that lowered them to the surface would power their return to lunar orbit, minus the habitat. Instead, the flimsiest of structures would strap the astronauts directly to the rocket motor, and blast them off the surface. Three tall jacks, with splayed tripod feet, held the bottom edges of the rocket nozzle.

Sellenya Laurencino was responsible for assembling the solar cells. Her normal duties for Protocol A would have been to prepare the payload launcher assembly, but each crew member had been cross-trained extensively. Whiteside's group had given Laurencino this job, despite the high level of exertion required. This normally would have been Lomish's work. It was slow going for her. First, she had to disassemble the cargo chord, and remove the solar cells, transformers, cable, and assembly kits. Then, she had to lift each fragile cell (no dragging), and carry it a dozen meters north of their landing site. Back and forth, she carried the components, beating down a path in the virgin moon dust. She kept her feelings to herself, and just about her only communications with the rest of the crew were during the hourly breaks back inside the spacecraft.

By the end of their 14 hour work day, Lomish had plowed east-to-west rows down to the basalt substrata, and piled the regolith close by the landing struts of soon-to-be Habitat A. The underlying rock was a darker grey, and would form the foundation of their new home. Their true home, Terra Firma, hung permanently in the sky directly above them. It projected the only distinct colors besides black, silver and dun in the whole scene.

Marcharelli and Barstow had moved the rocket motor off to one side, but the rest of the Ascent Spindle would be assembled later. Using the same jacks that had held the nozzle, they now lowered the entire habitat. The landing struts had been removed, laid aside, and were ready to be re-used on the next phase of the habitat preparation.

The day was finally over, too exhausting to be exhilarating. It was bright outside, but their Earth-learned circadian clocks expected darkness. Laurencino was exhausted and grouchy. She had completed the solar assembly, and they were now free to use electricity without restriction. Lomish was playing C&W music on the habitat sound system, and cooking double rations again. She despised the man, while still respecting his capabilities. His voluminous diet inevitably resulted in flatulence, despite the careful planning of the payload engineers. At his frequent eruptions, he would laugh, and the other men would make an exaggerated show of hating the stench. But she had tired of it immediately, and dreaded the aftershocks of his every bite. She would steal surreptitious glances at him while he ate, willing him with her mind to eat less. When he caught her glance, he would grin, open-mouthed, often with food visible inside. She was thoroughly disgusted with the man, considering him a necessary evil.

Something metal twanged. They all went stock still, listening. Micrometeorite hits are a constant danger in space, but their ship was protected against most of them. Larger bits of rock, all traveling at speeds far faster than any subsonic jetliner on Earth, can strike at any moment and puncture a hole in their hull. After a few moments, Dallas Barstow drew in a breath.

"John, Sellenya, run your systems checks. I'll call CAPCOM," he said, toggled the radio quickly, and waited for the welcome routine of making contact.

"What about me?" Marcharelli asked, petulant.

Barstow stared at him a moment. "Watch the cabin pressure gauge, Bob. And the air regulator. If we got a grazing hit, it could progress. Let me know if there's any sudden change."

'Roger!"

The voice of Jason Broderick, their night shift mission director, came from the squawk box over their heads.

"You guys all right?" he asked.

"Just finished dinner," Barstow told him. "About to start rec time, and we heard an impact. Repeating, heard an impact to the hull at about 24 hundred on the mission clock." He looked to his crew. Lomish and Laurencino shook their heads. "Bob?" he asked.

"Nothing."

"So far, no sign of any funnies. Your advice Mission?"

The signal delay conveyed the impression that the sender was thinking over the response. Yet Broderick had replied at once, following protocol.

"We see you're running systems checks. Set tighter limits on fluid and gas diagnostics for the next 5 hours. You're all due for sleep, and we'll do the external inspection after you've rested. Acknowledge."

"Got it. Thanks, Jason. Any news on the CME?"

"Heliomax has you 24 hours from first rise to your dosimeters. Duration should be 4 to 6 hours, possibly a little longer due to pulse broadening. You're partially in the magnetopause, and moving further into it, so that will help a little. With a good night's sleep, we don't see any problems at all."

"That's good to hear," Barstow said, with a distinct lowering of his vocal pitch.

"I'm assuming you want us to take a trank?"

"That's right. The doc advises taking a sleep aid mid-way through your 1 hour rec period, which is due to start now. We've got a fun surprise for y'all in the morning, but make sure you get the rest you need. Tomorrow will be busy."

Barstow paused a moment before signing off. "How's everything on Earth?"

The rest of the crew paused. Lomish and Marcharelli turned to the box on the ceiling. Laurencino looked at Barstow.

"Fine. Just fine. A little tropical storm headed up the Gulf, but nothing we can't handle. You can probably watch it from there in just a few hours when we turn to face you. Like you, we'll weather out this little storm. Nothing Nature can throw at us we haven't planned for. Now, relax, take your sleep-aid, and get some good rest. Mission Control out."

"Hyginus out," Barstow said. He turned back to his crew, giving a grimace of a smile. As ordered, and as they had trained for, they each forced themselves to relax. Marcharelli leaned casually against the wall of the narrow, tall cabin.

"Ouch!" he said.

"What?" Barstow snapped.

"It's cold! What is this thing?" He tapped a metal pipe gingerly with one fingertip.

"Routes outside," Lomish spoke. "Nawmuch insulation."

"Well, it burns," Marcharelli said. "So much for having planned for everything."

Cislunar space

Midway to the earth, an expanding lobe of elementary particles raced outward. On top of the ever-present solar wind, a coronal mass ejection is a huge increase in the concentration of particles thrown off by the sun. In addition to hydrogen, helium and electrons, these burps of stellar matter are laced with heavier elements like carbon and iron.

Despite the greater concentration of particles in a CME, they are still spread out thinner than the last remaining molecules in the best vacuum achievable on earth. Yet, they travel with enormous velocity, giving them a powerful punch in terms of energy. At their high speeds, most particles will pass directly through a human body, doing a minimum of cellular damage. However, when they strike metal, like the hull of a spacecraft, or the fittings of a space suit, there is a cascading spray of lower-energy particles that fan out from the impact point. These slower, and often heavier secondary radiations have a longer dwell time in human tissue. The disruption to cellular structures and genetic material in the cell nucleus causes radiation sickness. In a CME, this can lead to horrible, painful death over days. Higher doses can make you die in just a few agonizing hours, or in extreme cases, within minutes. Even at lower doses, such radiation causes cancers, leads to rapid aging, and can affect reproductive cells enough so that conceiving a child would be ill-advised. All astronauts know this risk before their first launch, as do their mission planners and directors. The crew of Habitat A, having swallowed a generous dose of antioxidant before the solar storm's arrival, and soon to be resting beneath a grave's depth of lunar soil, would be safe. So would their electronic circuits, which are also susceptible to ionizing radiation. Yet, the Return Module in orbit above them would not have this same level of protection.

News of the solar flare gave excuse for widespread media coverage of the many 'angles' taken. To many watchers in Europe terrified of the de-stabilizing practices of the 'Crazy Americans', this turn of events was 'just desserts'. Many Mohammedan America-haters saw this as a sign from Allah, a 'Finger of Fire' to strike at them. NASA and the White House both ballyhooed the peaceful nature of the mission; but that did not forestall inherent skepticism. Most believed that the US was really establishing a military base.

China greatly limited foreign journalism ever since the Three Gorges War with the US. External Sinophiles had to operate on rumors. The current trends suggested that the Party believed the worst about US intentions. They had been tipped off to something already suspected within the vast bureaucracy, that the ballistic transport device on the Moon could instantly be converted into a weapon of mass destruction. Minister of Science and Technology Wang Jiachi had been dismissed from his post within the year, and had not been heard from since.

Press releases were still comparatively rare. But behind their 'Silk Curtain' the Chinese government no longer believed the US would complete the project and deliver the energy free of charge as promised. So, in addition to their continued scornful rhetoric towards America, there was ample evidence of alternative initiatives.

Sakhalin Bay, north and east of China, represents one of Russia's key sources of crude oil. Yet, the Bay is a long way from Moscow. What were called 'fishing boats', but in reality were disguised cruisers and patrol boats, probed the resources and military might of the Russians, as well as the Japanese. Sounding rockets, which were actually missile tests, were conducted almost every week from the north-eastern tip of China. The embattled natural gas deposits along the disputed border between China and Japan now saw new Chinese drilling rigs installed every few months.

Spokespersons for the Chinese government used news of the CME as the right moment to suggest other, more conventional means to supply the 18 gigawatt generating capacity destroyed by NASA negligence. They demanded that the US pursue two courses in parallel, in case one should fail or become delayed. For example, if the solar flare sickened or killed the astronauts.

Anti-space factions in the US cited this as further evidence that man does not belong in space. Aside from these extremes, most people followed the news closely, curious to find out what might happen. Would they die? Would they become sick? The threat of mortal danger appealed to the vicarious thrill-seeker in almost everybody.

Administrator DeRosier and mission coordinator Colleen Whiteside were busy mapping scenarios. Anticipating that there might be further cuts from Congress, DeRosier ordered a speed-up of the schedule for the second manned mission. He redoubled the efforts to make that rocket loaded up and ready.

Hyginus Rille

At 3100 hours on their mission clock, Henry Cox hailed Habitat A with a song. It was a take-off from an old punk rock tune.

"Giant steps are what you take
Working on the Moon,
Henry, let us take a break
Working on the Moon.
We must work forever
Working on the Moon
Gonna make it better
Working on
Working on the Moon."

The crew stirred, but were groggy. A full day of strenuous physical exertion following three days transit in zero-g with no exercise had left their muscles tight and sore. Four hours of sleep in net hammocks was not enough for them to recover fully from the mental fatigue of steady concentration on challenging tasks. The interior of their cabin, already pungent with body odor from their trip over, was becoming acrid. There would be sponge baths when the habitat was positioned, but for now, they had to get accustomed to each other's body odors. But with Lomish testing the limits of their scrubbers and the activated carbon filters, it had already turned rank.

Marcharelli was first to rouse from bed; and dropped from his high bunk lightly down to the floor of the vessel. He grabbed a headset from the rack, and folded the wire behind his ear, adjusting the boom mike to the corner of his mouth. Cox was already starting his morning pep talk, and Marcharelli sat down.

"Good Morning, Hyginus," he intoned, mimicking the movie *Good Morning, Vietnam*. "Welcome to sunshine, stardust, and song. We have a beautiful morning in store for you, even though it began three days ago. Today, as a special feature, we have comedian extraordinaire Lenny Balazar with a show designed just for you. Are you ready up there, crew?"

"Ready as we'll ever be," Marcharelli said laconically. The other two men waited in their bunks while Laurencino moved to the toilet, grateful for the gravity assist provided by the Moon. All of them were moving slowly.

"Hello Moon Units!" said the high-pitched maniacal voice of the ubiquitous comedian. His annoying voice was part of his shtick. "Hey, did you hear the one about the French boy who asked his dad if the Moon was made of green cheese? Huh? His dad said 'No, son, it's Gruyère'! Ha ha ha."

Lomish and Barstow looked over at each other. Barstow grimaced.

"Okay, okay, here's one. What did the Man in the Moon say to the Goddess of Venus? He looked at her with his big round eyes. He says: 'If I told you you had a celestial body, would you hold it against me?' Yuck yuck yuck."

Laurencino finished at the toilet and quickly re-dressed. She sat down opposite Marcharelli, frowning, and rested her eyebrows and forehead in her hand.

"I got one for Doc Laurencino. What's up, doc? … You! That's right, Que-Mah. That's my acronym for Quarter Million Miles High, cause you're higher than any woman has ever been before. QMMH. Que-Mah. Okay? You remember you heard it here first, from Lenny Balazar. So, anyways, *you're* up doc. What's it like up there, huh? Hey! You know why they haven't put a woman on the Moon until now? What, no guesses? Because she'd have PMS for 29 days!"

Up in their hammocks, Lomish and Barstow giggled. One of Laurencino's grey eyes peeked out at them from between her fingers. They laughed harder.

"All right, I can tell you're a tough crowd. So, I've saved my best material for last. Ready? How does an astronaut hold up his space suit? Huh? With an asteroid belt."

Marcharelli made a Bronx cheer, which would probably arrive in Houston in the middle of Balazar's next line. He didn't much care.

"And how about this one? What do you use to ride a space horse? I know you astronaut types are smart. You can figure this one out, right? Give up? It's a saddle-lite! Get it? A saddle lite. Ugh ugh ugh. Oh, this is so easy. You should see them here in Mission Control. They're rolling in the aisles here, folks; rolling in 'em!"

The four astronauts looked around at each other. Visualizing the staid mission controllers actually *rolling* in the aisles like Balazar said was such a ridiculous thought that the juxtaposition tickled some internal funny bone. None of the four were actually laughing, but there was a levity to their mood which had not been there before. It was easy not to think of the approaching danger. That was exactly Cox's idea.

"Okay. You probably know I've had a wife or two in my time," Balazar went on. "Well, maybe three or four. But there's someone up there who's got even *me* beat. Who's that, you ask? Who? It's Old Man Saturn. And how do I know that? Well, look at *all* those rings! Ha Ha Ha! Oh, I give myself a side ache sometimes. Har har har.

"All righty, Mr. Cox here is signaling me to shut up; says you've got important work to do, but hey, all work and no play make Jack a… a what? An astronaut! No, just kidding folks. Here's one last joke for you. One more joke, and then, you've got to come see me headlining in Vegas at the… oh, sorry Henry, no plugs. I forgot. Okay, one more joke. Here goes. What's an astronaut's favorite meal?"

It was an age-old jibe in the space program, and all four of them responded in unison.

"Launch time!"

And they finally laughed. It was a belly laugh, and the relief of tension that came with it carried them through several rolling moments of blessed mirth, unifying them. The sound carried all the way to Earth, and set a tone for the rest of the day. While certainly anxious about the coming storm, there was an ineffable sense of confidence in their success.

*   *   *

After a moderate breakfast, the four astronauts went through the laborious process of donning their Manned Maneuvering Unit; their space suits. Each custom-fit suit was assembled in layers. Over a tight-fitting, breathable body suit was their thermal layer, with non-collapsible fluid ducts, to carry away the intense heat from the sun's rays. The thermal layer was connected into the backpack of the outer pressure suit, which would seal them off from the harsh vacuum of space. The pressure suit was the bane of working in space. On the one hand, human lungs could only oxygenate the blood properly when the pressure of oxygen in the suit was above 3.1 pounds per square inch. Yet, the more air crammed inside the suit, the harder it was to move around. The working pressure of the outer suit was a tradeoff between comfortable breathing and ease of moving. Each astronaut could adjust their pressure, within limits, depending on the work at hand. Gross motor skills required more oxygen, and modest bending of the pressure suit. Detailed work required finer control of limb movements, but less oxygen. Regardless of the duties, each astronaut was under firm orders to take rest, sip, and breathe breaks of 15 minutes after each 50 minutes of demanding work.

During the first three work sessions, Laurencino teamed with Barstow and Marcharelli to prepare their vertical habitat to become a horizontal one. Using simple gear, they hoisted the landing struts to the top, and affixed them to the east face, pointing nearly straight sideways. At the bottom, also on the east face, they attached pure-titanium "wicked claws" of several sizes to the edge of the structure, to arrest sliding when they turned the structure on its side. Before their lunch break, they used hand shovels to build up the pile of loose regolith that Lomish had amassed with the rover-dozer. Their technique would be the reverse of the method used to raise obelisks in ancient Egypt. It was Stone Age technology, but far simpler than bringing cranes and hoists all the way to the Moon.

Lunch was mostly quiet. Their suits were coated from top to bottom with the clingy dust of the lunar surface. The light powder, in the weak gravity, took forever to fall, and they created thick clouds of it as they worked. The pole-mounted ionizers to minimize the attractive power of the dust would have to wait until later. Already, well into their second day, they could taste the grit of moon dust in their food. It did not bode well for the messy job they still had to complete before the CME hit several hours hence. The astronauts were tired, but grimly determined, and struggled back into their pressure suits as soon as lunch was cleaned up.

When they crawled outside for the last time with the ship in the vertical position, the dust had settled, leaving a scene just like their simulations on Earth. Its familiarity was energizing. They had worked so close to the job during the morning, that the lunch break revealed how much they had accomplished. Despite having performed these tasks in simulators, underwater, and on the industrial-dust and talcum powder training fields, seeing it happen on the Moon was fresh and exciting. They worked quickly, each knowing what needed to be done.

Video cameras were mounted at several spots around the site, so that ground-based controllers could keep an extra set of eyes on the progress. Their communications channels were all open now. At no other time before or after, would their cooperation as a team be so critical.

"Okay, folks, we're ready for first tilt," announced Barstow in a steady voice. They had attached a thin cable from the rover-dozer to the top of the east face. Pairs of constant tension, retractable guy lines counterbalanced off the north and south faces, attached to stakes driven into the basalt substrata. Lomish and Marcharelli crowded in the cockpit of the dozer for weight, and began to tug the top edge ever so slowly eastward. Lomish had the electric motor torqued down as low as possible. Laurencino monitored tilt sensors and compared them against pre-made calculations.

The greatest risk on first tilt was a sudden release of the wicked-claws at the bottom, but they were designed to optimize their holding power to shifting granule size. These self-adaptive spiraled talons alternated with interlocked bent rods meant to give the claw greater structural integrity as it crushed in upon itself. This one invention, designed with the aid of an artificial intelligence optimization routine, was a breakthrough in giving the 'habitat tilt' concept a high-enough reliability. Still, there was considerable risk.

"How's the tilt?" Lomish asked.

"Nominal," Laurencino replied.

"How nominal?" Marcharelli asked, needlessly.

Laurencino ignored him.

Barstow eyeballed the progress from a point north-west. He could see the top edge tipping over gradually against the stars. As the tilt rose to 5 degrees from the static stability angle, they stopped, and checked the progress. He and Laurencino inspected the claws from a respectful distance, and gauged the distance the habitat would fall before being caught by the high mound of regolith. They had performed this many, many times underwater. The water slowed the gravity of Earth, but did not properly mimic the gravity of Luna. This was the one point where gut feel could be allowed to modify the careful calculations of the mission engineers. They would stop here, and consult with Mission Control before proceeding.

"I think we need more regolith," Barstow suggested.

Cox's voice came to them from Houston.

"Anybody here on the ground have a comment?"

There was a brief pause, but everyone was being very patient, despite the oncoming radiation.

"Hi, this is Aaron Washington, from the physics group. We feel that a taller mound would put a larger moment on the claws. More might pop out, instead of scraping through." He ended rather uncertainly.

"What do you think, C-note?" Barstow asked.

It took Laurencino a moment to realize he was talking to her. His gold-plated visor was turned towards her, and it dawned on her that he had just coined a nickname. She hoped it would not stick. She waited an appropriate length of time to be insulting, then responded as if he had said nothing.

"I'd leave it alone," she said simply, eyeing the mound closely. In her mind, she played the motion forward, as she'd seen in the virtual reality cave a thousand times. Estimation was her forte, and her razor-sharp mind factored in dozens of variables in considering how the leaning tower would topple. It was designed to withstand some torque and some bangs. The most important thing was to avoid jarring impacts. Sliding was better than bouncing. She could see the habitat tilt over in her mind. The fulcrum point would be above the center of mass, but not by much. It was perfect.

"It will work," she said, and the confidence in her voice carried weight with them all.

"C-note weighs in with a 'GO'. Houston, are you okay with this?"

There was a delay longer than that required for the radio signals to traverse the immense distance separating them.

"We are okay on this end," Cox told them. His voice was definite and sounded sure, but it seemed he was almost too certain. They knew he was deeply worried for them. The only other option they had was to find a cliff or overhang, and bury themselves and their suits under piles of regolith.

"John, take her over. Keep the tension on as she falls."

"O-K."

Lomish engaged the throttle, and the habitat leaned ever so slightly further. Then, without a noise, or any visual warning, the cable started to fall slack. It was falling. The motion was very slow, compared to Earth standards, almost giving the impression that it was a much larger structure, but farther away. The gap between the east face and the mound of Moon dust collapsed into a thin parallel space, and then the claws juttered back. As the bottom edge scuttled out, the tall, soft pile of powder caught the habitat. With a few shuddering shifts, the habitat settled down into the mound. The point of contact was obscured by a spray of gray dust, glinting in the bright sunlight. In just moments, it came to rest.

A loud cheer broke out on the Moon, followed by more cheering from the Earth moments later. Marcharelli was pumping his fist. Laurencino instinctively put her hands to her ears, but only found her helmet. She was dreading the next part. Second tilt came after hard labor, and there would be no break until they got the habitat level. This would be the longest stretch in the pressure suit for their entire mission.

As the jubilation faded away, they took stock of the lay. The habitat had pushed down into the regolith very much like the dust and talcum field studies, which had been weight-corrected for gravity. It gave them further confidence that the rest of the operation would proceed smoothly. Now that the habitat was leaning over the escarpment that Lomish had created, the spindly landing struts were pointed nearly straight down. These would cushion the fall of the second tilt. But before that could happen, they would need to dig out from under the habitat. When they had cleared enough to almost expose the center of mass, Lomish could tug the habitat down the shallow slope, and onto the basalt foundation. The collapsed claws would be the sled.

Using shovels, they went to work profiling the down-slope of the regolith mound. The habitat loomed above them, and they had to work carefully not to dislodge it. The work went slowly, hampered both above, by the low-lying structure, and on the sides, since only one astronaut could work either side at a time. Lomish and Marcharelli did most of the work. Barstow removed the guy lines. Laurencino had no tasks, and just watched the men digging. It seemed oddly ironic that she had seen groups of street workers do the same thing in Boston, and had always silently derided them for their laziness.

At last they were ready for second tilt.

Using the rover-dozer, Lomish and Marcharelli pulled the habitat slowly into the peak of the mound. Wafts of powder pushed out from underneath, as the habitat went slightly up, preparing for its slide down the far slope.

"Wait!" Laurencino cried.

Lomish stopped the dozer.

"What is it?" Barstow asked.

"I'm not sure. I think it will slide too much. The dust is slippery. We need something to slow it down."

"Won't the lower gravity slow it down?" Marcharelli asked.

"Yes. But it's a balance of forces. Houston, in the simulations, we assumed a coefficient of friction, based on return samples from Apollo, right?"

"Hi Dr. Laurencino. This is Aaron again. You're right. We tested the regolith in a lab. But we modified the Einstein relation for the viscosity based on gravity. It didn't amount to much."

"What about electrostatic charges?" she asked.

"The baseline comes from an average," he told her. "Then, we adjust for sunlight and regolith conductivity. We're in protocol B, so we've removed the effect of the charge suppressors since they're not yet installed."

"Could there be some early arrivers from the CME?"

"Um. Oh! Oh, my God, yes. That could be a significant factor."

"I think it will be a lot slipperier than we calculated," Laurencino said in a distant voice. "We need to make the slope shallower."

"We can't do that now, there's no room underneath," Marcharelli said quickly.

"He's right. It's too dangerous to work under there. Not enough room," echoed Barstow.

After they had all been silent for a long time, Lomish said: "I've got an idea."

"Whazzat?" Barstow drawled.

"I could move around to the backside with the rover-dozer and hold it from going downhill. The three of you could take this cable, and pull it down."

They all thought about that for a minute, sizing it up with what they saw before them. It seemed to Laurencino that their combined strength would be too small to overcome the static friction, and the habitat would not be able to crest the mound of regolith they had built. They would need to get the habitat tilted like a teeter-totter, right on the verge of falling, before Lomish's idea would work. Too risky. But she couldn't come up with anything else.

"Houston, we need some advice," Barstow said.

"We're working on a solution right now," came back Cox's voice. They could hear the stress in his voice, and it began to set them on edge. There were two clocks running out, the coming storm and the need to get into a stable airlock. They needed an answer fast.

"I've got an idea, too," Marcharelli said.

Nobody asked him, so he kept going.

"We need more regolith underneath, right? Why don't we build smaller mounds under where the landing struts will fall. That will keep the angle smaller, and so it won't slip."

"Then, what?" Laurencino asked him, already fast-forwarding the possibilities.

"Um. Well, then we keep pulling, and the struts come down off their smaller mounds. But since they only fell a small distance, they won't go very far. We can pull it a little at a time until the tail end is off the big mound, and then lower the rest down using the jacks." He ended sounding excited, and pleased with himself.

"That's a good idea..." Laurencino started.

"Thanks!" he replied quickly.

"But, I wonder if the struts can take the lateral loads. What if one breaks free. Then the habitat will rotate and come down on a corner. That's the worst possible way for it to fall."

"They won't. Those things supported our entire weight when we landed. And that was a nasty landing too, right Henry? They'll be fine."

Laurencino asked for ground support. "Houston, how do the lateral load limits compare to the vertical limits for those struts?"

The answer came back almost immediately.

"They are pretty stiff, and we may have design margin, but that's only a guess. They were not really optimized for lateral strength. Repeat, not. We could run a simulation, but it would take a while to set up and run, maybe 30 minutes."

There was another pause as they all considered what to do next. A moonquake shook the ground under their feet, and continued rumbling, building slightly in intensity. The regolith under the habitat began to shake and slump, and the whole structure started to move downward. The slippery regolith was being vibrated away from under their habitat. Despite her instinctive panic, Laurencino saw what was coming.

"Quick! Dallas!" she yelled. "Grab a guy line. Pull west as hard as you can! Now!"

With agonizing slowness, she and Barstow hopped towards the loose cables on either side of the habitat. It was already moving downslope, and the lines began to drag away from them through the rubble. Barstow reached his first, and pulled back on it, digging his heels in against the tension. Laurencino grabbed her line, wrapping it around the wrist of her pressure suit, and began a tug-of-war with the Moon's gravity. Marcharelli, too far away to help, nonetheless began bounding towards Laurencino.

The habitat tilted over. The Moonquake had died away already, but the movement it started had enough momentum to keep the structure falling. They had to keep tension in their lines to arrest the downward swinging motion so the struts would brace the fall. Despite his small weight, Barstow was like a rock, and held his side of the habitat fairly steady. Laurencino, on the other side, was being pulled slowly forward, and could not seem to catch her footing. She kept losing traction in the regolith, and her side of the habitat hit down first. She fell heavily onto her rump with the sudden loss of tension. The strut slammed into the ground, quavered a second, then buckled slightly. Barstow's side impacted just then, and the habitat caught, wobbling for a moment before coming to rest.

Marcharelli caught up with Laurencino, and helped her to her feet. The cable around her wrist had punctured the pressure suit, and a jet of small ice crystals was flowing out of the glove, freezing the moisture from the air inside. He worked on the sleeve of her suit, and removed an emergency patch. With fat-gloved fingers, he deftly ripped open a strip, and applied it carefully to the crease in her glove were the cable had bit.

"How's your pressure, Sellenya?" he said, sincere concern clear in his voice.

"It's good," she said after a moment to check her head up display. "It only dropped a little, but it's already back to nominal." She paused, then added kindly, "Good nominal." It was as close to a thank you as he was going to get.

Barstow came over, and then Lomish. After checking her out, they all stood, looking numbly at what had just happened.

"Well, that took care of *that* problem," Lomish said.

As they caught their breath, the feather of a shadow came over them. In the stark landscape of black and white, it was a most eerie effect. They all turned to look sunward.

"What the hell is that?" Marcharelli said.

Spreading out in a line, part-way to the nearby horizon, was a roiling wall of grey. Cirrus clouds of sparkling moon dust arched upward from a seething grey mass closer to the surface.

"Doesn't look good," Lomish said.

"Houston?" Barstow called. "I assume you can see this too. Are we in danger? Please advise."

"Ah, well," Cox began, "it appears to be a lava tube subsidence, folks."

"You mean the surface collapsed, Henry." Laurencino said. "Like a Texas sinkhole, right?"

"Um, yes, Sellenya. That's about the size of it."

"Lateral extent?" Barstow asked of Mission Control.

"Nearly a hundred klicks in a northwest-southwest line. Minimum range from your location is 9 klicks."

"How far are we from the fault line?" Laurencino asked.

"Horizontal projection based on best guess is about 3 klicks."

"We need to get out of here, right now," Marcharelli said, the pitch of his voice threatening to break.

The pause from Houston was longer than needed. Henry Cox explained in a atypically hushed voice. "Robert, I'm sure you remember the deadly radiation heading your way. Inside the RM, radiation shielding is minimal. You would all be subject to the full force of the storm."

A silence stretched on for almost ten seconds. Marcharelli finally answered; his voice now back to normal, sounding almost calm.

"Well, I guess we bivouac down here, then do what we came here to do."

They all shared a tight laugh of relief, but there was no time to relax. They had to jack the habitat the rest of the way down, and drag it fully onto the bedrock. Then, using the mound of regolith, push tons and tons of more regolith on top of the shelter. They had just five hours before the CME broke upon them, and, with breaks, it would take them nearly all of that time to finish.

* * *

The radiation alarm first sounded at 4733, half an hour after they had quit work, and come back inside for the storm. The work had been much slower than anticipated. With the sun's constant bombardment with ions, the lunar dust carried an electrostatic charge which repelled itself. Using the skittish dust for construction projects had not turned out exactly as planned. It took Lomish an additional work period to build the ramp of regolith high enough where he could begin pushing more dirt on top of their habitat with the dozer blade. After installing the eaves, Barstow and Marcharelli used hand shovels to toss the powdered rock up to the roof. Laurencino used her collapsible shovel in the half-bent position as a hoe, and scraped the covering layer evenly across. They would be covering just half the habitat for now, to the necessary 2 meters depth. The rest would have to wait for after the storm. This restricted them from the uncovered half the habitat, but they made sure the livable portion included the toilet and bath area.

After such a strenuous day, the oppressive odor of sweat and halitosis was thick in the cabin air. Despite mineral salts used to kill odor-causing bacteria, feet, crotches, and armpits were fetid. Laurencino was taking a sponge bath when the alarm hit. All of them instinctively looked up – in the direction of the Sun. But there was nothing to see, nothing to feel, nothing to hear but the warning tone. In their cramped half of Habitat A, privacy was an unaffordable indulgence. Laurencino had hung her skin suit and her fresh change of clothing from the low ceiling, to give at least some concealment. She was not body conscious, having grown up in a large family.

However, men being men, she knew they could not help but be brutes. Especially Marcharelli. Ever since she had shown him that one scrap of kindness, he had been looking at her with lascivious eyes.

Marcharelli attempted to busy himself with the painting wall, a large, touch-sensitive screen that could be painted with 16 million colors using a fat stylus. He had been painting a meadow landscape, and the clouds in his sky were taking the shapes of a female body. It drove him crazy that just a few steps away was a naked woman, sponging herself clean with a damp rag. He couldn't concentrate, so badly he wanted to offer to help. The thought of asking her to let him sponge her clean burned in him and pushed away all other thoughts. His fear channeled into lust.

Using the stylus, he made a mountain range, coloring it all black with the wide brush setting of the stylus. With that dark background, he could watch her reflection in the glass panel. God, she was beautiful. Every move she made, every way she bent, aroused him even further. The crude screen she had erected fueled his desire even more, adding fresh wisps of excitement every time the shapes and curves of her athletic body were revealed between the hanging garments. When she bent over to wash her feet, he couldn't help stealing a look around. Just one glance, because he knew she'd despise him forever if she caught him looking. Turning back around just in time, he closed his eyes, and relived the moment, stretching it into a fantasy that made him quiver and pant.

The other two men, ostensibly playing a game of chess, were suddenly next to the painting wall. One of them stood on either side of Marcharelli, their backs to the bathing woman. His view was blocked.

"Hi, guys," he said to them, rather breathlessly.

"Whatcha drawin'?" Lomish asked, in a loud voice.

"Picture."

"I think maybe I've been there before," Barstow said. "Yosemite?"

"Yeah, maybe Yosemite."

"Don't they have pine trees in Yosemite?" Lomish asked.

"Yeah, and eagles in the sky," Barstow pointed, where an eagle might be soaring.

"Oh, yeah. I think they have eagles. And pine trees too."

"Can't wait 'till it's finished," Barstow said, nodding.

"Yeah, me too," added Lomish.

Barstow patted him heavily on the shoulder, and the two of them walked the three light steps back to their chessboard, and resumed their game. For Marcharelli, the spell was broken. Laurencino had a towel wrapped around her, and was reaching up for her fresh clothes.

"Who's next?" she called out.

"Me!" Marcharelli said, a bit too hastily. Realizing he was betraying his emotions, he added, "I haven't been this smelly since Hell Week at my frat initiation."

While around them poured hundreds of thousands of high-energy impacts every second on every square meter of the Moon, being six feet under gave them plenty of protection. They had muted the radiation alarm, but the indicator dial in the wall panel flickered unsteadily throughout the yellow zone with frequent spikes into the red. The sensor mounted outside, on the edge of the eave, recorded the full force of the atomic storm pounding down on them from overhead.

* * *

In orbit above the Moon, nearly forgotten and largely taken for granted, flew the Return Module. This Spartan piece of engineering had one simple function: accelerate the four astronauts up to 8,000 kilometers per hour, and carry them to Low Earth Orbit. There, they would link up with the ill-fated ISS, and a shuttle would drop them back to the Earth.

The RM was built in the tradition of a big, dumb rocket. Plenty of brute force, and very little in the way of luxuries. Nonetheless, it was designed for reliability. The time-tested electronics, while not state-of-the-art, were radiation hardened. Silicon-on-Insulator process technology allowed fabrication of transistors in very thin layers of semiconductor material. Compared to conventional transistors, which are highly sensitive to ionizing radiation in their bulk layers of silicon, SOI transistors barely notice.

Fault tolerant circuit layout and generous linewidths gave an extra margin of safety to the mean-time-between-failure metric for these hardened electronics. The packaging material for the computer chips was made of low-alpha source plastic composites, and the cases were built with an integral layer of boron compounds to help absorb stray neutrons. Control software was extensively redundant, with thousands of fault diagnostics and error-correction routines embedded in the code. Virtually every conceivable failure mode was accounted for, with fail-safe default states hard-wired in. When the electronics modules were complete, they could outlast consumer electronics by a thousand to one, even in a high-rad environment.

The challenge in building any high-reliability product was knowing how much was enough. Reliability cost functions had to be derived, and an acceptable level of risk determined up-front. End users always demanded perfect functionality under all circumstances, and for all time to come. But such perfection was just not possible with the technology of the day. The design engineers for the Return Module accepted a probability of failure of one part per million under worst case conditions. With four manned missions, and worse case conditions happening only a few hours a year, such a level of risk was deemed quite satisfactory. Yet, when computing the cost functions, one assumed that all failure modes were known ahead of time. The technique of Failure Modes Effects and Analysis was used to anticipate all conceivable means by which a failure could occur. Each failure mode was then ranked with a risk priority number, and those with the highest risk were eliminated or accounted for, until the final risk remaining was better than fifty-to-one for the entire mission.

Where the FMEA method has a weakness is in interactions between two or more effects. To be sure, these were less likely to happen, but in a solar storm, the amount of ionizing radiation could compound the danger from multiple failure modes. In the power driver circuits of the Attitude Control Module, a high-current output transistor ignites the hydrazine thrust jets that hold the RM on-orbit. An orbit that passes through the Moon's lumpy gravity field is diddled by mysterious subsurface gravity anomalies which make navigation a constant affair. When the on-board navigation units detect a correction is needed, they apply roll, pitch, or yaw and X, Y or Z motion to the vehicle, nudging it's orbit back on true.

When the solar storm arrived, the interior of the Return Module was flooded with ionizing radiation each time it crossed the Moon's dayside. Static charges built up on the inside surfaces, pulling stray dust to the walls. Inside the electronics modules, oblivious to the radiation problem, a navigation algorithm determined that a small amount of roll was needed to position the pitch thrusters for an orbital correction burn. The roll thrusters would be held on for 120 milliseconds, the vehicle would be allowed to roll 40 degrees, then roll thrust in the opposite direction would be applied for 120 milliseconds, to stop the roll motion. Finally, the pitch thrusters would be used with the Y and Z thrusters to correct the orbit upward with a slight nudge, compensating for the sudden rise in pressure from the solar wind, fueled by the CME. This compound maneuver, more complex than most of the orbital correction burns, engaged the power systems at a proportionately higher draw.

Full-up operational testing under high-radiation environments is expensive, since parts cannot be re-used, and must be put into expensive test vaults. Further, testing of the entire RM system in a high-rad environment is nearly impossible, and was performed instead on individual sub-systems. Complicated cause-and-effect sequences are infinite in number, so with finite resources, some scenarios must remain untested.

During the last milliseconds of the burn, as software indices counted down to zero, an especially heavy dose of alpha particles and protons collided with the Return Module. The impact of sub-atomic particles with the metal hull of the spacecraft caused a cascade of ions and electrons which blasted through the cabin like a sawed-off shotgun. Many were stopped by the casing and packaging material of the power module, creating powerful electric fields. The statically-charged power cables, spiked with the inductive kick-back of the shut-down thruster current, became a broadcasting antenna. The signal penetrated the control bus, and discharged into the power transistor itself, rupturing the gate, but spraying the crater with molten copper that shorted the source to the drain.

The thruster fired full jets, dumping precious ounces of burned hydrazine, and spinning the RM like a rifle bullet. The control routine compensated with the reverse thruster, which it first fired with a trial burst. The shut-down of that burst fried the reverse power transistor in the same way as the first one failed, and burned the reverse thrusters full jets. This threw the RM into a tumble, a complex combination of roll, pitch, and yaw. In some cases, tumble exits maneuvers could be performed with two thrusters out. It mattered which two. If both forward and reverse thrusters for the same degree of freedom were lost, the craft was lost. The control system sent an error log to base through the omnidirectional antenna, and initiated a warm restart with diagnostics set to the highest level.

The craft had also vectored before the restart, and its orbit was no longer aligned with the equator.

Chapter 13

Hyginus Rille

"That's odd. We missed the Return Module state-of-health signal this pass," Barstow announced, after responding to the new alarm at their control console in Habitat A. He had a portable keyboard-monitor, and was busy typing commands.

"Where's the glitch?" Lomish asked.

Marcharelli, who was just finishing his sponge bath and was drying off with a flair trying to attract the attentions of Sellenya Laurencino, just barely caught the words.

"What? Something happened to the RM? What's going on, guys?"

"Hold on, Bob," Barstow said, not looking up from his monitor. "It's probably nothing. The CME must have just jiggled its comm link. It has a running reset function, and should come back online any second. The handshake signal just dropped out for a moment, okay?"

Marcharelli wrapped his towel around his waist and came by to huddle over Barstow. He noticed that every field held a zero.

"It crashed, right?"

"It just means we're not getting the signal right now, okay? Like I said, it should reset itself, and then all those fields will fill in with telemetry and orbital parameters."

They waited. Laurencino watched them from her perch with an ebook in her lap, but did not come over, not wanting to stand next to the almost naked Marcharelli.

"Well, we'll just have to wait until the next pass," said Barstow. Turning to Marcharelli, who would surely want to know all the details, he explained what the man had already been told over and over during their training. "We only have line-of-sight comms with the RM. Our on-board monitors have a handshake that engages each pass, and updates telemetry, recent automatic corrections, environmental conditions, and a ton of other stuff. The alarm just meant that this time around, there was no update. I sent an active ping before she was due to drop over our horizon, but there was no response. Like I said, it's probably nothing. Either way, Bob, there's nothing we can do for the next fifteen minutes. Until then, we just sit tight." He fixed a stare on Marcharelli, challenging him.

Marcharelli backed down, and sulked over to the wet corner and began dressing. Laurencino put her ebook down.

"Does Houston have a link to the RM also?"

"Yeah. I'll call them right up. Houston just gets a burst transmission from the forward antenna every time the RM comes around from behind the Moon. They should get their next update a few minutes before we do."

Barstow donned a headset, dialed in Houston, and raised Mission Control. The connection was crackling with static from the intense ionization in the van Allen radiation belts which surround the Earth, but their sigma-delta notch filters still allowed intelligible communications. He explained what they observed. The ground crews had received their last burst from the RM right on schedule, and had just received an error log from the omni. The COMM team was already working on it. The CAPCOM counseled the astronauts to sit tight, and congratulated them again on getting their shelter ready before the storm.

The next ten minutes passed with agonizing slowness. The only sounds in the United States' first permanent Moon base were the rustling of clothes, the whir of circulation fans, and the occasional pop and click of the habitat settling. Marcharelli had moved over to the south view port, and was looking across the lunar maria, hoping to see the orbiting RM fly past. They were supposed to be sleeping soon, and two of them were still in need of a bath. They waited.

The link with Mission Control was left open, but had been muted on the far end. When it clicked on, the voice of Jason Broderick addressed them.

"Hey, folks. We confirm failure to receive burst transmission from RM. We are initiating visual and radar location activities," he said formally. "We suggest active ping on your horizon window at 48 seconds; mark."

"Roger, Houston. We will active ping. I'm turning up antenna power to maximum."

Without an ionization layer like the Earth's which deflects radio waves and can bounce them over the horizon, and without the project expense of relay satellites, the astronauts could only contact the Return Module when it was overhead. With the curvature of the Moon being much tighter than the Earth's, the horizon was close by.

And with lunar orbits much lower than Earth orbits, because of the weaker gravity and without the inconvenience of atmospheric drag, this meant the RM was in communication with them for only a few seconds out of each overhead pass. It was an economical arrangement, but relied utterly on the automatic course correction systems to function properly. Small deviations could be corrected manually if needed, but if the vessel happened to drift quickly out of its nominal orbit, recovery could be much more difficult. With an estimated one part per million risk of failure, this was one of many automated systems the entire mission relied upon for success.

"It's gone, isn't it," said Marcharelli in a very soft voice. His breath fogged up the view port. He had seen nothing. No white cylinder whizzed across the starry sky.

"We don't know that," Barstow replied. "More than likely, the communications are fouled from this damn solar storm."

"So, what do we do?" Lomish asked. In the pause, he volunteered an answer. "How about if I take a trip on the Ascent Spindle, and pay the RM a visit. I can refuel the spindle from our reserve tanks, and come back to pick the three of you up." He looked ready to go, seeming to forget the deadly radiation still pouring down around them outside.

"What else could have gone wrong?" Laurencino asked them all.

"Comm overloads, meteorite strike, software glitch, power drop-out, and about a dozen other things," Lomish said.

She narrowed her eyes at him. Lomish shifted uncomfortably in his seat, and screwed his face up to concentrate.

"Could be a navigation malfunction. Or maybe something in the attitude controls. Maybe she's just drifted off course a little."

"Okay," Laurencino said, taking that. "If the RM is off course, but *not* crashed," she added emphatically, "how would we verify that?"

"Well, assuming it's navigation and not comms," Lomish answered, looking up and to his left, his eyes focused on where the orbiting Return Module might be. "Then, Houston should have no trouble finding it with radar. Once they've found it, then they should be able to send a signal to the omnidirectional antenna to correct the course." He paused, narrowing his eyes, apparently checking his reasoning. "Yeah," he said finally. He nodded his big head a few times. "Yeah."

Right on cue, the static-laden voice of their ground controller broke in with good news. "John, it looks like you were right again. We've just got word from the tracking station at Malindi in Kenya. They're right above you at the moment. Their lead tracker, Mr. Marcus Lafontaine, just called us to say they've identified a body circling Luna. They're doing telemetry and should have detailed orbital parameters to us after the next pass. Also, we're planning to reconfigure the omni to transmit telemetry and system status."

"Ya-hoo!" Marcharelli crowed.

The rest of them were encouraged, but more subdued. At least, the RM had not crashed. If the orbit was nominal, they could cheer too. It would be another agonizing fifteen minutes before they knew. Marcharelli returned to his painting, and began putting pine trees into his landscape. He hummed to himself, which would normally be annoying in such close quarters. However, with nothing to do but quell anxiety, the others just watched him numbly, unable to concentrate on anything of their own.

"Look, you guys might want to consider bathing," Laurencino suggested.

"Naw," Lomish said, sitting like a mountain on his tiny fold-out stool.

"They say a fox smells his own hole," she continued. "It's the one that doesn't smell – to him."

Barstow caught her meaning. He glanced at the chronometer.

"Right after the next pass, I promise, C-note."

She glared at him for using that stupid nickname again.

"You guys are like regulars at the local pub. Once they park on a bar stool, you can't get them off it until closing time." She paused for effect. "Maybe that's what I'll call you: Barstool."

Barstow drew in an audible breath, and let it out in an explosive laugh, surprised at her uncharacteristic humor. Lomish guffawed as well. They both continued to chuckle. Lomish turned towards his friend.

"Yeah, Barstool, you smell like bar cheese gone bad. Go take a shower."

Barstow grimaced, and hesitated, glancing again at the clock.

"Go ahead," said Lomish. "You got ten minutes. I'll cover. No problem."

With a look and a smirk at Laurencino, he got up slowly, and bounced over to the wet corner. In a flash he stripped his suit and was rubbing the cleansing foam on his smelly parts. He did not bother with any privacy, but he didn't need to.

"Okay, we got that update," Broderick announced over their speakers. "And let me welcome back that old smoothie, Dr. Henry Cox, who missed you guys so much, he couldn't sleep; and came in early."

"Hi folks," Cox said, and his voice sped towards them at 300,000 kilometers per second through heavy radiation.

A chorus of "Hey, Henry" echoed back across the inconceivably wide gulf between them.

"Hey to you too," he echoed. "Jason's going to focus on the RM tracking and retrieval, so I'll be your ground support. Is that okay with everybody?"

Lomish answered, as Barstow hastily struggled into his fresh change of clothing.

"Yeah, Hank, that's okay with us. Look, how bad is it?"

"It's too soon to tell, John. We've got a lock on the orbit, and it looks fairly stable. The only problem is that it's deviated. Now, don't get worried; it's not much. Just a degree or two. It's precessing slowly, but not too bad. I don't want you to get alarmed. Our flight experts say it's well within the capabilities of the attitude control to correct this orbit. Now, the thing we need to do next is lock in on the yaw, pitch and roll, and send the course correction to the omni. Piece of cake. Any questions?"

"What can we do on this end, Henry?" Lomish asked.

"At this point, there's nothing for you to do. And we need time here to work out the particulars. Let's not forget you've all had a tough day. I'm sure you are exhausted. It's well past the end of your shift, and you've still got three or four hours before the CME passes. Why don't you all take a trank and pack it in for the night. Hmm?"

"Yeah," Barstow said, resignedly. "Okay, Henry. Anything else?"

"Only that we're all really proud of the four of you. This has been a rough mission and you've pulled through every challenge with aplomb. I'm sure we'll all get through this one together. I'll just bet, by the time you awake in the morning, we'll have the RM passing right over your heads, and you can get on with the payload launcher assembly."

* * *

Sellenya Laurencino was the first to step outside the next ship's morning. She craned her neck, looking up towards the sun which had so recently besieged them. Through the dark glass of her helmet, it was still an impossibly bright circle, swimming at the edges due to the persistence-of-vision effect it had on her retina. Not far from it was the crescent Earth. The two celestial objects seemed to form the letters DO. It reminded her of the old adage, "Just Do It", and she decided that was appropriate. Her only reason for being on this mission at all was to install and start-up the payload launcher, designed and built by her team of colleagues at MIT. Her crewmates blustered on with ground control over the RM, but there was absolutely nothing they could do from the Moon; at least until Houston got it back to an equatorial orbit. So, we might as well just do what we came up here for, she thought.

She moved over to the cargo chords, using the skipping hop that seemed the most efficient means of locomotion. Now, out of the sight of the others, she allowed herself to play around a little, to satisfy her scientific curiosity without appearing unprofessional to the men. She kicked at the dust, watching it roll and boil away from the toe of her boot, and settle in slow motion back down again. Leaning over, she brought a handful of the stuff closer to her helmet. It was calico gray, she decided. Every color between black and white.

Much of it was dust that sifted down between the fat fingers of her glove, peppered with tiny rocks and glass globs that stayed behind. There was no sense to the size distribution, no coherence to the composition that she could tell. It spoke of the regolith being born of violence. All around her she could see crater impacts. Large ones, like Agrippa, were like a mountain range in the distance. But even here on the mare, she could see the characteristic circular impact craters dotted everywhere, and of every conceivable size. As she looked closer, she could imagine seeing even tiny impact points, caused either by micrometeorites, or perhaps by the gentler landings of ejecta from larger craters nearby. Each of the larger craters marked the arrival of new material to the Moon. Scattered throughout the regolith they could probably find any given element known to science, dumped on top of the native minerals containing oxygen, silicon, aluminum, calcium, iron, and magnesium.

She stood beside the storage holds for the launcher, and looked up at the Earth again. It seemed so small and alone. She knew that almost everywhere on its surface, life could be found, all interacting in complex ways. Despite the apparent conflict between many forms of life, in reality, she thought, they were all symbiotic. Standing on the Moon, looking back to her home planet, Laurencino felt oddly removed, like one who had died, and was leaving the Earth behind. It was so small from here. She imagined it becoming smaller and smaller, fading to a wan point of light, and then, falling into darkness. The thought made her melancholy, and she longed for home. She desperately missed the color of green and the familiarity of having a blue sky above her.

Lost in thought, she was aware only of her breathing, deep, even and relaxed. But a part of her wanted to cry. Although she would never let tears come out of her eyes, that slightly hot tingle in her chest was unmistakable. She felt like Gaia now, the Mother Earth, crying for her children. She cried for Man, burning so brightly, growing so quickly, hell-bent on progress and acquisition of power. They were each so busy doing their own little thing, they scarcely gave a thought to their home planet, so alone and precious in the wasteland of space. Government leaders jostled and struggled and killed one another over their particular piece of it, or because they could not agree on religion, or because their skin was a different color, or sometimes just to see things explode in flames. Corporate directors, driven by greed, took whatever expedient would boost profits, only regarding the consequences of their decisions if forced to do so by regulations or angry consumers. Their motto seemed to be: 'Better to ask forgiveness than to seek permission'.

She felt so small and helpless, standing on the Moon. There were so many humans up there, what could she do? Maybe like everyone else, just do what you can, and hope for the best. That seemed so lame. And yet that was why she had accepted the contract with the inventor, Parks Vambolent. It gave their academic group an opportunity to use electromagnetic launcher technology for peace instead of war. She had been skeptical at first, but later fascinated by his sweeping dreams of creating a cislunar economy, and using it to relieve the burden humanity placed on the Earth. It was a chance to make a difference.

She turned her head to look sideways, and there was another space-suited figure standing right next to her, also looking up at the Earth. So startled was she, her heart leapt into her throat and began pounding furiously. She drew in a few shuddering breaths, then quickly regained her characteristic control.

"Hey, John," she said, realizing from the size it was Lomish.

"Hey C-note," he said, turning his helmet to look at hers. She could hear the smile in his voice, and decided to stop protesting. If they insisted on giving her a nickname, fine then.

"You ready to get started?" she asked him.

"It sure is beautiful," he said after a second. "You know, I've always wanted to go to the Moon. Ever since I was little. Didn't figure it would ever happen. Sometimes I pinch myself to see if this is a dream. I want to remember *everything*."

They stood for a moment, looking at each others' opaque faceplate. Despite being an arms length apart, they were separated by the deadliest of vacuums. Their thoughts might be as far different as the Earth was from the Moon. But they weren't. She smiled, although he couldn't see it. She nodded her head slightly, which he could. His chest bounced up and down in a silent chuckle, which she could see in return. It was enough. They had made a connection.

"C'mon John, we got a job to do."

"Yes, ma'am," he said enthusiastically, and they bent down beside each other to unfasten the launcher components. Their suits rubbed together at the elbows.

* * *

As they worked, Barstow and Marcharelli came out and powered up the rover-dozer, outfitted with a sledge they would use to haul components over to the rille. As soon as they came through the airlock and were patched into the suit-only link, Laurencino and Lomish could hear them arguing about the RM status. Marcharelli kept asking questions Barstow had either already answered, or could not possibly know the answer to. Finally, Barstow's patience reached a limit.

"Enough already, Bob. Enough!" He paused, to let the menace of his voice sink in. "Right now, we have work to get done. I need you to focus on your tasks. Can you manage that?"

"Sure I can," he replied immediately, in a defensive tone. "But if we can't get out of here, what's the use?"

Lomish and Laurencino stood up to watch them approach. They saw Barstow stop the rover-dozer. Marcharelli stepped off it.

Barstow squared off in front of him, and flat-handed him in the shoulder. It was not a violent hit, but neither was it gentle. Marcharelli backpedaled awkwardly in the meager gravity.

"Hey! What's got into you?" Marcharelli shouted in all their ears. His voice registered panic, and they watched him take another step back.

"Listen, you worthless, no-good piece o' crap. The rest of us are here to do a job. There's three more missions coming that are counting on us to get this done. That's what the rest of us signed up for. What are *you* doing here?"

"Look, Barstow," he replied with a quavering voice. There were equal parts fear and defiance in his answer. "I'm here for the same reasons as the rest of you. But I didn't sign up for a suicide mission. If there's something we can do to get the RM back, we should do that *first. Then*, we can put this contraption together."

They could all hear each other's breathing. The hissing was probably Barstow breathing through his teeth in his rage.

"I've told you a hundred times. There's nothing we can do from here. Not until they restore an equatorial orbit. Now, if you can't accept that, and move on, then maybe you'd better just go back inside."

But Marcharelli didn't have the sense to shut up.

"We can raise the antenna higher, it will get us more range," he said quickly, grasping at straws. "We can send Lomish here up in the spindle, and he can go looking for it."

Barstow, stepped towards Marcharelli, curling his gloved fingers one by one into a fist, and then cocked it. The four ribbed knuckles were aiming right at the faceplate of Marcharelli's helmet.

"Hey Bob," called Lomish, in his loud, growling voice. "Why don't you give me and C-note a hand here. We can't get these lugs loose. They got dust in 'em or somethin'. Maybe from all the activity yesterday."

Marcharelli, whose main task was payload deployment, was responsible for the lugs, and a thousand other items in their inventory. The appeal to his expertise gave him a way out of the confrontation where he could save face, in more ways than one.

"Yeah, I'll be right there. You just need to whack it a few times with the zero-torque handle." Then, realizing what he said, pulled his head back from Barstow's still-cocked fist, and began backing up towards the cargo chords. He kept a pattering stream of details going in his nervousness.

"Remember John, it's 'righty-tighty, lefty-loosy'. And sometimes the whole thing can bind if it's not lying flat. Try picking up one corner, and whiz the lug nut as it falls."

The rest of that work-morning they cooperated efficiently and silently for the most part. Barstow, who was still fuming, was decidedly reticent. They dragged the entire payload launcher to the lip of the rille in three trips, bringing the assembly kits by hand on the final run. Lomish kept Marcharelli busy, by pretending to need help with things of which he was eminently capable. Whether Marcharelli saw through this, or just allowed it to happen, he didn't let on. He just stuck close to Lomish and was especially helpful.

After their break, Barstow and Lomish set up the winch, which would be used to raise and lower components, and themselves, down the steep slope of the rille. Marcharelli and Laurencino staged the payload pieces at the edge of the steep cliff, and then the two of them had to make the descent.

"Ladies first," he insisted, and she just looked at him for a moment, thinking what a coward he was. Cowards were more dangerous than heroes, she thought. Heroes die courageously, but cowards get everyone else killed.

She passed the point of the hook into her waist harness, double-checked the engagement, and stepped over, with her back to the drop-off. She turned around, to look down into the hole. It stretched for miles in either direction. The cliff wall was hidden in shadows, and the starkly-lit bottom seemed far away. Barstow, at the controls, took out the slack.

"Setting unwind at zero-point-three. Ready?"

"Ready," she said, and her normally low voice sounded girlishly high. She must be afraid, she thought to herself. The others had no doubt. Despite training for this over and over on Earth, you never knew what would happen up here, as they had found out all too often already.

With a slight jerk, the cable began paying out. She held her breath, and leaned back, putting tension between her feet and the waist harness. She leaned backwards, and began walking heel to heel down the 80 degree slope.

"Breathe, Sellenya," Barstow told her.

At least he dropped the stupid nickname, she thought, and began breathing deeply again. Soon, she got the rhythm right, and it was just like being in the pool with her leaded boots.

"Let's increase unwind to point-five, okay Barstool?"

Barstow let out a tight, controlled chuckle. It was his first sign of loosening up since the confrontation earlier. He sped up the payout rate to Laurencino as she rappelled down the slope. The two of them began to banter, helping to quell Laurencino's anxiety as she descended. Everyone relaxed a little. Marcharelli had seemed to finally figure out that keeping his big mouth shut would do them all some good.

The rest of that day, they completed the positioning and pre-assembly of the rail launcher. The pieces fit together easily, thanks to the excellent engineering work that went into their design. After several more breaks, they quit for the day. The next day they would begin drilling the support anchors. The third day of this task would be final assembly. On the fourth day they would make all the electrical and systems connections, and the fifth day, run power connection and initial checkout. By the sixth day, they expected to be operational. The rest of their time would be split between trial launches, and completion of the regolith covering of the habitat. Then, it would be time to go home. If the RM could be repaired.

Johnson Control Center

Jared Ventris walked fast to keep up with Henry Cox, as they made their way down the corridors to the Administrator's on-site office. There were floaters in his eyes that danced across the bright linoleum. They doubled the number of stains he could see on his shirt when he looked down. That motion made his neck burn. With just an electric shaver and no lotion at his desk, his necktie-tightened collar snagged and tugged with every movement of his head.

His navigation team had very sketchy data to work with, but they had finally figured out the problem. After hours of tumbling, they had lost absolute position data on the Return Module because of small but cumulative errors in the gyros. Without a stable craft, they could not locate navigation stars, so their on-board methods to relate attitude were lost. He coughed, his throat tight.

They stepped into the elevator, joining a small crowd. Henry leaned over. "What are you thinking about, Jared?"

After a quick glance to see who might overhear, he angled his head closer.

"Yaw, we got nailed down. Radar and telescope give us that. Roll, we can get mostly, but once we get it closer to zero, we'll lose that in the resolution. Pitch, we're completely blind to. You combine the uncertainties, and..." he shook his head slowly.

Out of the elevator, they turned right, and Cox resumed their brisk pace, speaking quietly but urgently.

"So, we've got the omni, you can take control every eight minutes as the RM comes around, or goes behind the limb, right?"

"Right."

"But you can't make big guesses, or, with the uncertainty..."

"Right. With the uncertainty, a bad guess could make it auger in."

They entered the waiting room. A receptionist recognized them and assured them the wait would be brief. Ventris stood, not wanting to make his unkept shirt un-tuck. His team had been working round-the-clock in 12 hour shifts to piece together the status of their craft. With a smaller workforce than was prudent, some of his people were reaching a breaking point. With media involvement, the pressure on DeRosier and all of NASA-lite was intense. Without new developments to report on, the media giants re-hashed past debacles and over-analyzed the current structure. Parks Vambolent's name cast a pall over any trouble that arose.

The inner door burst open, and a harried man in a pinstripe suit stalked out. His eyes flashed sharply about and his red-limned nostrils flared. He recognized Henry, and opened his mouth to speak, showing teeth tightly clenched together. Thinking better of it, he pressed his lips shut and yanked on the hallway door. Before it swung shut, they heard DeRosier.

"Next!"

Ventris and Cox stood before the large desk.

"Finally, you've brought me a solution, right? Except by the looks on your faces, I won't like it. If it's anything other than 'Yes, Administrator, the problem is fixed', then you're dead on." He sat back and folded his hands behind his head. "Talk to me."

Cox nodded to Ventris, who took a breath, tried to relax, and began speaking rapidly.

"We've got it, sir. With the help of the math department, we've solved a massive system of linear equations derived from repeated test bursts. They found an unstable eigenvalue on the roll thrust. We called in statisticians to help rank the most likely scenarios causing the instability. Experts in chaso theory and indeterminate analysis were brought in from the outside. System engineers from the assembly contractor have been here since the start.

"Using pair-wise problem solving, we narrowed down the root causes and began singleton experiments. The RM drifted, actually precessed, for three and a half days when we finally deduced that both clockwise *and* anticlockwise roll thrusters were not firing. When we re-analyzed the control equations under this hypothesis, everything fell into place."

"And the bottom line," DeRosier prompted.

"Without the ability to roll both directions, most classes of solutions are ruled out. The only conservative option is to make very small adjustments, waiting on radar and visual feedback after each one. We can repair the orbit in 8 to 12 more days, depending on our luck."

"Henry, when's sunset at Hyginus?" DeRosier asked.

"Seven more days, sir."

"And survivability after that?"

"Food, water and oxygen for at least 4 more days. But their power source is all solar with battery backup. To keep the habitat warm enough to avoid hypothermia, maybe a day and a half after sunset."

"That makes eight and a half days. Ventris, can you repair the orbit – guaranteed – in that time?"

"I can guarantee 12 days, sir. Eight and a half days is possible, but not a certainty."

DeRosier came forward, slapping his palms on the desktop and raising his voice.

"Not acceptable, Ventris. Not! Acceptable! Now, you get back down to your labs, and come up with a way to get them that damned Return Module in position by next Wednesday. I want a report on my desk by noon tomorrow."

Ventris stared at his boss's boss's boss and did not know what to say. They had already run the analysis every way they could conceive of. The eight day scenario was a best-case assumption, counting on them to learn more and more of the craft's new dynamics along the way. The only way to shorten the time was to increase the risk. Ventris blinked as he stared, wondering how to put this.

Henry Cox spoke up for him.

"Jared, the tradeoff between risk and expediency has some kind of curve to it, right?"

"Right."

"Maybe you can prepare for Mr. DeRosier the tradeoff curves for each scenario, or for your various assumptions. Then, as managers, we can make the appropriate choice."

"That sounds reasonable," Ventris said, addressing Cox. "I could have that in an hour."

"I'm not interested in your tradeoff curves, Ventris. And Cox, stop giving him an out. Listen up Ventris. I don't want any curves; I don't want any tradeoffs. I want a solution, and I want it tomorrow. Is that clear?"

Ventris was beginning to get angry with the way this man was treating him. He wondered if it was because of his skin color, and grew even angrier. It emboldened him.

"Mr. DeRosier, the answer is obvious to all of us on the team."

"Then let's have it?"

"Send a rescue crew."

DeRosier drew in a sharp breath to object, but Ventris wouldn't let him interrupt. "We have another bird on the launch pad. They can dock with the RM, and guide it manually. They could even repair the roll thrusters, and get them fully functional. That way, you guarantee the safe return of Crew 1."

DeRosier took in another breath, preparing to raise his voice. The message he had received that morning from the Senate Chairman was clear – no more Moon shots until further orders. Still, the remark by Jared Ventris gave him pause. He remembered the last words of Senator Mason Barnes: 'make sure those astronauts are safe.' However, he had not yet stopped the preparations. DeRosier was in a fix. And the only way out of a fix, he knew, was to pin the responsibility on someone else. His mind shifted gears, looking for a way to weasel out of this. He paused one, then two seconds, exhaling the breath he had taken in anger. A plan was forming in his head.

"Okay," he said, in his most reasonable voice. "Let's assume that you've convinced me. You said yourself that the only way to guarantee their survival is to make the second launch, right?"

"Yes, sir."

"Can you prove it? Mathematically or statistically?"

"Well," Ventris said, his intuitive senses sending alarm messages to his gut. "The solutions overlap. It's not like a proof. It's a statistical probability."

"But your best case scenarios – the ones that end up saving the RM – they are based on things you don't yet know, but hope to learn. Am I correct?"

"Yes, to an extent."

Cox began frowning, as he, too, watched DeRosier work. He had a deep, instinctive distrust of the man.

"So, you could run me scenarios based on everything you *know*, but nothing that you *don't* know, where you could guarantee the astronauts will die. Right?"

"I'm not sure if I'm following you, sir."

"Oh, I think you do Ventris. It's like this: if we try to save the RM and fail, we all fail. That's clearly no good. Too risky. But if we send up the next bird and they save Crew 1, then we are heroes. But if they also fail, then it is *they* who fail, and not us."

Ventris had a foul taste in his mouth, and swallowed a powerful urge to spit. The end result was what he wanted, but the means by which the decision was coming did not appeal to him at all.

"Let me see if I get you right, Mr. DeRosier. You want me to run scenarios which show that no matter what we do, the astronauts up there are already dead. Does that about sum it up?"

"That about sums it up, Ventris. Unless you can come up with a guarantee for me…"

Ventris pressed his lips together, teeth grinding and nostrils flaring. He wanted to save the RM as much as anyone, but they just didn't have the means to assure it. DeRosier smiled as if he had achieved what he wanted. He took the man's silence as an answer.

"Then, I suggest you prepare those scenarios. I want them on my desk by noon tomorrow. *Capisce*?"

"Yeah, I understand," Ventris said in a disgusted voice. "Anything else?"

"Only this. Your report, *and your records*, must all be in agreement. Is that clear to you Jared, or do I have to spell it out?"

"No, I understand you just fine, *Kevin*." Ventris spat out the man's first name. He looked to Cox, who seemed sympathetic, but had nothing to offer. "This is bullshit," he said, gathering his things to leave.

"No, Jared. This is real life! Get used to it."

Ventris left, followed by Cox, who could not stand being with DeRosier one second longer than he needed to. At least they were doing the right thing. But it was the wrong way to do it. That kind of bad karma could really turn on you, he believed.

Hyginus Rille

The sun set with agonizing slowness over the stark lunar horizon. From their tinted viewports, Crew 1 had watched the terminator drop down the eastern mountains. Now, the fuzzy edge between light and dark was advancing across the "sea". In just minutes, they would pass into night, and the temperature around their habitat would plummet hundreds of degrees.

Fortunately help was nearby. After Crew 2 had arrived in lunar equatorial orbit, to their great relief, the second spacecraft had vectored out of radio range to go rescue their Return Module. Henry Cox was updating them every fifteen minutes, providing a summary only, to save radio power on their batteries. Long before the sun went down, the solar cells began to provide weaker and weaker currents, as the angle of incidence of the sun's rays tilted over. Their batteries were nearly full, but their fuel cells were nearly empty. There was very little room for error in the modified mission profile being run by the new crew.

Barstow's crew had hunkered down for the 14 day-long winter. They shut down all non-essential systems, turned off most of the lights, and had dressed in several layers of dirty, smelly clothes. Laurencino had pulled down her hammock and was huddled against the galley on the floor in a tight ball. An ebook at her feet illuminated her nose and eyes, the only part of her visible from her hammock cocoon. Marcharelli had wrapped himself like a mummy, and was vigorously rubbing his upper arms, stomping his feet, and pacing the floor. His breath came out in white wisps.

Lomish and Barstow were playing chess on the electronic game board, under a make-shift pup tent they had erected on the floor. The habitat walls were to be avoided, since that was where their home would loose heat the quickest. The ceiling and the floor of the habitat were insulated by the rock and regolith outside. The floor still held its warmth.

Barstow's watch beeped, and he curled his knees under himself, and reached out of the pup tent to power-up the radio. He used listen-only mode, which took virtually no power. There was no need to expend their precious reserves of electrical energy to transmit a radio signal back to Earth.

"Good news, folks," came the upbeat voice of their capsule commander. "Crew 2 has matched orbits with your RM. They report it's in great shape, no visible damage anywhere on the hull. They're in the process now of guiding it with the omni. Boy, you just can't underestimate the value of being right there. From what I can tell, they can do in seconds what was taking our nav boys hours."

Marcharelli gave a small cheer. He had been more subdued since the announcement to launch the second Moon shot. Now that he was going to be rescued, he seemed ashamed of his earlier cowardice. He'd been a good soldier for Barstow, taking instructions without question, and working hard at his tasks. He didn't joke or hold forth, and had even stopped entirely his advances towards Laurencino. At long last they had all settled into a routine, and truly become a crew.

"We're working on projections for you," Cox continued. "I'm sure you want to know all the facts and figures Robert!" But Cox was out of the loop. The ground based teams had missed the confrontation a week prior. Marcharelli had changed, and Cox's assessment of him was now out of date. None of them spoke of it, but they no longer had the emotional need for what Henry Cox had been providing them.

Waiting to be rescued, Crew 1 passed the time as fully as possible, extending out simple tasks, and absorbing themselves fully in whatever activities they did. They had completed the assembly of the payload launcher, and before the sun had set too low to power it, had fired three small test cylinders into their black sky. The cylinders had been filled with regolith, and then accelerated vertically between the rails of the launcher, powered by heavy inductor-capacitor loops charged up by the solar cells. The burst of high current, passed from one rail to the other through a moveable armature, created a huge force tending to make the new loop of electricity larger. This flung the armature upward, carrying the payload bucket with it. At the top of the rille the payload passed through one final stage, a ring of electromagnets, which modified the trajectory of the projectile, taking out any small deviations from the initial launch. A deceleration rail, arcing high above the rille, then captured the armature for re-use, and the payload was flung into space like a pebble from a slingshot. Their three test shots had been flawless, sending several kilograms of compressed moon dust out into deep space. Now, with nothing left to accomplish, Barstow, Laurencino, Lomish and Marcharelli waited.

Barstow's watch beeped again. This time Lomish reached up to activate the radio.

"Okay, I'm sure you're getting tired of hearing my voice," Cox began, "but the good news keeps rollin' in. DeSoto has stabilized the attitude of the RM, and Mittenberg is going EVA to go fix the roll thrusters." They knew, without needing to be told, that extra-vehicular activity was slow and careful work. It would probably take several passes around the Moon before Dishon Mittenberg was aboard their RM. Then, who knew how much longer it would take him to repair the thrusters.

"Hey, I've been thinking," Lomish said, turning down the volume on the radio. "What if we get them to fix the orbit first, and then *I* go up and fix the thrusters."

"You just don't trust that Marine to make the repairs as good as you, John. Relax, he's one of our best," Barstow assured him.

"No, hear me out. Mittenberg has a solar cell drop planned at Vasco de Gama, right? We could use my spindle to make that drop, and just leave it there. I would stay aboard our RM, and the three of you could use their spindle to come up and rendezvous. That's still just three trips."

Laurencino saw the flaw in his plan, but waited a moment to see if he'd figure it out for himself. Finally she had to speak, but cleared her throat first.

"Ahem. You forgot one thing, John."

"What's that? Oh, they still need their spindle back, don't they?"

She nodded, and her whole spherical mass moved slightly with the motion. She went back to her reading.

With a huge sigh, Lomish forced his attention back to the chess game. He knew he should just go ahead and sleep, but for some reason, he had lost his ability to doze off easily. Ever since he had stood shoulder to shoulder with Laurencino, looking up at the Earth, something inside of him had shifted. He had known many women, and greatly enjoyed their company, but his relationships were all shallow. Not that he minded – he actually preferred that. Common interests, vigorous sex, and sharing food and drink endeared him to any woman who didn't look half-bad. Over the years the attitude his father raised him with had been reinforced: women were pretty to look at, but didn't have a straight thought anywhere in their heads. They were fun to be with at first, but sooner or later, all that craziness inside their minds began to leak out, and it was time to head out on the highway.

But Laurencino was different. She didn't fit into his neat and tidy explanation of womankind. When he had snuck up on her that morning, she had been thinking, not just looking. Time and again, she saw situations clearly the rest of them struggled through. All of them, ever since the near-disaster of tilting their habitat, had sought her advice on any number of things. She never got condescending, never got cocky. She just listened, thought things out, and then told you what she thought. Ice-cold, but dead-on. Lomish had known men like that – assholes all. But not Laurencino. Just watching her handle this Moon mission had shattered his whole conception of the fairer sex. It unsettled him, this new emotion, and kept him up at night. He shifted his position lying prone on the floor, and tried to focus on his chess game. He was losing badly.

After seven more passes, Dishon Mittenberg had boarded their RM and replaced the driver circuits with spares brought from Earth. Using manual controls, guided and cross-checked by navigation computers on board and back home, Mittenberg nudged the orbital precession back towards zero. As the orbital angle came true, he applied counter thrust to bring the drift rate also to zero – a steady, stable orbit. After re-activating the autonav computer, and waiting through the first several corrections, he passed back through the airlock, and returned to join the rest of his crew.

With two spacecraft now in equatorial orbit around the Moon, mission commander Richard DeSoto parked his craft 180 degrees behind the original RM. Crew 1 could now leave whenever they wished. The rescue was complete.

Low Lunar Orbit

Ned Austin sat fuming in his sling. Finally their module was descending. They tilted over and now the reaction force of the rocket motor gave weight to his body at last after three nauseating days without it. Space sickness had hit Austin hard. As one of the twenty percent who experience severe nausea and disorientation in zero gravity, he was incapable of functioning all three days. He felt humiliated. Those military jar heads had no respect for him, and he was cut out of their camaraderie. His was the most important job for their mission – the start-up of an isotope separation machine. With this machine, they could manufacture the pure, simple, raw materials needed for semiconductor fabrication. Feeding these raw materials into a solar concentrator driven float-zone recrystallization furnace would produce excellent device-quality silicon. Using small quantities of boron and phosphorus from their well-trimmed cargo allotment, they could build pre-fabricated solar cells. With the payload launcher, they could then ship the cells directly to geosynchronous orbit.

This was his baby. Austin had developed the initial design of the isotope separator from the simple concepts invented by his long-time friend and now boss, Parks Vambolent. Using his intimate knowledge of physio-chemical reactions, Austin adapted the basic concept into a super-efficient dynamo of creativity and energy. In short, he made it work.

Now, they were landing on the Moon, carrying what could arguably be called the most valuable cargo ever to reach the Earth's natural satellite. Beyond the necessary expedient of supplying energy to China, Austin was in full agreement with the plans Vambolent kept from his congressional masters – the plans he shared with just a few of his closest associates. With their isotope separator they could create an entire thriving economy within cislunar space, that circle traced out by the Moon's orbit. A grand plan, indeed, and Austin was dead-set and determined to make it happen. Without such an outcome, why come to the Moon?

Yet they treated him like a flunky; especially Mittenberg, probably the ugliest man alive, in Austin's opinion. What a waste of skin! The guy was stronger than anyone he'd met in person, and more single-minded than a rattlesnake poised to strike. No wonder world opinion was rising against Crew 2 as a decided military presence in space. That fact solidified the growing perception of the activities on the Moon. Some people were looking at this as an act of aggression. To fathom that the USA was establishing a permanent base that would circle above their homes once every 25 hours was more than many nations could stomach. Some very powerful individuals, co-ops, keiretsus, chaebols, mafias and corporate interest groups were making some very unfriendly-type remarks about the likes of DeSoto and Mittenberg. DeSoto was alright, thought Austin, serious but sympathetic. But he could see why people objected to such obviously combat-oriented astros as Mittenberg. His very presence on their mission spoke of undercurrents of power and control that were not shared with the astronaut crews during their accelerated training. The two regular pilots that he and Juneau had trained with were replaced as soon as the hasty launch was announced. They had supposedly taken ill with the flu and would miss valuable training time. But Austin knew them both to be in perfect health. He was told to keep quiet. No explanation.

Get that, gag orders for the nauseated. It almost made him sick again to think of it. Three days of utter hell. He had lost a lot of mass. The docs in Houston were all worried about him. There was a heart arrhythmia. *Horsefeathers*!, he thought. *I'll show them*. Now that they were going down to the Moon, the retro-gravity was a comfortable 1.2 gs. It felt great, and he wouldn't even mind when they plunked down hard in the dark lunar dirt. The sooner they could get that over-sized goon out to the edge of the Near Side to hook up their solar cell arrays, the better off everyone was going to be.

* * *

Franklin Juneau's insides re-adjusted to the apparent gravity of their retro burn. He felt organs settle back into their accustomed places, and his stomach took a much-welcome journey from the back of his throat to the center of his trunk. His eyes returned to their non-bugged-out normal state, and were scanning the faces of his crewmates, wondering what they were each thinking about. Inside his own mind, he was doing a rapid free-association on their mission profile, thinking through threads of possible futures. Corners had been cut to get them to the Moon so quickly. The overlap with Crew 1 had to be timed right, and all their tasks had to go down like clockwork for success. Too many aspects were interdependent. Although he thrived with complexity, he hated it, considering it his enemy, his challenge. Chaos was Juneau's own personal demon, and he exerted his fine mind against its forces of dissociation, dissolution, and disrepair with a nearly religious fervor. For this mission, Chaos waited patiently at the fringes of each task, knowing that eventually, it would always triumph. His duty, his calling, was to carve out a slice of the Universe and make it more Ordered. Juneau drew great satisfaction in creating structure from the unstructured, sequence from the random, and coherence from the general trend of the Universe to run down and become cold, dead, and lifeless.

Pressed from behind by rocket engines, feeling their thrum of power through his very spine, he fast-forwarded through the consequences of an engine failure. Not good. In his mind's eye, he could see the extent of the wreckage; imagine the most probable status of the occupants: broken from the impact; burned by the vacuum, suffocating painfully as their cabin air leaked out into space. He shuddered.

Franklin Juneau was secretly terrified of dying in space. He kept up his false bravado for the sake of DeSoto and Mittenberg, but Austin's predicament affected him more than he cared to admit. For three agonizing days, Juneau had watched his good friend puke his guts out. The guy looked like a skeleton stretched over with sallow skin. It made the specter of death all too real and personal.

Then they were down.

Juneau felt such a wave of release that his bowels signaled their readiness to be emptied. A deep cramp formed just above his left thigh. He looked over at Austin who was having a private moment of re-adjustment, eyes-closed, looking utterly exhausted. Mittenberg was already unstrapped, and moving to open a storage cabinet. DeSoto pulled an electronic notepad from its niche, and started running system checks. The two soldiers practically ignored the two engineers. Juneau unstrapped and moved to the toilet, not even pausing to savor the new gravity. He unfastened the lid, and positioned himself over the apparatus.

Juneau thought of the moonquakes, bringing to mind the picture he had of open channels under the lunar surface, waiting to cave in, collapsing under the new weight of their ship. The new, mile-long trench just eastward of their landing site was a stark reminder of the risks. The mission planners had been forced to accept the Hyginus Rille site. If Crew 1 had been able to leave as scheduled, they might have decided to abandon Habitat A, and set B and C somewhere that made more sense, like the Aitken Basin. But with Crew 1 stranded, Crew 2 had to land nearby. The lunar base would be at Hyginus. They would just need to take their chances. Another opportunity for Chaos to strike. He imagined the Moon as a soft, gentle mistress, luring them in with her ginger gravity field. But the image turned hostile, as an angered Moon swallowed them down in a sink-hole of broken basalt. As he anthropomorphized, he began to see the Moon in a more living, more dynamic way than he ever had before.

Luna, he would call it. Juneau finished his business, and suddenly decided that he would start calling the Moon by it's truest, loveliest name. He spoke up, and his voice seemed oddly out of place in the busy quiet.

"Luna! We're here!" he said. "I can't remember when I've been so happy to *get* somewhere. Calling it 'The Moon' seems too impersonal, too distant, don't you think? 'The Moon' is fine when it's way out in space. But now that we're here, we need to give this planetissimal a more human name, a feminine name: Luna."

DeSoto looked up briefly, but his face was emotionless. Mittenberg didn't break stride, unpacking his special MMU.

"Sounds okay to me," Austin said. "Luna firma. Glad to have somewhere to call down. Sorry, guys, for … you know," he added, addressing Mittenberg and DeSoto mainly, referring obviously to his un-manly display of nausea and retching.

"S'aright," DeSoto said, glancing up again briefly.

"No problem," Mittenberg said in his machine-like monotone.

Austin rolled his eyes towards Juneau, who shrugged in return. The two of them began their own duties. They heard DeSoto speak in a near whisper, communicating with the ground controllers. Mittenberg had laid out an experimental space suit, and pieces of it were spread across the floor. Austin and Juneau paused to watch him.

The new suit was black, actually a charcoal gray, but striped with removable reflectors for visibility. Instead of being a single unit that one struggles into, like a Houdini in reverse, Mittenberg's hard suit was in several pieces. And instead of being a partly-inflated balloon, like the suit he had recently taken on EVA, this one seemed articulated, but in a way that was not obvious. As they watched Mittenberg go over the suit, inspecting it for any damage, they were mesmerized by the way the suit pieces moved. They seemed to flow in motion, but when they came to rest, appeared rigid. Juneau longed to touch the material, but Mittenberg looked busy.

Sensing them watching him, Mittenberg looked up. It was like his eyes were staring at them from a different universe, and the thoughts behind them were so alien as to be inconceivable. Then, the sensation passed, and Mittenberg gave a lopsided grin, seeming to say '*Watch this*'.

DeSoto paused to watch now too. Beginning with a heavy carapace, the big man swung the largest component of his suit up onto his lodge pole shoulders. Powerpacks for the suit heater, compressed air, electronics and communications were all contained in a sleek bulge that rested between his shoulder blades. He then bent down and picked up the abdominal unit. It was sculpted like a cupped hand, from an over-sized codpiece at one end, to a curving, graceful edge that married perfectly to the carapace as he fitted it in place.

Juneau and Austin watched in amazement as the suit sealed its own seams. Mittenberg merely jiggled the whole assembly slightly. As they peered closely, they could see the two sides of the seam stick and begin to pull together, adjusting their own alignment as they drew the gap closed automatically. When the process was barely half-complete, Mittenberg was already reaching for the next component.

"I think I've heard of that material," Juneau said. "What's it called?"

"It's classified."

"I see."

Mittenberg wrapped a large butterfly-shaped piece around his hips, and he jiggled this one into place as well, letting the amazing self-sealing process do it's work for him. The leg pieces were next, and he stepped into them like jack boots, pulling their edges up to the rest of the suit. He stood, completely covered except for his hideously-ugly head, and shimmied. The black seams between charcoal suit material were vanishing, and it had already become impossible to see where the seam had been in most places. Mittenberg walked the two steps back to the cabin, and the way the suit moved with him was almost organic. The soldier looked over his shoulder at the two civilians with a look of self-satisfaction at the jealousy he sensed from them.

From the cabinet, he removed a helmet. It was so black they could see no features in it, as if the light in the room went into it, but could not escape. He lifted the headpiece almost reverently, and wiggled it down over his blocky ears, then pushed it firmly against the seal ring that was already formed around his neck. They could hear a hiss of air pressurizing the inside of the suit as the seal formed itself in that strangely disquieting way, like mitosis in reverse. The completely suited figure no longer looked human. It was an alien robot, of giant proportions, and reached a black-gloved hand into the cabinet, pulling out a toolbox.

Mittenberg gave a slow salute, and moved to the airlock door. DeSoto had already pressurized it. Mittenberg undogged the wheel, and swung the door open. He ducked his head and stepped inside. DeSoto set down his pad and came over to seal it back up, giving a thumbs up through the small portal when the seal was made. They could all hear the rush of the pumps evacuating the airlock, in preparation for his trip outside. The whole process had taken less than five minutes.

"I don't suppose any of the rest of us have a suit like that," said Austin.

"It's experimental," DeSoto said. "You're safer in a pressure suit."

Austin simply nodded. His normal banter and need to have the last word seemed to have been lost since his embarrassment on the trip across. He went back to work preparing the meal. "Is he coming back for breakfast?"

"No," said DeSoto simply.

"Oh."

* * *

Outside, Mittenberg felt drunk with strength. His apparent weight here was just 40 pounds. On Earth, he could bench-press his own weight, and then some, meaning he could lift almost three-quarters of a ton on the Moon. Like Hercules, he could hurl large boulders without breaking a sweat. And once he accomplished his 12 immense labors, he too, would achieve immortality of sorts – even though no one would know of his accomplishments except a select few. That was okay with him - he was going to enjoy himself immensely.

Mittenberg set to the task of removing the rocket engine from the underside of their spacecraft cum habitat, a job normally requiring two. He was re-assembling the engine into an ascent spindle when John Lomish came to join him on the surface. The autolanders had placed the second vessel within 150 meters from Habitat A.

"Hey, welcome to the Moon. It's Dishon, right?" Lomish called over the suit radio frequency.

"No need for you to be EVA," Mittenberg responded, standing up. Then, apparently as an after-thought, like he had been poorly coached in human relations, he added. "Lomish, right?"

"Yeah. Nice to meet you," Lomish said, not really sounding like he meant it. "Hey, thanks for coming so quick."

"De nada," Mittenberg answered, and bent back down to his task.

"Can I help you with that?" offered Lomish.

"Nope. Hey, you're gonna need to stand back."

He was done already, and did one last walk around to inspect his work. Lomish, a big man in his own right, was daunted by the even greater size of this new arrival to the Moon. The other man's manner invited no easy conversation. Lomish stood watching as Mittenberg, satisfied, sat down near the simple control armrest on the outside of the exocraft and strapped on the retracting 5-point harness. Mittenberg's body movement froze, and the infinite-black helmet turned towards Lomish.

Lomish skipped back a few paces. He knew Mittenberg needed to fly over to Vasco de Gama and hook up their remote solar cells, or in another day, his own habitat would be too cold to support life. But the man was as cold as the frigid dust sucking the heat from the soles of his boots.

The Ascent Spindle rose up on an invisible fire of burning propellant which blasted down noiselessly. A spray of microscopic regolith fragments shot out all around, some hitting Lomish's pressure suit and sounding like a sandstorm. Lomish instinctively shielded his face with his hand, and watched as, in the light of his helmet lamp, Mittenberg rose right up into the starry black sky, arching west across the Moon, headed for sunlight. As Mittenberg crossed the zenith, his silhouette passed over the half-Earth, the only bright object in the sky. The shadow was ominous and black, and Lomish felt an uneasiness arise within his own being. As the spindle moved off towards the horizon, he lost it in the stars, and it was gone. He turned slowly, and hopped over to the newly-arrived vessel, to offer help in the lowering process.

*   *   *

Forty minutes later, Juneau, Marcharelli and Lomish were working on the jacks and legs, lowering the unit onto the lunar surface in the first phase of its transformation from spacecraft to habitat. Laurencino and Austin were out checklisting the electrical transformer and junction box where solar power from Vasco de Gama would soon be routed to their waning power systems via the parallel iron cables laid down by the lunar gleaners since their own landing the year before. Inside Habitat A, Commander Richard DeSoto consulted with Dallas Barstow, his counterpart from Crew 1.

"Thanks, Dick," Barstow said, as he helped DeSoto out of his pressure suit.

"Not a problem. Glad to help out. Everything okay here?"

"It is now. We were a little worried for a while."

"Look, Dallas, I've got word passed down from the highest sources. They want me to go back. Something about the military presence."

"I'm confused. Are you talking about leaving just a 3-man team here? Or are you asking for volunteers to stay behind?" Barstow asked.

"Well, the General left it up to you and me."

"Oh. Well, wasn't that kind of him."

"Look, it's out of my hands, Dallas. I don't like this any more than you do."

"Why don't *both* of you fly back." Barstow was fuming, as long pent-up emotions flowed through him. "And what's with that suit we saw him wearing?"

"It's experimental. There are a few things we're trying out. You've got to realize that this is a prime opportunity to test out some new systems that our boys have been developing for years and years. Think of your country, Dallas."

Barstow was deeply patriotic, and the appeal to his beliefs stopped him for a moment. It sickened him to see the USA piss away their lead in space. As a former serviceman, he knew the value of good technology. However, at the JSC, he'd been asked to remove a bumper sticker from his pickup truck which read 'I love my country, but I fear my government'.

"So, what are you proposing," he asked, more calmly.

"We ask Laurencino to stay on. She's vital to the commercial part, and she could really help out Juneau and Austin."

"We'd need to ask her, of course," Barstow said.

"Of course."

Watching DeSoto, Barstow moved over to where he could hail Laurencino. He dialed in a private encryption.

"Hey," he said.

"Hey," she replied, knowing his voice.

"DeSoto was told to cave in to public pressure, and he wants to go back early."

"So, you're asking me to stay?"

Barstow smiled. She was so smart, he felt like a fool around her. No sense sugar-coating it.

"Yeah, that's one idea."

"Why don't both of them go back?" she asked.

DeSoto stepped up to the mike.

"Hiya Doc, this is Dick. We still have a few things for Dishon to do, as long as we've got him here."

There was silence on the channel. The two men inside waited.

"I'll stay," she said simply. "But I have conditions."

"What's that?" DeSoto asked immediately.

"Lomish stays also. And you send G.I. Joe back right now."

"We'd have to ask John, of course," Barstow said.

"We've already talked about it," she replied. "He's agreed."

Barstow and DeSoto looked at each other significantly. DeSoto wondered if there had been a security leak. Barstow knew that Laurencino had somehow figured this all out ahead of time. He never ceased to be amazed by her.

"I think that might work out okay," DeSoto said after a suitable pause.

"You gave in too easily," she replied over the radio.

"Pardon me?"

"Just what *is* Mittenberg doing over at Vasco de Gama crater?"

"He's hooking up the solar cells," DeSoto answered.

"And what else?"

DeSoto hesitated just a split second. "Well, he'll be putting his new hard suit through it's paces. It's never been tested in lunar conditions. Dust and heat extremes are hard to emulate dirtside."

"Right," she said.

"Thanks, Sellenya," Barstow said. "See ya for lunch."

"Yeah."

Barstow signed off, and turned to DeSoto.

"What else *is* he doing out there – for our country?"

"It's classified," said DeSoto.

"Right."

Vasco de Gama Crater

Mittenberg descended on a flat plateau in the gradual rise to the ridge wall of crater Vasco de Gama, nearly on the edge of the Moon as seen from the Earth. It was a point almost exactly 45 degrees around from Hyginus Rille, and together with a similar station planned for Schmidt Mare, would enable almost perpetual power to the station at the Rille. Since 50% of the Moon's equator is always in sunlight, at least one of the three stations would be in direct sunlight except for a few hours every New Moon. With battery power to bridge the gap in time, the power flowing back to Hyginus would be enough to run continuous operations.

Ultra-wideband radar from his own RM swept the landing site, and located the exact path of the iron rails, plus the precise location of the lunar gleaner at their ends. In a radical new concept of flexible launch vehicles, the moonshot engineers had devised the launch spindle as an efficient way to cover large distances, and effect lunar orbit rendezvous for modest cargos. He touched down with the aid of autolanding software tied in automatically with his RM, and with Houston. He landed with a bump, and quickly dismounted from the ascent spindle. He turned, and without hesitation began to remove the solar panels from the interior compartment.

Laying out the folded assemblies one pace apart on the regolith, he then came back down their ranks and unfolded each bundle. Fresh solar cells began immediately to drink up the full-strength solar energy streaming down from overhead. At the crater named after the Portuguese explorer who opened up trade with India, and lying on the very limb of the Moon as seen from Earth, it was high noon.

The last piece of commercial hardware was an inverter-transformer assembly that first converted the DC power of the solar cells to an alternating current signal, and then boosted the voltage for more efficient transmission down the parallel iron conductors. Power transmission by high-voltage AC is the best choice for covering long distances, such as from Vasco de Gama to Hyginus Rille. The high-efficiency solar cells that Mittenberg now connected to the power connections were over-sized. They would need to serve three habitats if the final mission proceeded.

He then activated a power-up sequence, and waited. Indicator lights flashed quickly on the front panel of the inverter-transformer, and then, one-by-one, the red lights all gave way to green. Two indicator bars showed forward and reflected power, both near the upper limit, and well above the nominal design levels. Mittenberg personally thought this system crude. Civilian hardware, in his opinion, was designed sloppily and built by unmotivated individuals. He hated the thought that his life would depend on two thousand miles of barfed-up crude iron connected to a solar array that they could not afford to visit if something went wrong. If they had asked him? He would have sent a small nuclear power plant – it made the most obvious sense. Maybe a little more work to maintain, maybe a little more mass to launch, but then, plenty of power right at your fingertips. Damned civilians wouldn't even think of it. Damned UN treaties. Damned politicians.

Mittenberg sent an uplink to his RM as it passed over the next time, and sent a burst signal to Houston. His civil responsibilities now complete, he rubbed his gloved palms together, eager to move onto the next phase of his assignment – the one almost nobody knew of.

* * *

Inside the cargo hold of his ascent spindle were several silver cases, each secured with electronic locks. He removed the largest one. On Earth, in practice, it had been quite heavy. Here, though, he held it one-handed out at arm's length from his body, and felt like Charles Atlas. He walked the package a healthy distance away from his ride home. Using his boot, he scraped a smooth spot in the moon dust and set the case down next to it. He bent over it and entered a 10-digit string of numbers, then touched the latch release. A white puff of air escaped into the vacuum as the lid popped up an inch. With a deft movement, he raised the lid fully, and pulled out a stout tripod. Using a twisting motion, he jammed the tripod down into the loose regolith until he felt the teeth dig into the underlying bedrock. Next he removed a stout barrel, and fitted it onto the tripod base.

Buried even deeper in the foam lining, Mittenberg then carefully extracted two mortar shells. Each shell was shaped like a pointy mushroom atop a soup can. A two-part movement at the base of the shell allowed him to unclip the safety, exposing the firing mechanism. Standing off to the side of the small weapon, Mittenberg activated his helmet camera to record the festivities, and checked his chronometer. He then dropped the first shell into the tube. With a violent burst of smoke, the round shot up. Mittenberg watched its travel, which was marked by a tracer. It moved much more slowly than a mortar on Earth. This was intentional, since the purpose of this weapon was for ground-to-ground action. Too much muzzle velocity, and the charge would go into orbit. The motion was graceful, like watching slow motion, or seeing something really large and far away. Mittenberg's heartbeat stepped up in anticipation.

The soldier's eyes narrowed as he watched the perfect parabola finish its sky-writing back down to the lunar surface. He imagined it like a lunar rainbow, and cracked a smile at the thought of the pot of gold. The tip of the charge was designed to penetrate up to 2 meters of regolith before detonating. It would be a useful offensive weapon against habitats or control stations using the piled-dirt method of radiation protection. Mittenberg knew there would be no noise here on the Moon, so he made up his own sound effects. He whistled a descending tone as the shell dropped lower and lower. He actually enjoyed the slower pace of its fall – it gave him more time to enjoy the thrill of anticipation. "Ka-bsh!" he said inside his helmet, in time to the explosion that formed on his visual horizon. The high-explosives created a bright flash. But the fireball was all wrong, compared to what he was accustomed to in explosives training dirtside. Instead of a billowing cottonball of flames, smoke and dust, the fires of the detonation seemed to spread outward in an expanding, diffuse starburst. The light flared, but died quickly. It was not very impressive, he thought.

But then the shock pulse reached his feet, giving a tantalizing jolt to his skeleton. For the seismic wave to be so strong even at this distance must mean the pounding at ground zero would be lethal. His RM was timed to pass nearby the impact point, and would be photographing the damage from overhead.

With the safing removed on the second shell, he dropped it into the mortar. His test would hopefully show that in the absence of wind, the repeatability of this simple weapon's aim would be sufficient to make quick work of an unfriendly target on the lunar surface. Again, the shell rose noiselessly through the sky. Dropping gracefully in the weak gravity of the Moon, it arched downward and deposited its concentrated punishment to the already pulverized lunar soil. Mittenberg activated an integral binocular lens in the optical material of his helmet, and focused on the spot where the last shell had landed. The starburst was much clearer, and he could see the blast wave this time, tearing through the still-settling debris from the first blast. He smiled, a hard smile. He wished there were more fireworks to play with. This was all going to be over way too soon.

When the mortar was disassembled and stowed away, Mittenberg opened a second case. From the gray foam insert, he lifted a mobile charge. This was a miniature of the rover-dozer, capable of traveling long distances and then delivering its charge to a preset destination. It could be operated remotely, guided by vision, radar, or infrared; or it could be set on autonomous mode. Future models would have a search and destroy vision routine. Struggling with the fat fingers of his glove, he finally managed to connect the battery cable to the motor, and then powered up the control unit. This model would follow a pre-programmed course, tracking its own motion with an on-board inertial nav unit, and report its accuracy remotely to a receiving station built into the case. Mittenberg verified the linkup and activated the routine.

The mobile charge accelerated quickly, bumping along the soft regolith, leaving parallel tracks in the dust. Its tall whip antenna waved back and forth. The on-board steering system would allow the mobile charge to avoid obstacles, back up out of blind canyons, and avoid driving off a precipice. Within minutes it was out of Mittenberg's sight. Telemetry from the nav sensors allowed him to track the progress on a small display screen. The polar map showed his location at the center, and he watched as the green line of the mobile charge's path extended farther and farther from him. It stopped, and the green path began flashing red. It would blink ten times, then detonate. He looked up.

A tremendous geyser of flame erupted straight up from the lunar surface. Flashing bits of metal and rock flew outward from the blast, creating a nimbus of sparkles. The shock wave from the mobile charge was much sharper than from the mortar. Mittenberg was bounced up by a centimeter, but the motion of the ripple pushed his feet backward. He tilted forward, but with highly-trained reflexes, hopped his feet forward to catch his fall. But his instincts were trained on the 1 g gravity of Earth. The hop significantly overcompensated for the kick of the mobile charge, and he bounded up rapidly. Finding himself suddenly a meter above the ground disoriented him, and he flailed, expecting to hit hard. Again, his reflexes betrayed him. With the extra time to fall, his twisting caused him to rotate faster, and he began coming down on his side. He braced for the fall with time to spare, and landed almost casually on his side. In a heartbeat he was back on his feet, already correcting for the low gravity. It was a lesson he would not forget. Others would need to be re-trained as well. It could make all the difference in a combat situation.

The final box contained a laser pulse weapon. Arrayed in small niches in the foam were several deflated balloons of various sizes. Each balloon was equipped with a tiny rocket motor and a simple sensor that detected the direction of gravity. When the gravity sensor indicated the motor was pointed mostly down, it would activate a small burst of thrust. The net effect was that the balloons, as they spun about, would rise up, away from Mittenberg. But their path would be erratic and hard to predict. The test was to see if he could pop all the balloons before they escaped him. Of course, balloons did not float in a vacuum. With no atmosphere, there was no buoyancy. The targets had balloons for two reasons: it made for a larger, more reasonably-sized target (about the size of a helmet); and they emulated the pressure suits that most astronauts wore, which were basically just an inflated balloon.

Mittenberg removed the top layer of foam, and set it gently on the ground. Inside was his pride and joy – a laser pistol. A laser pistol wasn't of much use on Earth, needing a heavy power supply, and being nearly useless in rain or fog. But here in the vacuum of space, a laser had infinite range and a ready power source. Covering nearly every surface of the largish weapon were photodiodes. Made of an indium-phosphide compound, they were much more efficient than silicon, so they could absorb more power for the same area. Mittenberg gripped the pistol, and held it up in the sunlight, turning it over both to admire it as well as to charge it more quickly.

It was a beautiful piece of military engineering. Unlike the fanciful weapons envisioned in movies, this gadget was sleek, smooth, and deadly. Mittenberg had broken it down hundreds of times, and could see, in his mind's eye, all the internal components through the ultra-black skin. Behind the solar cell skin was a layer of stacked ultracapacitors. The weapon could be powered from a space suit battery, or be recharged from the sun. Electric current from the solar cells charged up the capacitors. When he pressed the trigger, this stored charge would dump into the flash lamp, which pumped up the laser cavity. The solid state laser crystal was specially tuned to the lamp wavelength, but had two sources of stimulated emission. Each of the electron traps was designed to emit a different wavelength, so the output laser beam contained both a high-energy ultraviolet ray, as well as a long-wavelength heat ray of infrared light. The length of the Fabry-Perot cavity was a multiple of both wavelengths, making the dual-frequency laser possible.

In the center of the weapon was a magnetic slip gyroscope. By pressing the trigger to the detent, the gyro would spin up and hold the gun steady. Once spinning, the magnetic slip would allow the shooter to nudge the gun a few degrees to make final adjustments to the aim. In space, especially in the free-fall of orbit, it could be very difficult to track a target at long distances. Without reference points, it was difficult to keep from drifting and twisting. The slip gyro gave the user an extra edge in aiming the laser shots. With a recharging time required between them, each shot had to count.

The powerful soldier plucked the tiny balloons from their niches. He broke a glass vial inside each one, and a compressed gas immediately inflated the bladder. Holding each rocket upside down, he twisted a ring on the exhaust nozzle to activate the electronics and the micromachined pumps and valves that operated these tiny motors. Once they were hot, he tossed them away from himself. On their own, whenever the rocket motor swung down, the balloons would jet upward at odd angles. The off-balance rotations caused them to spin wildly, making it impossible for him to judge when exactly they may move next. Just like a smart target might do in an actual orbital combat situation.

The gun was powered, and the balloons deployed. Mittenberg waited though, watching the balloons to get some sense of their motion. His initial throw imparted a small lateral velocity, but there was no air resistance to slow it down. As a result the balloons were all moving upward and away from him. He would not have much time to get them all.

Using his other hand, he dialed the pulse power up to 50%. If he spaced his shots out, he could keep the capacitors charged enough to make 9 shots in four minutes, as long as he was careful to hold the gun in full sunlight. After those first 9 shots exhausted the capacitors, any further shots would take a bit longer to charge. He had 6 balloons.

Mittenberg took aim at the farthest balloon. He pointed slightly high and activated the slip gyro by a detent in the trigger. The gun bucked in his hand as the gyro spun up, but he held it firm. In free-fall, the gyro could only be used when the user was anchored. Otherwise, it would spin the bearer. Mittenberg watched, and waited until the next time the balloon's jets fired. The balloon came right into his sight, and he squeezed harder on the trigger. Nothing seemed to happen – no recoil, no flash of light, no smoke. But the balloon vaporized in a puff of gas and sparks. The rocket motor began to fire erratically, and spun crazily against the starry background. He turned to the next target.

One by one, Mittenberg nailed them, missing only once when a jet had fired unexpectedly. It was almost too easy, and he felt unsatisfied with the lack of challenge. But he had two good shots left, and nothing to shoot at. He looked around him. At the edge of the horizon, almost unnoticed before, was the crescent Earth. *Earth!* he thought, and aimed the laser at it. He wondered what would happen to the laser beam if he fired directly at the Earth. Across space, the beam would widen slightly. The laser cavity inside his gun was long, but not long enough to keep the beam aligned over such a huge distance. Once the beam hit the atmosphere, it would get absorbed somewhat. If there were clouds, nobody would ever see it. But if there weren't clouds, *and* if some dumb fool happened to be looking straight up at the Moon, the laser beam would probably fry his retina in a microsecond. And nobody would ever know what really happened. It would be one of those unsolved mysteries, an urban legend perhaps. There might even be a television documentary on the phenomenon! He grinned at the thought, and pulled the trigger.

Then, he did it again, for the other eye.

He chuckled, thinking of that blind idiot who would never see the Moon again, and began packing up his gear. The video of his target practice would have uploaded to the RM by now, and it was time to evac. For him, the fun was over. He was not looking forward to another week of assembling civilian gear and enduring their boring engineering talk. If only he could go home with Crew 1! He had lots of ideas how to improve these weapons even more. Hi picked up a good-sized moon rock, and strapped himself back onto the Ascent Spindle. With a little extra vectoring, he could pass right over the Chinese lunar base and drop it.

*   *   *

Under a black sky, with the Sun above and the Earth on the horizon, the lunar gleaner cooled down at the end of its long, successful mission. Month by month it had crawled, just a few inches every second, from its landing site near Hyginus Rille, to the very edge of the Near Side. In 16 day stretches of lunar day, the gleaner had gulped down the powdered regolith, and extracted iron particles. The mini car-sized factory had an articulated scoop in front, and an inchworm conveyor to carry the minerals under the solar-powered electromagnet. Small bits of elemental iron rose up from the sifting pyroclastic rock, and were directed into the crucible of the smelter unit.

Raw sunlight was concentrated by three motor-controlled arrays of adaptive mirror elements, and directed into the crucible. At 1160 degrees Celsius, the iron particles softened and then melted into a liquid. The lunar gravity drew them down through extrusion tubes lined with a protective layer of diamond-like carbon. The extrusion tubes were also heated to prevent re-solidification, and were surrounded with a traveling inductive field to precisely control the flow rate of molten iron out the orifice. From the rear of the gleaner, orange-hot metal rails were carefully laid down atop the sifted-through regolith disgorged from the lower region of the crawling factory.

Slowly, patiently, erratically, the unit was guided every minute of every lunar day by teleoperation from the Space Manufacture team sequestered in the Johnson Space Center. Their strategies combined creativity, finesse, and brute-force to lay down an unbroken pair of iron cables stretching one quarter the way around the Moon. Deep chasms and torturous fields of broken rock had presented many challenges to a continuous cable. Rather than a straight line 2000 miles long, the East gleaner had wound a serpentine trail across the face of the Moon as the telescopic path planners steered them around difficult obstacles.

Simplicity of design, redundancy of features, and robustness to variation were principles that enabled the success of the gleaners. The iron rails, which would weigh thousands of tons on Earth, were far, far too heavy to bring up by rocket. But by building them *in situ*, with local materials and free energy from sunlight, they could now transmit power from two widely-spaced points on the Moon. The extra sunlight they gathered allowed the installation at Hyginus Rille to extend their stays from 14 days to more than 20. And by the time

Habitat C arrived, if it was not cancelled, a final power station would be installed at Mare Smythii and connected up to the rails laid by the West gleaner. With the habitat and recycling equipment planned for Habitat C, the site could be occupied for two months or more without re-supply, assuming the isotope separator could generate their oxygen from the moon rocks.

As soon as Mittenberg had energized the circuits, electric power raced westward, filling the space between the two rails. A few tens of microseconds later, AC power surged into the junction box at Hyginus Rille. This connection had originally been planned to occur at the end of Crew 2's mission, which was supposed to occur during a lunar day. However, the rescue had forced them to re-arrange the mission activities so that Crew 2 could complete their own mission, even in the dark of night.

Hyginus Rille

Monitoring the power from inside Habitat A, Richard DeSoto watched his oscilloscope monitor for fluctuations and noise that may have corrupted the signal on its translunar trip. Solar radiation, conductive particles in the regolith, and imperfections in the iron rails could all contribute to noisy power. Few things are worse than noisy power when operating electronics and precision equipment. Glitches, ripples, drift and spikes on a power line can jumble algorithms, force inadvertent resets, or even fry sensitive components. The line conditioner in their step-down transformer contained active noise reduction circuitry, but considerations of weight limited the size of capacitors and inductors. Normal ripple could be evened-out, but they were relatively exposed to sudden surges of electromagnetic bursts. So, for example, during any future CME events, they would disconnect the line and run off battery.

DeSoto was disappointed that the received power was at the low end of the acceptable range. In space, large margins for error were *de rigueur* to account for acts of nature, foolishness of man, or just bad luck. Still, he put on a confident face to deliver the news, turning to Barstow. Laurencino and Marcharelli had come back inside, bringing Ned Austin, who was cross-trained with DeSoto on power management and distribution, or PMAD.

"Looks good. Everything is nominal."

"I hate that word," Marcharelli said, but softly, almost to himself. Nobody paid attention.

"So, are we ready to patch it in?" Barstow asked.

DeSoto nodded. "Ready. Ned, do you agree."

Austin was already hovering nearby, and gave the console one last intense stare.

"Yeah. Looks okay to me."

Dallas Barstow grunted softly as a prelude to interjecting his own opinion.

"Is there any risk in turning it on suddenly? We've been on battery power for a while, and many systems are on standby. Should we power them up first? Or maybe we should power them down? Jason, are you on-line?"

Jason Broderick, their CAPCOM for this shift, spoke up.

"The manual says we should be able to swap power from batteries to solar cells without any change in state," Broderick told them. "I'm calling in our experts for a second opinion."

"Naw, you don't need to wake them, Jason," Barstow replied. "I'm just a little worried about the transmitted power. It's not as clean as the solar cells here, or the battery of course. It's probably just because I wasn't cross-trained in PMAD. I suppose I'm a little edgy. Been here too long, I guess."

"Richard?" Broderick asked, delayed. "I've got a download of the power metrics in front of me. But you're right there. How does the power look to you?"

"We're ready to go. All metrics are within tolerance limits." DeSoto said formally. Then he seemed to soften a little. "But I have been saved many times by following hunches. Maybe Dallas has a point."

Barstow looked at the soldier, grateful for the accord.

"Okay," Broderick told them. "All our material says a hot swap is okay. We're well into the safe zones on everything. However, it is also okay to power up from a total shutdown, like if the habitats had been empty for a while. So, I'll leave that up to you folks. You can hot swap, or do a safe shutdown on all non-essential systems, then a normal re-boot sequence after the relay is activated."

"How about the life support and comm systems?" Laurencino asked, startling them.

"Do we take that risk? What if they don't come back on?" Marcharelli said.

"If the logic is that a power surge could damage the electronics," she said, "then wouldn't it stand to reason that the essential systems should be the one we are most protective of?"

"Well, yes that does follow logically," said DeSoto.

"And shouldn't we tell John and Franklin? So they don't cycle the airlock just as we patch in?

"Um. Very good point," Barstow said, and switched his circuit over to tell the two men still working outside.

Lomish's reply came back over the cabin speakers.

"Hey, if y'all are *that* worried, why don't you suit up and c'mon out?"

His joviality broke the anxiety that had grown around the group inside.

DeSoto was about to speak and make a decision for the group. But in his peripheral vision, he noticed a sudden, large blip appear on the power monitor oscilloscope. He faced the screen, allowing the others to continue their debate. If a spike big enough for him to see in the corner of his eye had reached their electronics, almost certainly something would have been fried. He wondered if it had anything to do with Mittenberg, and his experiments at the other end of their transmission line.

DeSoto turned to look around. Laurencino was staring right at him, locked onto his eyes. Had she seen the blip? Had she heard him gasp? He gave her an easy smile. She did not return it. Damn, that bitch was cold, he thought.

Feeling like he had been put under a microscope, DeSoto finally broke off his gaze and turned back to the power monitor. He was careful to position himself so that Laurencino could not see it. He thought through his options, and decided to stall them, if necessary, until Mittenberg had time to complete his tests. However, it was most likely the laser weapon that had caused the power spike. The mortars would have been too far away from the solar cells. And the mobile charge didn't have a big enough EMP to strobe the transformer. As he watched, another blip raced across his monitor. There would be at least four more, but there was no limit really. He had to find some way to contact Mittenberg. It would have to go through Houston, he realized, and watched yet another pulse ripple across the power curve.

DeSoto turned, but in a way that still blocked Laurencino's view of the scope. She was an electrical engineering professor, and would have no trouble interpreting the trace if she saw it. He was startled to find her standing just a step behind him. She whispered quietly, so that the others could not hear her voice.

"How much longer?"

"Beg your pardon?" DeSoto whispered back.

"Until he's done." She folded her arms across her chest, and continued to lock him with her stare.

DeSoto swallowed. It rankled him to face off with a woman, but he was trained not to show it. A dozen options scrolled through his head. He finally decided to trust her. He mouthed his reply, holding up three fingers.

'Three minutes.'

Before their exchange could attract attention, Laurencino walked to Barstow and Marcharelli, and looked over their shoulder at the clipboard. The checks were done.

"Okay, that was good," Barstow said. "Everybody ready?"

"If this works, great," Laurencino said. "But if it doesn't, do we have time for everyone to suit up and get out to the Ascent Spindle?"

They thought, calculating, recalling performance tables they had all memorized. A complete loss of power left them with just enough air for four people to suit up and get outside. But today, nearing the end of their resources, it might be close. She watched them all thinking this through, and then made a suggestion.

"Why don't we turn up the nitrogen inside to reduce the fire risk, and then each use the breather with pure oxygen. That way, we're better prepped to go EVA if we need to, plus it gives us a little longer inside – just in case. I'd suggest we use it for, say, a minute each. Sound good to you, Robert?"

Marcharelli nodded silently. Knowing the biophysics, they were all variously aware that a single minute would not make a great deal of difference to their chances of surviving a glitch. But somehow, they all just played along. She had earned it, and they did not really think to question her.

"Good thinking, C-note," Barstow said jovially, setting a positive mood. He got up slowly, and moved down the habitat towards the exercise station, where the oxygen breather was hung.

Laurencino stepped back to her perch, looking pointedly at DeSoto. He smiled, and this time, it was genuine.

*   *   *

For two and a half days, the combined workforce of Crew 1 and Crew 2 accelerated through the mission tasks for the second Moon shot. They tilted Habitat B, but this time the landing struts had been lengthened before they left, to account for the slippery dust on Luna. Juneau had badgered them to use the new name, and now everyone played along. Their ground controllers in Houston took up the name enthusiastically.

The rover-dozer, with Lomish at the controls and Marcharelli aboard for ballast, dragged Habitat B across the lunar sea towards Habitat A. They all aligned the new module with crowbars and cables, settling it into the scraped-off area that Lomish had prepared two weeks earlier. With eight pairs of hands, the work went very quickly, and they soon mated the two habitats together. That night, after testing the seals, they had officially opened the hatch between the two sides. DeSoto and Barstow shook hands across the gap, and onboard cameras recorded the whole event for broadcast back to Earth.

In a flurry of effusive press releases, NASA-lite Administrator Kevin DeRosier proudly announced to the world that there was a sudden change in the crew schedule. Two members of Crew 1 were to stay behind to expedite the lunar factory process. From Crew 2, the military personnel, assigned at the last minute in place of the regularly-scheduled astronauts, would be brought immediately back to Earth. The United States did not want to give the impression that these habitats were a military base, he told the press corps.

Chapter 14

Ballistic transport.

Ballistic. What a great word. To 'go ballistic' means to erupt in anger. Cop shows mention ballistics when they wanted to know where the fatal bullet came from. Plus, it's the middle name of an ICBM.

Perfect.

With so many negative connotations to the word, the NASA-lite techno-namers paved the way to the technology's obsolescence.

Writing for the Harvard Business Review on Space as himself, and for numerous op-ed pieces in the Wall Street Journal – Asia Edition under a variety of pseudonyms, Jenks laid the foundation for Vambolent's undoing. His milder writings, using the name Jolanda Vasquez, appealed to the amorphous middle who liked the idea, but weren't ready for it.

> *Space solar power can benefit all of us, not just China, once the problems of ballistic transport and power beams are successfully resolved. Generations yet to come will thank us.*

In his more vociferous missives, Jenks' carefully-constructed alternate persona, Henry L. Williams takes a hard stance.

> *Giving the Chinese access to our space installations is tantamount to treason. What's to stop them from another Oasis Massacre using Gigawatts this time instead of Megawatts. No place on Earth would be safe. If I were Japanese, or Indian, or South Korean, I'd be learning Cantonese about now.*

The fact that the US had no intentions of ceding control of the GEO power stations to China did not hinder Henry L. Williams from railing against it. The force of his argument invited counterpoints, and like a tar-baby he ensnared them and escalated the debate.

The Williams character was a short-lived shock tactic, intended to set up an extreme to be pulled back from. It made the opinions he published under his own name seem well-balanced, moderate, and considered. It gave him a foil with which to compare himself, and to be perceived as having a more refined, intellectual approach.

> *While it is true that the US* could *mount an attack from the Moon using ballistic transport, there are many reasons why they never would. First, it takes two days for the missiles to arrive – plenty of time for a countermeasure. Second, the base on Luna is very expensive and totally exposed, vulnerable to many forms of attack. Finally, precision aiming from 224,000 miles away is just not possible with today's available technology.*

In the last few pieces by Williams, Jenks attacks his own arguments.

> *Sure, Winnimer Jenks of Hah-vad assures us that everything space is Perfectly Safe. Just ask the families of those who died aboard the ISS. He says we can't aim a missile, yet NASA can fly a million miles and land on a dime. Like we'd have all that power in our hands, and never use it when push comes to shove.*

All of his articles could be searched and found on the internet. Jenks paid for a service which posted the articles on thousands of servers around the world, and hyperlinked them all to each other so that web crawlers would bring them to the top of almost any broad search on the subject.

Then, he sat back and watched the responses. Once he got the opinions flowing, he knew, they would become self-sustaining. Subtle points he had merely hinted at found fuller expression through other voices. Deemed "The Ultimate Weapon," Jenks read articles and editorials in major news magazines telling the world that power beams from space were accurate to tens of meters, could burn holes through just about anything, vaporize people like paper dolls in a bonfire, and leave not a curie of radioactive material behind. Occupation troops waiting just outside the circle of destruction could walk right onto the still-hot ground and declare total victory over their foes.

None of this was lost on the Chinese, whose aggressive acquisition of neighboring lands was a constant source of international criticism. Jenks knew what they must be thinking: *Is the US really setting up a counter-offensive, under the guise of helping us?*

Neither was this lost upon the President's ailing Administration, the cash-strapped Congress, a more circumspect NASA and the malicious US media; or upon the Europeans and Middle East citizens who hated them all. World opinion was polarizing quickly, headed for a breaking point.

Test shots from Hyginus Rille to GEO had all worked perfectly. The first production shipments of solar cells were now scheduled to begin in just a few more days. More and more people were determined to stop it.

Jenks smiled to himself. Everybody was going ballistic.

Hyginus Rille

In fast motion, the work on the lunar factory looked like an ant hill must look to the untrained eye. Fixed cameras mounted on the periphery of the settlement captured one frame every 5 seconds as the various components were installed, tested, and started up. Played back as a movie, one could watch the factory coming steadily together as space-suited workers blinked in and out of existence.

Central to the work was the isotope separator, the invention by Parks Vambolent with just over a year remaining before the patent expired. Tall cylinders of aluminized plastic film stretched across frameworks made of amorphous metal and aerogel. These cylindrical mirrors tracked the orderly migration of the sun across the face of the Moon. Their combined focal line produced a segmented vertical tunnel of exceedingly dense light with a temperature equal to the surface of the sun.

Into this containerless tunnel, a steady stream of sifted regolith was shot from above, where it quickly vaporized from the heat. Radio-frequency energy ionized the stream into a plasma, and octupole magnetic fields at intervals kept the beam contained. A pair of opposing capacitor plates supported a powerful electric field which bent the charged particles an amount proportional to their state of electrical charge divided by their atomic mass. Heavy elements like iron were bent very little, while lighter elements like calcium were separated much further from the central beam. In between were oxygen, for breathing, aluminum for conductors, and silicon – the main product needed for making solar cells.

Despite Ned Austin's optimization during start-up, much of the beam never ionized, but fell to the naked surface of the Moon as a molten slag. There was plenty of slag, which had to be hauled away.

The solar cell factory, fed purified silicon from the isotope separator, was also surrounded by cylindrical concentrating mirrors. There, at temperatures slightly above the 1415 degree melting point of silicon, the precious semiconductor was formed into thin wide slabs, and extruded downward in a continuous ribbon. Leading the extrusion was a bar of low-defect single-crystal silicon, to serve as a seed for recrystallizing the slab. A moving slit of concentrated sunlight passed from the seed bar into the extruded slab, heating it just enough to scramble the silicon atoms without losing the surface tension. As the slab re-solidified, it took on the exact crystal orientation of the seed bar, forming a ribbon of single-crystal silicon, suitable for moderate-efficiency solar cells. By using a slice off the end of that ribbon, a new seed bar was made, so that the recipe regenerated itself like sourdough bread.

A scanning ion implanter was designed to propel dopant atoms into the silicon ribbon, forming a p-n junction capable of converting sunlight into electrical power. Evaporated aluminum contacts completed the circuit, with automated lasers annealing out any defects found during testing. In the highly-automated process, these 10x30 centimeter solar cells were gently stacked by robotic arms, and packaged for shipment.

Pig iron from the isotope separator was funnelled into the smelter. This part of the factory produced cylindrical iron canisters, capable of accepting one of three payloads. First and most important were the rectangular solar cells. Half-width cells, split by cleaving along crystal planes, filled in the odd spaces between the full-size cells and the cylindrical canister. Second, aluminum struts and antenna pieces, poured and milled by an automated processing machine, were designed to pack tightly inside, and when retrieved, to assemble easily. Finally, some canisters were filled with slag from the isotope separator. A clever re-use of this waste material would avoid the need of Earth re-supply of fuel for the the station-keeping thrusters.

The packed canisters were loaded by hand into the rail launcher assembled along the cliff wall of Hyginus Rille. Each Earth day, there was a launch window of approximately 40 minutes. Into this launch window, they would fire a regular sequence of payload canisters on a one-way trip to GEO. There, through the reverse technology of inductive capture, the canisters would be brought to a relative rest, ready for assembly. With a modest amount of materials from Earth, such as control electronics, microwave generator and PMAD infrastructure, crews of technicians could assemble the pieces into a solar power farm and transmitting antenna.

The movie stopped. The factory was fully-tested, and the systems integration complete. With minimal ceremony, the astronauts declared the facility operational. Luminous plasmas and the glow of molten metals illuminated the stark shadows of Hyginus Rille. The first solar cells were connected to a test fixture, then turned to the Sun for testing. Power flowed into the test stand and the astronauts cheered. In just hours, the precious payload would be packaged into the canisters, ready for the first production deliveries to GEO.

They awaited the okay from Houston for ballistic launch.

Geosynchronous Earth Orbit

Gordon Washmick squinted into the distance, trying to see the incoming canister. He never could. But that didn't stop him from trying. Waiting, he imagined what an off-course projectile could do if it struck his work pod. As a kid, he'd seen a high-speed video of a rifle bullet piercing a green apple. That reminded him of William Tell, which reminded him of oppression by tyrants, which reminded him of the government, and he got a new idea for an illustration. Politicians sending missives (missiles), that traveled so fast that woman (Eve, the apple) didn't even know she'd been cored (her seeds destroyed). Gordon was deeply disturbed that the abortion debate seemed to be lost now that the rate of mutations had been growing at an increasing rate almost everywhere in the world for over a decade. Fewer and fewer people pushed back, as more and more of them came face to face with a mal-formed child. His brother's wife had just created a perfect baby niece for him, Elina Washmick. Gordon had fallen in love with her from the moment she wrapped her surprisingly-powerful infant grip around his index finger.

He watched the inductive decelerators begin to dance along the line. The incoming canister was watched by countless sensors, and the attitude control algorithms were making last-second adjustments. Suddenly they lurched backwards.

Beep.

The chime sounded, indicating that the payload was captured by the series of coils arranged on a curved path. It was over that fast. He zoomed in for visual confirmation. The canister was a gray pig iron cylinder, and it floated, turning slowly, against the backdrop of stars. Keying in the confirmation, his jaw tightened again. All of his tasks were being automated. Dozens of engineers in Houston monitored his every decision and every action. These were fed into a decision-making expert system and a control routine for a waldo. It reminded him of the old book *Player Piano* by Kurt Vonnegut, Jr., where the movements of master craftsmen are recorded on punch cards, and then the men are sent home to a life of leisure. But without purpose or usefulness in their lives, as Abraham Maslow would have predicted, the men went stir-crazy. Mobs formed and marched into the factories to demolish the machines. It also reminded him of the classic movie, *The Stepford Wives*, in which husbands replace their wives with robots. These automatons serviced their every need, never had mood swings, and never acted irrationally, basically solving the 'problem' that men have had with women since time began. He felt an urgent need to sketch and draw.

To retrieve the canister, a lifetime of intuition told him to go straight towards it. *The shortest distance between two points is a straight line.* Yet his training for the last four years had taught him to override that impulse. *To catch up, dive down to a faster orbit, then drift back up to rendezvous.* Moving in orbit is like trying to cut your hair on the back of your head using mirrors. Now, however, he had on-screen controls. He had merely to point a stylus to his destination, and automated guidance systems took him there on the most fuel-efficient route.

The laser thrust jets on his pod fired their beams at the ablation targets situated on long girders sticking out away from his work pod. Once aligned, a burst of precious liquid fuel accelerated him in a direction askance to the canister. It would take a while for him to get there, so he called his co-worker, Benza Loray.

"Hey."

"Hey, Gordo."

"Y'on the graveyard?"
"Stackin' an' rackin'. You?"
"Vectorin'."
"Huh. Last one for the day?"
"Yep. See ya in a while, okay."
"'kay."

For Gordon, the most frustrating part of his work was also the most rewarding. Twisting in his cramped work pod, he fished out the e-pad and stylus. It would take him 40 minutes to get to the fresh canister, then just a few minutes to attach it to the tug waldo. Waldo Central, which Gordon called the WC, would then automate the delivery of the canister to the collection assembly. Gordon would follow the tug, since it would lead him back to the tin can they called home.

Northwestern USA

Wackos are easy to find. Push their particular hot button, and you can get them to do almost anything. And if you find the right wacko, they would act anonymously, and be completely untraceable to the button-pusher.

The plan was dreadfully simple. Among the thousands who still sent fan mail to Henry L. Williams, Jenks sifted through their letters over an anonymous link. He used a parsing routine of his own design which sorted the senders according to three criteria: paranoia, hatred, and single-mindedness. Starting from the top of his ranked list, he conducted a deeper search. Those passing his further requirements, such as being: unmarried, convicted felons, and having some electronics background, he studied further. To watch the few which remained, Jenks sent his favorite brand of private investigators, hard-luck ex-cops who quit for ideological reasons, or were too greedy for their department's level of tolerance.

A month of observation had culled his candidates to two. Under his *nom de plume*, Jenks wrote to the first one, a resident of rural northern California, a hater of the New Age community around nearby Mount Shasta, and a grandson of an MIA from the Vietnam war, who supported his rustic lifestyle repairing small appliances.

Darron,

Thank you for your letters. Your views on world events are right on target. You strike me as a straight-shooter who cares deeply for his country, and the American Way Of Life. When civilization starts to fall apart, it is America which must lead the way. We must not turn aside from this duty – any of us. And when threats to America are not taken seriously by our Government, then We, The People, must take matters into our own hands. That is why our Founding Fathers granted us the right to Bear Arms and form Citizens Militia. It's in the Constitution, the Second Amendment, right after Freedom of Speech.

I am writing to warn you. My Freedom of Speech has been cut off. The Establishment has pressured newspapers, magazines, ezines and blogs from publishing my letters. That is why you have not seen them lately. They do not want the world to know of the most serious threat to our Country ever.

Agents of the Chinese government are walking freely throughout our land, entering under student visas. Our own Government knows of them, but refuses to do anything until these Agents perform a hostile Act. But we cannot wait for that to happen. There are hundreds of them, all centrally coordinated, and they have plans for destruction on July 4th of this year!. We cannot afford to let them strike first. It will be too late.

I need your help. We need to stop this secret aggression before our Way Of Life is lost forever. You are among a select group of my readers. I am appealing to you to help America in her time of need. The need for Heroes is great. Won't you consider volunteering for our Militia?

Eagerly, I await your reply.

Sincerely Yours,

Henry L. Williams

Jenks received a reply within hours. Darron volunteered. Jenks sent him a name, and asked Darron to follow the supposed Chinese agent from a careful distance to determine his patterns. He taught his man how to delete all records of communication with Henry Williams. Jenks told him that new materials would be coming soon: schematic drawings, detailed action plans, and specific timing. He assured Darron the legacy of a hero when the rightness of their actions were finally revealed.

Jenks shut down the link and erased his traces. The packet he had prepared from several international sources would be delivered, along with a final note of encouragement to his cat's paw. This left him free to pursue his next course of business.

Picking up his packed things, he left for the airport. His first stop would be Boston for a meeting with Robert Marcharelli, Sr., who would soon have a heroic son who could start his own space solar power company *without* using ballistic transport. They could deliver power to relay satellites in low Earth orbit using a wavelength which does not penetrate the atmosphere, eliminating the risk of power beam weapons. Jenks would, of course, act as consultant.

Then, he was off to Switzerland to support Bruno D'Argeste's quarterly meeting with the plutocrats. There would be plenty of good news to report. Any time now, they were sure to invite him as a full-fledged member.

Geosynchronous Earth Orbit

Benza Loray, self-appointed space monkey, self-named man-in-a-can, self-nominated best looking man in space, operated a laser soldering tool through the arms of his work pod. His five-armed egg-shaped pod featured three independent, articulated, AI-guided robot arms on the back to manage payloads. Tipped with van der Waals grippers, the same technology used by spiders to walk across ceilings, the three arms autonomously unloaded the payload canisters, and prepped the solar cells for assembly. Loray's pod had two more arms on front, under the window, waldoed to his VR gloves inside.

As soon as he welded one solar panel into its niche in the aluminum superstructure, another slab was brought into place from over his head. Like a factory worker on a stick-lead line, he oriented the next piece, spot-welded the four corners, and then did the next one.

Down one row and back another he came, repeating the whole process across a wide empty array. He and Gordon built the aluminum lattice from the first test shipments coming in from Luna. The less-expendable solar cells had just started to arrive during this work period, and Loray was assembling them as quickly as possible. Washmick would help too, as soon as he finished vectoring. Both of them on-station, plus everyone at JSC, knew that their work might be halted at any moment.

The mission control team tried to filter the news to the station at GEO. More crew members were on standby to help with the work, but the President put them on ice. Loray knew there were plenty of people downstairs sending waves of hatred his way. Didn't bother him. So long as they didn't send missiles.

A dual-purpose Expendable Tank borrowed from the old Space Shuttle design was home to just Washmick and Loray now. They rattled around inside, but enjoyed the freedom of playing in zero-gee. In the quiet moments before retiring the night before, they had talked about their unique position in history.

"Did you hear what the Egyptian government said?" Washmick had asked him.

"Nah. What?"

"They actually admitted they couldn't control their *jihadists*. Hababek and his team of crazies are stirring up the whole country. Satan's Sword, they're calling it."

"The microwave beam?"

"Yeah. They're making it a religious argument. Saying it's the fire that will destroy the Earth. They blame all of Western civilization. And they're tired of trying to explain it to us. Somehow, we don't seem to get it."

Of more immediate concern to the two of them, the Chinese government had announced the intention to send up Chinese taikonauts to help with the assembly process. China was no longer at war officially with the US, but both countries were at high states of alert. The US had objected. The Chinese complained that they were still waiting for their energy supply, and the whole issue was becoming a political hot potato.

Kevin DeRosier, their ultimate boss as Administrator of NASA-lite, had personally instructed them to get a beam, any measurable beam, up and running ASAP. If they got stopped before the first microwaves reached the receiving antenna waiting patiently at the southern edge of the Gobi desert, they might never get to finish.

Colorado Springs

Parks Vambolent breathed heavily, hands on knees, at the top of the third floor landing. In an effort to conserve energy, DeRosier had asked everyone above-ground at HQ to forgo the elevators in favor of the stairs.

He and Eric had been eating poorly since Elise took Mindy to Africa. They were touring sub-Sahara countries interested to adopt the Kenyan Education System. Afren Kwapili was gaining in popularity, and with Aunty Lise, they were becoming mythical characters in a saga that grew upon its own popularity. Countries fractured by ethnic warfare and despotic governments crowded the streets along their itinerary, calling him the Osagyefo, and inviting him to take control of their country. Tensions were high, and he feared for his wife's safety.

She and Mindy had hopped through Schiphol just before a dyke broke and flooded tens of thousands of acres of lowland with more salt water. The airport had been shut down shortly after their departure. With all the re-routing needed for international travel, Elise's return trip had still not been re-booked. Then, their motorcade had been attacked in Zimbabwe, their very first stop. True, it was only rocks, bottle, and cans, but Parks knew all too well the kind of violence that could erupt.

Straightening up, leaning on the handrail for support, Parks sucked in an extra heavy breath against resistance in his chest. He coughed, and had to find a hankerchief for a hard wad of phlegm he brought up. Leaning into the door, he entered the executive suite, where administrative assistants behind low-walled cubicles of mahogany wood worked keyboards and headsets. Several stations were empty as Congress had shuffled more NASA-lite money to the CDC in anticipation of this year's SARS variant.

Walking across the carpet to his office, Parks felt now more than ever the incredible downward force of gravity. He longed to be on Luna with Ned and Frank, or even in orbit with his 'new kids', Gordon and Benza where there was no sense of weight at all.

Opening his e-mail, sorted by priority, he noticed further delays in repairs to the rocket factory in New Nawlins. Damage to the restored levees around the Port of Louisiana from this seasons' category fives had further complicated delivery schedules. DeRosier was planning to shift all subcontractor facilities to interior locations. Even Houston, 50 miles from the ocean, was increasingly at risk.

Parks felt the presures as a physical sensation. Winters in Colorado were noticeably warmer, as the ski resorts kept complaining on the evening news. Melting yedoma permafrost in northern Sibera expanded lakes there. The darker lakes captured even more of the sun's heat, leading to the decay of ancient vegetative matter. Bacteria on the lakebeds turned the organic sludge into food and methane. Some Siberian lakes actually boiled from the released waste gasses. And with methane twenty times as effective a greenhouse gas as carbon dioxide, global warming was starting to accelerate.

China's continuing growth had pitted them squarely against Japan in tapping the oil and natural gas along the meridian line between the two countries. The East China Sea was the lastest hotbed of contention as both sides circled each other in stealth subs. Energy for both countries was now clearly the limiter to future growth, and the supply of easy power grew ever shorter.

Parks let his head fall into his hands and moaned unconsciously. His work was going so slowly. They were just days away from the first power transmission to the Gobi, but it would be almost nothing in comparison to the 18 gigwatts he had promised. They needed far more people in space to scale up the facilities. They needed fleets of waldos and banks of lunar factories.

The world needed his technology *now*. As his breathing quickened, he began to despair he would make it in time. The 'Hurricane Mother' his father had predicted so many years ago was now a household word. Climatologist and meteorologists expected it to start any year now – almost exactly when his late father had predicted it would occur. Parks pulled an inhaler from his breast pocket and sucked, pulling the mist hard into his lungs.

Waves of nausea spread through his head and his shoulders felt tight. All the bad press on 'balllistic transport' and 'power beams' in the light of so much strife worldwide had swung his personal barometer of public support to lows he hadn't experienced since the Oasis Massacre. DeRosier had already informed him that

Habitat C was on permanent hold; that with the CEV they could staff the Lunar Energy Facility for a few more months, but after that…

A hiccough spasmed his chest and his pectorals siezed in a cramp. Parks brought a hand to his chest, clutching the muscle clawlike as his vision sparkled with a hundred flashbulbs. His admin stepped into the doorway as he began to topple sideways from his chair. Her pealing scream echoed in his head as the last thing he remembered.

Klamath Falls, OR

Pre-dawn, the growing GEO installation hovering over mainland China was barely visible as a white speck just above the southwest horizon. It faded with the other minor stars as the glow from the east began to brighten the atmosphere.

Forest Zhang sipped tea at the Finer Diner on Main Street. Forest was the American name he adopted while taking English lessons in Guangzhou. After completing his college program in China, he began taking graduate courses in Forestry Science at UO in Eugene. His goal was to bring forest management skills to the badly-damaged woodlands of his home country. This summer, he was interning with the NFS, working in the Fremont National Forest, in southern Oregon. The date was July 2nd, and Forest had been on the job almost four weeks.

Filling the booth were the three Rangers with whom he shared a truck, Dan, Geoff, and Randy. Forest took the seat closest to the window, as usual, and the others packed in around him. They quaffed coffee from heavy, oversized mugs, and ate sticky pastries the size of dinner plates. Forest, a slight man, ordered the same thing every morning: two scrambled eggs and a slice of toast. Plus his tea, which he drank unsweetened.

Outside the Finer Diner, an orange highway cone had been placed in the parking space the Rangers typically used as the first customers of the day. As Forest watched, a panel truck backed into the spot. The orange cone scuttled back, then bent over and folded under the rear axle.

"Hey," said Dan, their captain, "That's our spot."

As they munched, they watched. They could hear the sound of the still-running engine through the window, but nobody exited the vehicle. Soon, their attention turned to other matters. The veterans told their same old stories of great fires, and of stupid tourists being attacked by bears. When the rear door of the panel truck rumbled up, they all looked over.

Aimed at them from inside the panel truck was a densely-packed matrix of square metal horns, inside-out pyramids surrounded by colored wires. The engine rumble revved up, rattling the window. Without any other warning, the men were suddenly baked with microwaves.

Forest, directly in front of the array, burst into flames on his right side. He began to flail, torn between protecting himself and his struggle to escape. Dan, sitting beside him, was burning also, and scrambled to get out of the booth.

Randy had the presence of mind to grasp the situation and devise a solution. Hefting his mug of coffee, he dashed the contents against the window. The liquid absorbed the microwaves, stopping them wherever the window was wet.

Geoff caught on, and threw his coffee also, aiming to cover more of the glass. The thin layer of brown liquid turned to super-heated steam. The plate glass window cracked loudly, and then shattered. By then, Dan was out, and was already dragging Forest away from the killer beam and the flying shards. He rolled the student in the same motion, smothering the flames on his skin and clothing.

A second later, the panel truck pulled out of the parking spot and accelerated down the street.

"Get me the First Aid kit from the truck." Dan croaked. One of his eyes was blistered shut. "Radio the EMTs to get here pronto. Then, go follow that truck. Call the Sheriff. Call the Troopers. Hell, go call the F.B.I.!"

As he tended to a smoking Forest, Dan seethed with rage at the audacity of the act. Deeper down, he knew this would make an even bigger story than either the fires or the mauled tourists. A story this bizarre was bound to go nationwide, and fast. They would probably be on the news within an hour. This was going to make one hell of a great story!

Chapter 14

Colorado Springs

Kevin DeRosier held Parks Vambolent's "retirement party" down the street from the faithfully re-created White House at the New Renaissance Mayflower Hotel. He waited until Parks was fully recovered, timing it cleverly to invite key people rumored to be under consideration for the sure-to-be-elected President's Cabinet and key Administration appointments. DeRosier was under heavy criticism already from space advocates decrying the sacrifice of exploration and scientific research in favor of the now-defunct Lunar Energy Facility. He was also criticized by peaceniks and alarmists accusing NASA of becoming the sixth side of the Pentagon, because the lunar base at Hyginus Rille maintained a constant military presence ever since the first visit by Chinese taikonauts. Then, after the scandal in Oregon resulted in the summary abandonment of his major program, it was widely expected that DeRosier would be replaced in the coming transition months. During the cocktail hour, DeRosier circulated among the political guests, boasting of the positive accomplishments that had been made under his leadership.

"The yields on Mr. Vambolent's invention have been raised from just ten percent when we started to over thirty!" he explained to a former Fortune 50 CEO hoping for a Cabinet position.

"Thirty percent is an atrocious yield," came the reply. The man had a substantial stake in Japanese banks. The near-collapse of the government there was probably making him grouchy. "Why can't it be 100?"

DeRosier mentally marked him off as a loss, and moved on.

"We've got one astronaut who stayed at the LEF for almost two months!" he exclaimed to a Federal Circuit judge. The judge was rumored to be next in line for the Ambassadorship to Russia. "He tells me he'll stay as long as we let him."

"How long before he declares Independence?" joked the small-statured judge.

DeRosier threw his head back and laughed, using the movement to scan the crowd for his next conversation. He spotted the Chairman of the Energy Alliance, who was virtually guaranteed to be the next Secretary of Energy.

“Vince!” he called out, breaking into the Chairman’s conversation. “How nice of you to come.”

“Hello, Kevin. Excellent canapés.” Turning to his small group, he excused himself, and led DeRosier by the elbow to a balcony alcove.

“Look, Kevin, what’s the skinny on all that solar cell capacity on the Moon. Gonna sell it, or just let it rust?” he asked in a hushed voice.

“Good thing about space,” DeRosier said, laughing, “without water or air, nothing rusts!”

“I heard that Marcharelli Sr. was making a bid for it.”

“Oh, I doubt anyone will be allowed to touch the place. It’s as good as condemned,” DeRosier said, trying not to get too deep into this particular conversation.

“But listen,” the big man said, “what’s to stop them damned Chinese from just marching right up to our base and taking it over? They’ve gotta be sore about us canceling the power project.”

“That’s why we’ve got the Aerospace Force up there, Vince.” DeRosier dropped his voice even lower. “I hear they’ve got some pretty potent little greeting cards for any unwanted visitors.”

The Chairman pursed his lips, thinking.

“Still, it’s a pretty tempting target, Kevin. Oil prices have gone through the roof since your project got the axe. The new President states publicly his intention to honor our commitment – but I have no idea how we can manage. The PRC already has drilling platforms well into the South and East China Seas, and I hear whisperings they are mounting an offensive to take over Sakhalin Island from the Russians. The whole thing is a mess. I say we dismantle the whole shootin’ match. Burn the bridges behind us, if you know what I mean.”

DeRosier nodded, giving himself time to formulate a reply. Most in the Energy Alliance opposed space solar power. The popular opinion was decidedly against it: environmentalists cited disruption to migratory bird routes, peaceniks recalled the Oasis Massacre, and citizen groups said: ‘not in my backyard’. Their collective cause had been clinched when Forest Zhang was fried in a public diner by a home-made microwave weapon. After such an incident, people would not even look at their kitchen appliances the same way again. Microwaves from space were already a dead issue. The LEF crew was down to four charged with decommissioning the facility, staying another 2 weeks. The final two astronauts at the GEO installation were due to return in just a few more days. And besides, when oil became scarce and prices skyrocketed those in control of energy stood to make windfall profits.

"Yes, I can see the wisdom in that," DeRosier answered. "But there are other uses we could still get out of the LEF. I'd like to see us direct our lunar manufacturing prowess into building components for ships and orbiting hotels. Grand hotels, not the cramped tin cans or inflated balloons we have in LEO. What do you think of that?"

"Capital idea! How soon can you book me a room?" the Chairman said.

"You'll be our first guest," DeRosier boomed, clapping the man on the back. The Chairman walked away with a smile, angling towards a waitress with a tray of hors d'oeuvres. DeRosier intercepted his next target.

"So, what exactly do they *do* up there when they are not working?" asked a woman whom DeRosier suspected might be named Chief of Staff.

"They have video recreation, of course," he told her. "But one of our more restive fellows began tinkering with the slag that comes out the bottom of the element separator. Starting with fist-sized rocks he picks up from the surface, he built a fixture that holds it in the hot waste stream. The slag builds up on the rock, and he turns it this way and that. When he's done, there's a ceramic brick in almost any shape you can imagine. Now, get this, he has actually built an igloo on the Moon! Can you believe it?"

"An igloo? You must be joking!" she said, laughing.

"Honestly! Of course, it's not airtight, so at this point it's nothing more than a hobby project. But his little experiment has opened our eyes. He wants to use molten aluminum to seal the joints! We're looking into using this building material to make more permanent structures on Luna."

He touched her arm and excused himself, gliding through the crowd towards Vambolent. Along the way, he shook hands, clapped backs, and tossed pithy words of wit left and right. Surrounding the inventor was a collection of people DeRosier felt were in Vambolent's 'inner circle'. He had ordered background checks on all of them, and felt like he knew each one intimately.

"Parks, how *are* you!" he called, walking up and pumping Vambolent's hand.

"Hello, Kevin." Parks had never before seen DeRosier happy to see him.

"Elise, how nice to see you," DeRosier said. She nodded back.

"Ned, Franklin, hello gentlemen," he said to Austin and Juneau, shaking hands. "And Dajwal. You must be very proud. It was you who single-handedly brought Parks into our fold in the first place."

"I suppose the Chinese Ambassador had a bit more of an important role than me," said Chellapilla.

"And you must be the inestimable Afren Kwapili," DeRosier said, extending his hand to the tall Kenyan.

"A pleasure to make your acquaintance, Mr. DeRosier," Afren said.

"Call me Kevin. I feel like I know you already," DeRosier said. "In fact, I hear that you have become quite a legend throughout much of Africa. There's some name for you. The Osagyefo, right?"

"I certainly don't call myself that," Afren replied.

"No, of course not. But your education program has been so successful that graduates from the Kenyan schools are being hired all through the sub-Saharan continent to start similar programs. There is even talk, from what I've been told, of forming some sort of pan-African coalition. With you in charge."

"The Self-Schooling program is actually Elise's, not mine," said Kwapili. "She and Parks developed the idea together."

"Is that a fact?" DeRosier said, pushing up his lower lip and giving the Vambolents an appraising glance. "Well, quite a group we have here. Two men who've walked on the Moon, a world-class inventor, a revolutionary teacher, and the Savior of Africa."

His flattery fell flat, and his smile faded. Clearing his throat to end the moment, he pointed to the dinner hall. "Gentlemen, Lady, I think we are about to begin the feast. I do hope you enjoy it. We even have a special vegetarian plate for you, Elise."

"Thank you," she said.

Sellenya Laurencino approached the group, flanked by Dallas Barstow on one side, and holding hands with John Lomish on the other. DeRosier folded them into the circle, raising his voice in greeting. Nearby heads turned to see him smiling benevolently over a group of mostly astronauts, obviously in command. He talked them all up, laughing and gesticulating to beat the band. It was a great image, and a few discrete photos were taken from several angles around the room. Catching the eye of his head photographer, he received a confirming gesture that they got some good shots.

With a smile full of white teeth, DeRosier moved back out into the crowd, calling out names and herding clutches of people towards the next room.

"That guy gives me the creeps," confided Ned Austin. He looked at Laurencino, but she didn't seem to object. Barstow nodded and made a noise of disgusted agreement. Then he and Lomish headed to the bar to refresh their drinks before dinner. Elise and Sellenya put their heads together. Elise wanted to know all about the romance between her and John Lomish.

"He is merely practicing political Darwinism," observed Dajwal, as the four core members of Space Manufacture watched the receding back of the lame-duck Administrator. "The new President is surely going to fire him. He thinks he can survive by throwing this party. No offense, Parks."

"None taken, Dodge," said Parks. "Elise and I suspected all along that this project would only last long enough to appease the Chinese. It lasted longer than I hoped. I was expecting a far more ignominious sendoff."

"I saw him talking with Vince Liddyard," Juneau said. "Do you suppose they were talking about the gigawatts of solar panel production capacity lying wasted on the Moon?"

"I doubt it," said Ned. "Probably talking about a way to get rid of it. Hey Parks, is there a way we could get control of it before they do something stupid?"

"If by we, you mean Space Manufacture," Parks said, "I don't think so. Only a government or an aerospace conglomerate could gain control of the facility. And with public opinion towards space solar power more negative than even nuclear power, it's a hot potato that nobody wants to touch."

"So, what *do* you have planned next to do, Parks?" Dajwal asked.

"I do have one more ace up my sleeve," Parks said.

"Do tell," Dajwal said.

"Not now. Not here. But I can tell you one thing," Parks said. A hand bell rung out a trill. Around them, slow-moving guests gravitated towards the food. The group tightened around him, leaning in to listen. "I'm thinking beyond the Moon," he said in a hush. "It was Elise's idea to go there first, and except for the course of events, it was a great idea. But my new goal is to go after a comet!"

Elise heard her name, and turned to listen.

"You gotta be kidding me," Ned Austin said.

Elise caught the word comet, and started moving towards her husband, frowning.

"I figure that people aren't yet ready for anything from space to come down to them. But, there are people who want to go up there. With a comet, you have plenty of water to drink, and plenty of nitrogen so people can breathe without fire risk. Together with slag-brick habitats and solar cells from the separator, we can make space a place for permanent habitation."

"And how do you plan to pull this off?" asked Austin.

"You just watch. But for now, let's eat."

Parks moved off with his old friends. Elise stood planted to the floor, a look of deep concern on her face. Sellenya hooked elbows with her, and the crowd moved them towards the next room.

*　　*　　*

"What are you thinking?" Elise asked Parks after the limosine dropped them off, and they were alone in their bedroom.

"What?"

"A comet?" Her question was an accusation. She tugged jewelry from her ears and wrists, slapping them down one by one on the marble vanity.

"I just thought…"

"You thought nothing," she spat. "Do you have any idea how crazy that sounds?"

"But a comet has all the elements we need for a self-sustaining colony in space."

"A colony? A colony! Parks, you just got 'retired' from NASA. They put you out to pasture. You had a great idea, and you gave it your best shot. But it didn't work. It nearly killed you."

"That hurts, Elise."

"There you go always thinking of yourself. Do you ever stop to think how *I* feel? Do you realize how much *I* hurt when you collapse in your office from overwork? What is this, your second heart attack?"

"First."

"What about that day in West Lafayette, huh? Are you just going to call that 'angina'?"

"Minor," he said.

She smirked. "Don't you see? These dreams of yours, the pressure you put on yourself, the problems these technologies create – they're tearing you apart, Parks. They're tearing *us* apart."

"What about you? You're always flying over to Africa, braving even greater dangers, following your own dreams. Being away from me when I need you. What's so different there?"

"The difference is…" she hesitated only a moment, then lowered her voice and hardened her face, "is that we're being successful."

Parks sat heavily on the bed, as if punched in the gut. He rubbed his face with both hands. It was hard to think.

"What are you really telling me?" he asked.

Elise came over and sat next to him, but not touching.

"Look, Parks, I'm not saying what you're doing is wrong, or not any good, or anything like that. It's just, well, you have to judge a tree by its fruit. So far, there hasn't been much good fruit from your work."

He started to object, but she interrupted.

"No, wait, I know there have been positives, but lets not forget all the negatives. The point I'm trying to make is that there are ways, and there are ways. You've tried to be a builder. But maybe the world's not ready for it."

"Then it may be too late for everyone anyway," Parks interjected.

Elise looked away, thinking, then turned back and grasped both of his hands. "Listen, Parks, you need to try a different way. I mean really different, not going after a comet, for God's sake. There are plenty of ways, less-stressful ways, that you can make a difference. You always told me it didn't matter if it was *you* who made things happen, as long as you contributed, right? Do you still feel that way? Parks?"

"Yes," he said, downcast.

"I realize that may be a disappointment. You like to be a builder, but maybe the time has passed for that, and it's time for someone else to do the building. What we all need is an architect, a visionary. Do you know what I'm saying?"

"Yes."

"Do you, really? What I mean, Parks, is go back to inventing. You have enough networking that any ideas you come up with would be heard, would be considered carefully by people who can make them happen. Isn't that what you set out to do almost twenty years ago? Now, you can make it happen."

"Hmm".

"Or, you could write, Parks. Write books that describe the big picture. Get the word out there to people. Make a fiction novel that everyone can read. Capture their attention, and show them the way through."

"Hmm," Parks mumbled again. "Kind of like Arthur C. Clarke."

Elise put an arm around his shoulders. She kissed him on the cheek.

"That's the man I want to stay married to," she said kindly.

Parks stiffened, but she seemed not to notice.

"That's the kind of man who can be there for his family, to see his children grow up. Stop pushing on the rope, Parks. Go with the flow, but help divert the stream. Your voice will be heard now. Stay home. Invent new things. Write books. Low-stress. Stay alive for me, Parks. I don't think I can take any more chasing around the solar system. Do you hear me?"

"Loud and clear," he said, now looking into her eyes. He felt the sparkle, the chemistry, but it was weakened. He shook his head slowly, wondering.

"Good. Then it's all settled?" she asked.

"What's all settled?"

"That you're going to be a homebody now, and work on a computer instead of in a lab or a boardroom. That you'll take your retirement seriously. Okay?"

He was silent for a long, long moment.

"There's just one more thing I have to do," he said quietly.

Elise's arm dropped from his shoulders. She stood, walked towards the closet, and reached back to unzip her own dress. Parks noticed how trim and fit she still was. In the mirror beyond her, he saw his own apparition, dim and slumped. Her reflection was vibrant and beautiful.

"It's just one last business meeting," he said, a bit too loudly. "One last chance to see if I can make it happen. If this doesn't work out, I promise, I *promise*, that will be the end of it. Then I'll be an inventor, a writer, a house-husband, whatever. Okay?"

She whirled. Folded her arms across her chest, pushing the cups of her brazier together, but the look she gave him was anything but sexy. Her eyes were distant now, and she spoke as if she had not even heard him.

"I'll be speaking at the local Boys and Girls Club tomorrow. Peter Lausen is the host. He's really involved with the organization, and they are interested to start the Self-Learning program in their facilities around the country."

"Oh."

"And I've got a meeting with Winnimer Jenks in Boston next week. He's brokering a deal with Marcharelli Space for some computers and facilities for our various projects in Africa."

"I see," Parks said, and his face darkened.

"Then, Afren wants me to come back for the Independence Day festival. They'll be celebrating the achievement of the goals President Mtongwe set out years ago."

"Those were my goals! I'm not invited?"

"Our goals, Parks. You heard the doctor. No travel."

Parks thought of the arrangements he had made without Elise knowing about it, and decided to say nothing. She disappeared in the closet and re-emerged in a long nightgown.

He watched her brush her teeth with fast, angry strokes. She spat forcefully, and gargled loudly. Running her fingers through her hair, he saw a flicker of a smile as she looked at herself. Then, she swept past him without a touch and got under the covers.

"G'night."

Parks breathed a barely-audible reply, and sat on the edge of the bed for a long time. When he finally heard her breathing deepen in sleep, he got up carefully and went to his office to sleep in the chair.

Southern California

Parks Vambolent squinted at a stab of reflected sunlight. His fingers ran along the edges of a ragged business card which had been handed to him in Kenya. Shielding his sight with the rectangle of cardstock, he stepped down from the small jet at the Santa Barbara airstrip. The hot Santa Anna wind blew away the soft chill of the air-conditioned cabin. The sudden change made his heart race wildly.

A glossy black limousine stood a distance away on the tarmac. Bruno D'Argeste would be waiting inside, per instructions he had given Vambolent. Against doctor's orders and his wife's wishes, Parks had made the call to his angel investor. He timed their meeting to the return touchdown of the last crew from the China Array. D'Argeste sent his Gulfstream VI to make Parks' trip as comfortable as possible.

As he walked to the limo a rear door opened. He bent over to look inside. From the darkness a familiar voice hailed him.

"Welcome Parks, please, step in."

Vambolent stooped down and moved into the vehicle. He could see two figures, but no faces. He sat down in the rear-facing seat, and set his case on the floor in front of him. The door closed by itself, and the vehicle began moving at once. Opening his eyes wide to let in more light, the saturated photoreceptors in Vambolent's eyes finally sloughed off their surplus, and he could begin to make out more details.

"Hello Bruno," he said. "Hello," he said to the other man, who appeared ancient and Asian.

"Ah, I hope you are doing well, my friend," said D'Argeste. "I know you have been very, very busy! On my right here is a very close associate of mine. I am sure you have heard of him. Parks Vambolent, allow me to present Mr. Hiyate Kindoh."

Vambolent bowed his head, and held it down in a gesture he hoped conveyed great respect. He had heard only vague details from Mifune and Hanazawa, but he knew that Kindoh had somehow assembled the consortium which had allowed them to finish the Energy Project in Kenya. The reminder of his late friend saddened him.

"Hello, it is very nice to meet you," Parks said formally.

"The honor is mine," Kindoh replied. His voice was soft, heavily-accented, but clearly enunciated. "Mifune-san spoke very highly of you."

"He was a good man," Parks said.

Kindoh nodded and they observed a moment of respectful silence.

At last Kindoh spoke softly: "I think he would be very proud of all you have accomplished."

“I thank you, sir” said Parks, “but the future we envisioned together has proven more difficult to realize than we ever anticipated.”

“Do not sell short on your progress,” D’Argeste said. “You have shown the possibilities to all future generations. The history books will not forget you. Today, when we drop you off at Edwards Air Force Base, your astronauts will shine with glory.”

“Thank you, Bruno, but the astronauts are not mine anymore. If there is any glory left, it will all belong to them. I was just retired.”

D’Argeste mad a soupy sound of mock disgust. “You are sometimes too humble, my friend. You do not require that affectation with us. It is good for the general public, and it is good for the media, but both of us know your ‘retirement’ had more to do with politics than with your work. Among us eagles, there is no need to act like one of the flock.”

Vambolent chose not to reply. He was keenly aware of the minor role he was playing in events now. True, he was the idea man – the problem solver – and he guided each of his space projects to their technical success. But then, they had failed. Now, he was out of work.

Again.

Were any of the accomplishments actually his?

He considered that he might have been a puppet for forces beyond his awareness. Yes, he had supplied the technical talents, and yes, he had been blessed with good people to help him. Yet, perhaps he had been used or manipulated by others to achieve their own goals. Maybe even the two men in front of him. Vambolent suddenly felt disoriented, riding backwards as the car negotiated a turn.

Kindoh watched him closely, eyes narrowed, as if trying to guess his thoughts. His eyes fixed onto Vambolent’s. When he spoke, it was almost as if his whisper, barely audible, was supplemented by a sudden intuitive impression.

“You are perhaps thinking that you do not control events, and therefore you are powerless,” he said.

Vambolent nodded, wondering if this was a good guess, or something more. His mind began to race with the possibilities. He felt like he had just stepped from a flat world into a 3-D universe, where suddenly everything was infinitely more complicated than he ever imagined. Unsure how to respond, he spoke cautiously. “I have never truly felt like I was in command. Not in Kenya, and not with

NASA. What you say has some truth." Parks let out the nervous breath his lungs were jealously guarding. He drew in a deep breath, and exhaled up towards the window. Wasteland rushed by outside.

Waiting for the old man to speak, his mind zipped here and there in search of solid mental ground. How much of the amazing course of events over the last five years had been orchestrated behind the scenes?

"And, it has some untruth," Kindoh said, in seamless congruity with Parks' thoughts. "A man of vision is worth ten thousand men with strength alone. Perhaps you do not yet see this. Those with power and control are often seen as the leaders. On the surface, this is so. But it is the man of vision who sets the course. Some might say it is the vision itself. I do not agree. It is the man."

Vambolent was not quite sure what to make of this speech. Was he, Vambolent, the man of vision? Or, were they? He bobbed his head in acknowledgement, still feeling disoriented. The elder man returned the nod, and sat back in his seat, apparently having spoken his peace.

D'Argeste leaned forward with hands on his knees, and his elbows out to the side. "Parks, we have great hope for you and your dreams. What you have been able to accomplish is no small feat. Fifteen years ago, who could have guessed? We – Kindoh-san, myself and our colleagues – are quite expert at predicting economic outcomes. Quite expert indeed. However, as I explained to you before, we did not fully anticipate how quickly, or how well, space solar power could be accomplished. Now, your work for the US government is complete. A success by many measures, but stopped by public ignorance of what you are truly doing. At least in the US."

D'Argeste cleared his throat as if that were his opinion of the whole affair, then sat back and continued.

"But now, what comes next? For some time, we have anticipated that Space Manufacture, LLC might become a going concern. You might say we have been waiting on your call! So, it is with great pleasure and anticipation that we are here to be with you, to listen to your proposal. So!"

D'Argeste opened his hands towards Vambolent, his face expectant.

Vambolent reeled from the implications. He wondered what was really going on. Despite their speaking to him in plain language, Vambolent had the feeling that they had nonetheless told him very little. Or, perhaps they had told him more than he realized. The possibilities were at the same time frightening and exciting. He swallowed a dry spot in his throat, and began to reach down for his case.

"Please," D'Argeste said, holding out an over-sized palm. "No presentation materials are necessary. You may just speak with us directly."

Vambolent reached for the glass bottle of mineral water set in the wall of the limo. He twisted off the lid as he gathered his thoughts. Looking over at Kindoh, he could see a small, bright spark in each of the man's moist eyeballs, reflecting the sunlight coming through the front windscreen. Vambolent had not even seen the driver, but fought against his sudden curiosity. Their time was short, and his need was great. 'Vision', they had said. Start from there.

"Our vision, the guiding principle behind Space Manufacture, is this: Humanity is the Imperative. I believe in the spirit of Man. Within each one of us is a power and a purpose. Our innermost natures are grand, though sometimes hidden or dis-used. It is my greatest desire to see the glorious future that is our birthright unfold for all of mankind. Yet, through our collective ignorance and stupidity and haste, we have brought this planet, our only home, to the brink of disaster. My dream for humanity as a whole is to foster freedom, opportunity and peace, and to do so without injustice, exploitation, and destruction. My goal for each individual is that they enjoy life, liberty and the pursuit of happiness without causing suffering, hardship or sacrifice on the part of any other. As a key enabling technology to such a world, my ultimate objective is to build an entire economy based on principles of growth plus sustainability."

He paused to see the effect of his introduction. Both men were smiling slightly and were perfectly attentive. He pressed on, aware that the air base was not far off, but he still might miss the New Shuttle landing.

"What we've been able to accomplish so far demonstrates the feasibility of a space-based economy. There is a factory on Luna which can build components of its own duplicate. By multiplying the number of these factories geometrically, we can form a manufacturing base capable of tremendous output. Powered almost exclusively by the Sun, these factories can form the foundation of a cislunar economy, an economy that can return value to the Earth that is almost incalculable."

Even as he said that, he thought that these two capitalists almost certainly *could* calculate such numbers, and probably already *had.* In fact, they probably had worked out more details than he had been able to. Maybe they knew exactly what he was here to propose, and were just testing him; watching to see if he really knew the score. He took a small sip of water and continued.

"Without subsidies from the government and launches from NASA, the current manufacturing base on Luna does not achieve economic breakeven. All it would take is one failed launch, one inadvertent catastrophe, and the whole enterprise could collapse."

He paused and looked for a reaction. His analysis did not seem to surprise the two men. This was a relief for Vambolent. So far, he hadn't shared the depth of his concern on this point with anyone, even Elise. He had been working furiously on a solution. When he finally came up with an answer, he realized it was an idea he had considered years ago, in Kenya. But he had set the idea aside in the effort to establish the base on Luna and the facilities at GEO. Then, coming back to the cocktail napkin from the SMER espresso bar, he put together a proposal in one hard day, right after moving out of NASA HQ. Then, he had called D'Argeste. That was yesterday.

"The basic technologies for manufacturing are established. The remaining obstacle has to do with Luna herself: her gravity field, and her lack of rocket fuel. Every two months we need a Moon shot to re-supply the habitats, and to rotate the astronauts. But landing in Luna's gravity, and then boosting back up into orbit costs a lot of fuel at the midpoint of a very long trip. It's wasteful, it's expensive, and it's risky."

"And your solution is…?" D'Argeste prompted.

"A comet. An old one with the volatiles mostly locked inside," Vambolent said in a rush. "Get one into a convenient orbit, and transport is much cheaper than going to the lunar surface. Comets have a plentiful supply of vital ingredients that are totally missing on Luna. Water, obviously, but also carbon for fuels, structural elements, industrial diamonds and many other uses. Ammonia for fertilizer, and nitrogen for air. Old comets have plenty of minerals, and they are already finely divided, making it very easy on my isotope separator. And by operating in free-fall, many of the logistics of sun-tracking and transport become radically easier."

Vambolent was winding up, growing more animated and focused as he concentrated on his plans. The content of the information packets in his case were still fresh in his mind, and the words flowed out of him.

"The economics are radically different with the concept of geometric manufacturing – using a machine's output to create new copies of itself. The payback curve is steeper, so we achieve breakeven in months versus years. The long-term gains are far, far greater. In fact, the economics are so compelling that even the risk of two, possibly three failed missions still exceed the risk/benefit balance compared to lunar mining." He paused for a much-needed breath. "But there is a downside. The initial debt is much larger. In other words, to reap the greater rewards of a comet-based cislunar economy, the up-front investment is higher. Substantially higher. That is why I have contacted you."

Vambolent paused to let his message sink in. Had they already anticipated the comet ploy? How much did they really know? Did they have his office bugged, and his keystrokes monitored? He let the moment stretch on, inviting them by his silence to make reply.

"A comet," mused Kindoh. "Very interesting. Historical events have shown that capturing free bodies in space is difficult and dangerous. You have some better idea?"

Vambolent nodded his head gently. "Yes, sir, I do. We need to select a dead comet, one which was once much larger, and has lost much of the ice through repeated trips near the Sun. Some of the opaque debris that was once embedded in the lost ice remained, until it covered the remainder of the volatiles in the core. This will make it safer to work with. To move it, we simply expose the volatiles, and use a laser or reflected sunlight to vaporize them. The reaction force, applied over a long-enough period of time, will modify the orbit of the dead comet.

"Then, we guide our quarry to a safe orbit, not around the Earth, obviously. I have learned that you can't fight public opinion. For many reasons, a heliocentric orbit is best because the sunlight is stronger, so our installation can be smaller, and it is never eclipsed, so we get more up-time. All of the processing methods adapt easily to zero-gee, and zero-g space stations are already an established technology."

They had reached the gate of Edwards Air Force Base, and the limousine passed through, barely slowing. Vambolent wondered again at the power and influence of these two men, and made a mental note to take nothing for granted with them.

"Very compelling!" D'Argeste said. "Now, I think we are to the last point." They were nearing the landing strip, cars on one side of the road and grandstands on the other. "What do you need from us?"

Vambolent quoted a figure, a large one, and then watched them carefully. Neither man responded, so Vambolent spoke on quickly. "We'll use one of NASA's alternate designs for a Mars rocket that uses the Expendable Fuel Tank from the New Shuttle with strap-on solid rocket boosters pre-launched into low earth orbit. We load drilling rigs, reflective foils and high-power lasers, plus a habitable environment for a small crew, and in a little over a year and a half, we have a comet."

"Compelling!" D'Argeste exclaimed. "So, now the question for us to answer is: yes, or no."

He looked over at Kindoh, receiving a tiny nod in return. It was like agreement to something arranged prior to Vambolent joining them. D'Argeste turned back to Vambolent.

"We will agree," he said expansively. "But there are certain conditions."

"Of course," Vambolent said, bracing himself mentally. He recalled the 'conditions' imposed by Kenya's President, and all the unforeseen consequences they had driven.

"I am sure you do not need to hear me tell you what a large sum of money this is you ask for," D'Argeste said. "And I do not have to tell you how important a first try success will be. All these things you must surely comprehend. But what I need to impress on you is how this arrangement will affect myself and my friend Kindoh-san. You see, Parks, we are both members of a small society of interested citizens. We do what we can to help all of humanity, just like you said of yourself. Yet, among these noble persons, the two of us alone are in favor of developing an economy in space. In fact, a majority of us are decidedly against it, at least in the short-term. Kindoh-san and I run a grave risk to even meet with you today." To Vambolent's disbelieving look, he said more urgently, "It is true! There are wheels within wheels."

Vambolent almost gasped. It was the same phrase used by his supposed friend Mason Barnes. Were these men somehow connected? Were they on the same 'wheel', or was one group controlling the other. Vambolent burned with curiosity and alarm. Whose puppet was he? But D'Argeste was still speaking.

"We are much more concerned with our anonymity than with our own economic return. Kindoh-san and I both believe the benefit to humanity will far outweigh any risk to our own personal fortunes. This may be a once-in-a-lifetime opportunity for us, and we do not intend to miss it."

"But surely there is something you want in return," Vambolent asked him.

"Yes, but first, you must understand that it will require great discretion on your part to hide the source of your funding. I wonder if you are up to it? Any small slip on your part could cost us most dearly – most dearly!"

"I think I understand," Vambolent told him, not quite understanding anything except the importance D'Argeste was stressing.

"It is critical that you do understand – and not only for our sake. For yours as well. There are those who would stand in your way, Parks, especially if they knew we were associated. They are very clever at learning associations. Trust me, you do not want these people for your enemies. Everything must be handled with great care and circumspection."

"Okay," Vambolent said, now feeling a sense of anxiety and even fear creep up from his gonads.

"But enough of the cloak and dagger. We will work out all arrangements for you, and give you very specific instructions on how to handle the money and the publicity. It will be very important for you to follow these instructions to the letter. Even if they do not make sense to you, you must follow them exactly. *Comprend tu*?"

"Yes, Bruno. I understand." The way they were speaking, Vambolent had the distinct sense that they had just engaged in a verbal contract.

"Good," D'Argeste said, nodding. "Then, enough of this talk." His entire demeanor changed at once, and he was again the expansive, gesticulating host. "Now, as for what we want in return. We are not looking for repayment, Parks. We have something else in mind."

"Yes?"

D'Argeste came to the point as the driver slowed and pulled to the side. "There will be certain events come to pass in Egypt in the near-term which will change that entire continent forever. It is an unprecedented opportunity. We would like you to build an SPS cluster over Africa." To Vambolent's shocked reaction, he said, "It will not be a repeat of your earlier experience. We would like you to establish this electrical power for African and Eastern European nations. Our projections show that a source of low-cost power to these regions will have a stabilizing influence on the entire geo-political world economy."

D'Argeste paused, and smiled broadly.

"That's all?" Vambolent asked.

"Well, perhaps not all," he replied. "But that is foremost. You must keep our connection hidden, and you must establish an SPS over Africa. If there are any other small favors we ask, I assure you they will not be painful. Do you agree with our terms?"

They had come to a stop. Vambolent needed to step out and join the welcoming crowd for the return of Washmick and Loray. Should he accept? Could he refuse? The money he needed was surely not going to come from anywhere else he knew. He thought briefly of his wife. The door was opened for him. He stole a glance towards the drivers seat. It was empty!

"I agree," he said, awed.

"Very good!" D'Argeste shouted, reaching for Vambolent's hand. "Very good, indeed! We will be in touch."

Vambolent gave a small bow to Kindoh, then grabbed his case and stepped out into the searing heat. He began sweating at once, and realized his legs were quavering. He bent down to say goodbye, but the door was already closing behind him. The whole scene felt unreal. He had just asked for billions of dollars, and got the assurance that he would receive them. In his mind's eye he saw the first comet chaser, racing away from the Sun to lasso its prize, and bring it back home. The limo pulled away, nearly silent, and was quickly hidden behind a cloud of dust.

From behind him came the sudden blare of horns. He whirled at the unexpected noise. Headed directly towards him was a small radio remote-controlled car, bouncing over the rough road. Behind it by a hundred yards drove a Humvee, emergency lights flashing, and bearing down fast, honking the horn. The toy car swerved around Vambolent with inches to spare, and continued down the road, disappearing in the settling dust. He had a moment's impression of it, but it looked less like a sports car model, and more like a light tank.

At the last instant, Vambolent stepped back as the Humvee swept past him, narrowly avoiding a collision. He watched a man with a rifle stand up through the roof, aiming it forward into the dust cloud the limousine had just left behind.

Reeling from the shock and surprise, Vambolent was thrown further off-center when a sudden flash of lightning blazed through the tan dust. The white light changed to red and grew quickly bigger. A double percussion boom followed immediately after. Vambolent saw a geyser of flame lift up from the dust like a rocket launch. As the flames licked up, he saw pieces of the shattered limousine hurled into the air.

Vambolent fell onto his buttocks, hard, more from the shock than the sound. The dust of all three vehicles choked his breath, which came in ragged gasps. He could hear women screaming and men shouting. The frightened voices of children wanting to know what was going on pierced the air and tugged at his heart.

A sudden cramp pierced his side, growing sharper and deeper with each rasp of his breath. The pain rose through his neck and dug into his jawbone. A cold sweat popped out from the skin of his forehead. With eyes tightened in pain, he looked up at the rising roar from the sky. As his vision narrowed down to a tunnel, he watched the New Shuttle sweeping down. He watched the wheels hit as if looking through a telescope. Blue smoke was torn back under the body. A moment later he heard the squeal of their impact rip through the screams of the crowd. As dizziness began to black him out, he looked down the road.

The Humvee was parked at an angle, soldiers swarming around the twisted metal shape. It was rent in the center and smoke gutted from the exposed interior. A nauseated bile rose in his throat as he thought of the burnt and blasted bodies of D'Argeste and Kindoh. His last hope was now gone.

Suddenly, his whole speech of minutes ago rang false. He clutched at his chest and spat dirty saliva onto the ground. It absorbed at once. A pinching pain seized his heart.

Not again!

He sat there, gasping for every breath, feeling like each one would be the last. Earth's oppressive gravity sucked him inward, and all his dreams and aspirations drained out. Frightened crowds moved all around him, leaving him inside a space which filled with dust, swallowing him up and covering the traces. He didn't care. Ten thousand years away, when some distant progeny came on an archeological dig, they would find his bones.

He could almost feel them scraping at him now.

Chapter 15

Base Medical Hospital

"I need to leave for Africa Saturday."

"But I need you here."

"Afren has to leave the continent for a while. The Independence Day celebration in Nairobi is featuring our education system. I *have* to be there, Parks."

He rolled in silence, pushed by an orderly whistling tunelessly through the glaring hallways of the base hospital. The click of her pumps on the tile floor was sharp, snappish.

"I don't want Mindy to take care of me, Elise. She needs to start college."

Elise paused just a moment, but her reply was not softened.

"Well that's just one more thing you screwed up, isn't it."

Parks turned to look at her, but she only stared straight ahead. Each jarring step caused the flesh under her chin to jostle slightly – one of the very few signs of her advancing age. She looked extraordinarily tall to him, straight-backed, striding purposefully. He noted a charisma in her bearing that he had not quite noticed before.

"When will you be back?" he asked.

"When the tour is over, Parks. This is really important to our work. We have corporate sponsors now, a timeline. It's real now, Parks. I've got to be there."

"But when will you be home?"

"In a few months," she said, and stepped in front of the door sensor. Hot air flowed in around her, carrying the scent of a new perfume into Parks' nostrils. She strode out in front of them and climbed in behind the wheel of their rented hybrid.

Parks pushed himself up even before the chair came to a stop. The orderly pushed him back with a firm hand, and Parks semi-collapsed in the wheel chair. Stepping around him, the orderly opened the door, and gave a smile.

"You take care now, sir."

Grumbling under his breath, Parks stood, unable to straighten fully, and guided his fall into the passenger seat. The door was closed for him, and the seat belt wound around his trunk on a motorized track.

“Who are you traveling with?” he asked, as she pulled out of the eave.

“Ambassador Washington will be our guide for the first two weeks of the tour,” she said, and turned hard onto the access road. “Winnimer Jenks will join us after Zimbabwe.”

Parks winced instinctively.

“What is Jenks doing there? I thought he was in Boston working with the Marcharellis.”

“Marcharelli Space donated all their legacy computers to the Fund. Winnimer arranged matching contributions from 50 other Fortune 500 companies. This will be the biggest disbursement since we started. Like I said Parks: it’s real now.”

She stopped at the light, but didn’t turn to look at him.

“Well, I’m happy for you,” he said.

“Thank you,” she said, brightening briefly. Then her face became stone again.

“Can I at least get a nurse, to help Mindy?”

“No, Parks,” she said. “First of all, we can’t afford a nurse. We can barely afford to fly you back with a medical assistant, for God’s sake. Secondly, I don’t trust you not to do something stupid again. I *trust* Mindy. She’s already agreed.”

Elise accelerated hard onto the entrance ramp of the freeway. A truck was in the slow lane. Elise tried to beat the truck, but the hydrid topped out. The merge lane became smaller than their car.

“Watch out!” Parks called.

Elise shot him a look of molten lava. Rumble strips chattered their tires. Loose gravel spit from their wheels. The hybrid lost traction and began to skitter. Parks watched the huge front wheels of the truck loom directly ahead of him. It’s horn blared.

“Horse!” he called.

Elise turned the wheel out of the skid, and the rear wheels yawed out from behind them. They were traveling sideways down the berm, jackhammered by the rumble strips. Parks watched deadly-hard pavement threatening to slap up into his face.

The rear wheels swung into soft dirt, and they were snapped forward. His head whipped left, right, then whacked the window. Anti-lock brakes battled with the grooves in the asphalt and they shuddered and bounced to a suddenly silent stop.

"You idiot!" she yelled at him. "You incomparable idiot! Why in God's name did you have to say that! What were you thinking?"

"Me? You blame this on me?" Parks felt dazed.

"You screw up everything, Parks. You distract me right when I need to concentrate." She slapped the wheel, and turned full to face him. "You cost Mindy her chance to go to college with her classmates. You betray my trust in you by flying out here to make another damn space deal. You lose at everything you ever try. I'm sick of it. I'm sick and tired of dealing of the way you ruin *everything*!"

She turned back forward, caught her breath, and prepared to pull onto the highway. "You disgust me," she muttered. "I might never come back". She hit the gas and cranked the satellite radio full blast.

Parks slumped back in his seat, breathless and speechless. A droplet of blood from the split in his temple rolled down his face and came to rest in the corner of his lip. He tasted it, and another part of his heart died. It felt worse than the heart attack.

Cimmaron Hills, CO

Mindy Vambolent pushed open the screen door to greet Afren Kwapili. Several men were standing in the yard. She stopped, blinking white eyelashes. He moved forward, and they hugged awkwardly, aware of many eyes on them. As Afren released her, he caught her hand by the fingers and kissed over them, just as he had done to her mother years before.

She became a little girl again, and threw her arms around his neck. They stood there, embraced, as he swung her lightly back and forth. All tension left her, and she sighed mightily as they sat down together on the porch swing.

"Thanks, so much, for coming, Uncle Afren," she said, beaming pure energy at him.

"It is my great pleasure, young miss. And you have Mr. Chellapilla coming also, is that true?"

"Yes, but just for tonight. Uncle Yoshi told me he couldn't make it," she said, and tilted her head. "Are you sure this is okay for you? I know you're missing Independence Day."

"It's quite alright. Your mother is covering for me. We will just set off a few fireworks in the back yard… if my security men will permit me."

"How *is* my mom?"

Afren stared at the nearby mountains silhouetted black against a fiery sunset. These young upthrusts made his homeland's greatest peak seem lonely and out-dated. Yet, his land was the cradle of all civilization, and now, perhaps, a hope for the future of a continent.

"Your mother is doing well, young miss. She serves a great purpose, and she does it well."

"When she gets back from the tour, is she going to divorce Dad?"

"I don't know, Mindy," he said, and plucked her hand to his forearm, then covered it gently with his own. "I suppose it depends on what he does until then."

The look of pain in his face was so gut-wrenching to Mindy that she became lost in it. Seeing him and reading his face let her know, with utter certainty, that her father's dearest friend was hurt by this separation as much as she. Maybe, in some ways, even more hurt.

"Oh, Uncle Afren!" She hugged him again.

"Now," he said, his voice a little thick. "Now, about you, Mindy. Your college studies should have started. Are you at school?"

"No," she said, hanging her head, wagging it. "Dad takes too much time. But it's really hard on me," she said, tears welling now and throat tightening. Her voice squeaked upward. "I could really use a break."

"Then, please, consider it my gift to you."

Her eyes were perfectly clear, and a crystalline drop of moisture surged up from each duct. Twin globs rolled down her cheeks. They met on the tip of her chin, teared, then fell onto her bosom. She blinked, white over white. When she opened them again, her eyes were red.

Gulping and sniffling, Mindy stood and began talking rapidly. She explained about raisins and nutmeats, oatmeal and flax seed oil, broccoli and kale, carrots and yams. She listed on the tips of her fingers each kind of pill: Co-Q-10, B-plex, Cal-Mag-Zinc, Red Rice Yeast, psillium husks, and a telomere-replacement cocktail from overseas.

"Everything will be fine. You go now," he told her.

They both turned at a scuffing sound. At the end of the hallway, oddly illuminated by the fading light of the sky outside, Parks Vambolent seemed to float in darkness at the end of the hall. He stepped forward, and they both noticed the heaviness to his footfalls.

"Well, if it isn't the Osagyefo."

"Hello, Parks," Afren said. The two men embraced quickly, and Parks waved them into the den.

"Actually," Afren said, "it is not a good idea to call me that name."

"Why's that?"

"That is part of why I am here. Some disapprove of my calls to unify Africa's economy. Some would kill to prevent it."

"So, you're hiding out here. And I thought you were just helping out your favorite niece."

"That too. But most importantly, to see you, my old friend. How have you been?"

"Been worse, but not by much. The only thing I hate more than not having enough time is having too much. I want to do something – anything – but this drill sargeant has me doing nothing but swallowing pills that are too big for my throat, eating rabbit food that doesn't fill my stomach, and parading me around the neighborhood so everyone can see this tired old sot."

"It's for your own good, Dad."

"Well, a lot of good my own good is doing me. If only this…"

"Don't start, Dad."

"What?"

"You know what I mean, Dad. Don't go there!"

"You see, Afren? Drill sargeant."

"I suppose I can see why you two need a little break from one another. Mindy, why don't you show me what I need to know in the kitchen. Then you can be going."

They left the room. Parks raised his hand so the television could see it. An on-screen menu showed his options, and he waggled his wrist and fingers to scroll through his options. A news vignette showed an anhydrous hydrofluoric facility in Los Angeles that had just been ruptured by 50 caliber rifle shots coming from somewhere in the overlooking hills. The plume of vapor was plainly visible as a white gas of acid, spreading slowly under the temperature inversion. Helicopters above the leak televised pictures of people lying dead in the residential streets surrounding the plant. One helicopter had already flown too close and now lay on its side, burning, in a tree.

The doorbell rang, and he gratefully killed the feed.

"Dajwal!" he called, rounding the corner.

Mindy came out of the kitchen. She ran to the front door, edging out her father.

"Uncle Dodge!" she cried, flinging the door and giving him a 2-point shoulder hug with elbows akimbo. "How are you?"

"Just fine, thank you, Mindy. Just fine. I have something for you. But first, I must greet my old friend and favorite inventor. Hello, Parks. Good to see you!"

"Likewise," Parks said, shaking hands. "Now, you keep her busy for a few minutes so I can have a glass of wine."

"You wouldn't dare!" she said, and glared at the back of his head as he walked into the kitchen. Then, in a heartbeat, her frown vanished, and her voice crescendoed. "You said you have something for me?"

From a pocket, Dajwal produced a small jeweled box, with colored crystals inset to a small red treasure chest.

"Ooh, thank you, Uncle Dodge," she said, then narrowed her eyes. "No bugs?"

"No bugs," he said, completing their ritual.

She opened the box, and inside was a ring. A generous-sized ruby in a filigreed silver setting held warm soft light like the last fresh bloom on a rose petal.

"It's beautiful," she cooed, and slipped it onto her middle finger. It slid over the knuckle perfectly. She held it up in the last light of day, and the glow got stronger. Each facet showed a different geometric pattern, which shifted in a pleasing way as she tilted her hand this way and that.

She gave him another hug, and a kiss on the cheek. Holding hands, they entered the kitchen with Afren and Parks. True to his word, Parks had a glass of red wine poured. Mindy stalked over, plucked the glass, and dumped the contents into the sink.

"Time for you to go," Parks told her.

She looked at Dajwal and Afren, forehead steepled in concern. "He used to be smart," she told them. "Now, he's just sneaky. Don't let him push you around."

"We will take good care of him, won't we 'Uncle' Dajwal?" Afren said.

"It was me who *taught* your father to be sneaky," Dajwal said. "He won't be pulling any fast ones with me and 'Uncle' Afren."

They all laughed, and Mindy crossed her arms in front of her chest. She stood up to her full height and stared them all down until their laughing petered out. "I'm going back to Indiana," she announced. "My phone number is on the 'fridge. The trash has to go out Sunday night because they pick up early."

"We will be fine," Afren said, laughing again. He placed an avuncular arm on her shoulders and guided her out of the kitchen. "Everything will be all right, young miss. You go now, before it grows dark. We will call you if there is *anything* at all untoward. Okay?"

Her shoulders softened, and she stared at the poorly-shaven cleft in his chin, dusted now with short, white hairs. Looking up at his warm brown eyes, she drew a deep breath and nodded, exhaling. "Okay. Thanks."

At the door, she pulled out the handle on her overnight bag and dragged it over the threshold. It banged down the stairs of the stoop. In moments, the door to her fuel cell car clicked shut. Afren waved from the door as she backed out of the drive. Her departure made nary a sound.

Addis Ababa

Elise Vambolent took the podium at the square on the end of Churchill Road where the Lion of Judah overlooks the train station. Near the end of a long tour of stops in nation capitals, it was her turn to deliver the keynote address, bringing the message of the African Union to their member states. A transparent bulletproof teleprompter in front of her showed her speaking points. Behind her sat members of the Executive Council of the African Union, members of the AU Commission, and selected industrial partners. The schooling system Elise and Afren were introducing was being used as a vehicle to change the next generation of African citizens, should reform take that long. Before her, tens of thousands choked the modest square, some sitting on shoulders, other on top of cars. Banners of red, green and yellow flew everywhere.

"Gentle people of Ethiopia and citizens of Africa," she said. "In this city, who's name means 'New Flower', we have seen the beginning of a new age for this great continent. As before, in 1974, when this country was the first to attain freedom from colonial powers, Ethiopia is once again a leader."

"Aunty Lise!" A child's cry filled the silence between Elise's points. She stopped to smile, and a chorus of children's voices chanted her African name over and over. Other voices joined, and the sound swelled to an echoing din in the square. Elise stepped back, overwhelmed, and raised her hands to get them to calm down.

"Thank you. People. Thank you," she said into the microphone. She turned and pointed to the sculpture. The noise subsided.

"In the year 1940, an Ethiopian solider named Zerai Deres was invited to Rome on the occasion of Hilter's visit to Mussolini's new Empire. When Zerai Deres saw this very statue, stolen from Africa and planted in Italy, he drew his sword and killed many Italian soldiers in his anger. It took more than 20 years, but in the end, the Lion of Judah was returned to its rightful home: here in Addis Ababa."

Shouts of 'Zerai Deres' sprung up in various places around the square. The voices seemed older. The chant never caught hold.

"In Africa today," Elise continued, "there is a new spirit of unification. The esteemed members of the Executive Council are here to remove the last remnants of colonization and arpartheid. They are here to promote solidarity among African States. They are here to intensify cooperation and economic development. And what better place could they be than here, where it all began."

Healthy cheering rounded the space. Happy faces shouted joyfully into a clear blue sky in a perfect 22 degree day in the highlands of the Weina Degga region. Elise relaxed and resumed her speech in perfect synchronization with the ebb of the cheers.

"Since 1993, Ethiopia has been a country divided. This year, with the province of Eritrea re-joining the nation, Ethiopia once again leads the continent in progress."

Hisses, though barely audible, were instantly sensed. Security personnel surrounding the platform stiffened. Several fanned out into the crowds.

"And now we all stand on the threshold of an even grander unification. The African Union, including the Assembly and the Pan-

African Parliament have voted to create a new position in the Union. This new position will be called…"

A lone voice cried out, loud and crystal-clear. "Osagyefo."

Elise hiccupped involuntarily. The momentary delay gave just enough pause that her words could not overwhelm the echo from the crowd: "Osagyefo!"

"Speaker of the Union!" she tried to shout.

Cacohpony broke out. Several members of the Council rose to their feet, raising palms in placating gestures. From the crowd, calls came from every direction. Police were raising blackjacks overhead, bringing them down forcefully.

"Please, people, please!" Elise pleaded. Not even she could hear her words.

The AU Commission Chairman stepped up to join her at the podium. Elise fell back, overwhelmed at the pandemonium that seemed to engulf the entire city square. The crack of a gun made the din break, but only for a heartbeat. Then, the noise began in earnest. More gunfire followed. People began to run.

The teleprompter screen suddenly flexed inward, snapping back with a splatter of white surrounding a circle of damage. Elise's eye drew the line from the teleprompter inward, and stared directly down at her own sternum. After missing one of its regularly-scheduled beats, her heart jumped in her chest, visible through the fabric of her blouse. A woosiness fell over her, and she had to sit down.

More shots were fired, closer now. The Chairman's shoulder whipped sideways as a bullet struck bone. Elise dove under the chairs. Those sitting joined her until there was a pile of officials low to the platform.

Elise was under the hot, fear-scenting bodies, laboring to breathe, but safe. Her faintness passed. The noises in the square became remote as people fled the scene. Left behind were those trying so hard for African Unity, now speechless. Elise sat up, and noticed it was Winnimer Jenks who had been protecting her back. They looked at each other, dazed, for several long moments.

"I'm ready to go home," she told him.

Cimmaron Hills, CO

"How's Africa?" Parks asked.

"It has never been the same without you there, my friend. I miss working with you."

Parks took a sip from his water jug. The flippant arrogance he effected around Mindy would not stand up for a minute with Afren.

Dajwal, uncomfortable with silence, sat forward on his seat.

"I hear the newer HIV vaccines are having good success. How does that affect things in your continent?"

"My *home* continent, not *my* continent," Afren explained needlessly. "It is hard to give simple answers that apply to every nation. However, we are very delighted that so many nations are subsidizing innoculation programs. Even with the US reducing its support, the program has grown year over year."

"That's great news!" Parks said.

"As with so many things we've discussed over the years, Parks, there always seems to be a negative consequence to any new technical development."

"Are you getting ready to make a zero sum argument?"

"No, not at all. Only consider that this groundswell of humanitarian aide has also generated over-crowding in the cities. The crowding engenders violence. In the pays, the countryside, our deserts grow ever larger. The cycles of weather, always severe, are even more so now. Crops are stressed, and do not yield well. As a result, we need even more aid, just to feed people."

"Your continent is in dire need of infrastructure," Dajwal said, uncharacteristically quiet.

"Indeed. Once we have a continent-wide network of communications, transport and energy, the new generation of entrepreneurs and skilled labor coming out of our schools will foster an era of unprecedented growth."

"Can you wait that long?" Parks asked him, noting wryly to himself that their roles had surely reversed. Afren was now the dreamer, and he the realist.

"No. I don't think so. But I don't know any other alternative. In the meanwhile, we are doing all we can. We have the Education System, of course, but this is a slow route to success. While I am out of the country, I will be meeting with World Bank representatives. They have offered us in-kind work, which they pay for and manage, but we need more. We have the Aunty Lise Space School, but no one will hire our graduates."

"The what?"

"You mean you do not know?" Afren said, amazed. "Elise *always* talks about space, Parks. The new school at the Overberg range outside of Pretoria is named after her."

"Oh."

"They have a WETF facility, a space-plane rated runway of 10,000 feet length, and a long history of polar launches. Over sixty students have been enrolled, and our first graduating class completed last year."

"I didn't even know," Parks said, half to himself. He was looking at Elise's chair, where Dajwal was sitting.

"Everyone wants a moon base: China wants to return in numbers. Russia wants to mine Helium-3. Sweden has been working towards a manufacturing base for over twenty years. But none of them will fly our graduates. Even NASA-lite will not consider them, even though your president is of African descent."

They both looked at Dajwal, who shrugged his shoulders and raised his palms, placating. "I am just the Associate Administrator for Science," he told them. "I do little more than push paper from one big pile into other big piles."

Parks sat up, and smiled broadly, a little too enthusiastic to be genuine. "Well, aren't we something. From Nairobi to Colorado Springs. So much has happened. And so little has changed."

Neither of his friends responded. Instead, they sat stock still.

"Except, of course, me," Parks said, deflating. "My wife has left. My heart is on it's last hurrah. My first patent ran out this week, the rest are close behind, and I'm on the verge of personal bankruptcy. My daughter has fled the state because she can't stand me, and my son went to Africa with his mother instead of going to school with me here. I've been vilified by the media to the point where the country gentleman Henry L. Williams calls my works the 'Terrors of Vambolent', and nobody dissents. Just last month, the headlines reported 'Space Inventor Suffers Heart Attack at Shuttle Landing Assassination'. A Grand Jury dismissed charges brought against me in a class action suit, but there are still more people who wish me stone cold dead than those who care if I live or die. Okay, so I have changed a little."

"Oh, Parks, I do not think it is all that bad," Dajwal said, speculatively.

"Dodge, we parted ways at NASA HQ years ago. Now, I find myself a has-been. You, on the other hand, are an Associate Administrator."

"For Science," Dajwal said.

"For Science," Parks echoed. "And you, Afren. You are the great hope of the African continent. People call you the Osagyefo, yet you deny the term. Somehow, I feel it could not be otherwise. Someone who *wanted* to be Osagyefo would never become it."

"I do not seek accolades," Afren said.

"And here we all sit, more than 10 years after our first meeting together, talking about the exact same issues we faced early in the century. Today, tar sands and gas hydrates have given us several more years of not-too-expensive energy. Sure, the climate has been hammering people more than ever. But people adapt. They do what is cheap first, and secondarily, what is perceived safe. We've seen the last cheap oil. We're using up natural gas faster and faster. When that goes short, we'll be backed into nuclear power, but with such a burden for safe use and disposal that our economies will wither for lack of energy.

"If, by some incredible act of sanity, we may finally start using solar power satellites and storing our power with hydrogen. But by then, we'll live in domed cities, eating hydroponic plants, worrying over the ever-present threat of a nuclear missile that penetrates our defense shield, and exposes us to the hell which will be our ruined environment. What kind of life is that?"

There, in the Colorado household of the Vambolents, three leaders sat speechless. As Afren had expected, the world had gotten worse off. As Dajwal had surmised, people made expedient choices at the expense of what was best. And as Parks had hoped would somehow change, the inertia of the status quo remained the greatest obstacle to improving life on earth. All of his efforts had been worse than failures. They had been turned against him, growing into unmitigated disasters. The Vambolent home was expressly quite for several long minutes.

"Do you have any good news, Dodge?" Parks asked at last.

"Well, Parks, I suppose that depends upon your definition of 'good'."

"Okay, then. Tell us the bad news."

"Well, I think this is actually good news. Yes, Parks, I believe I actually have some good news for you."

"Alright, then. Let's hear it."

"Well, you know the LEF has been idle now for more than a month, yes?"

"Yes." Images of his 'retirement' party played in Parks' mind. Vacuous speeches proclaiming victory when in reality they had just given up and come home.

"Well, there is talk now of selling that facility. It might get some further use."

"Is that right? Who wants to buy it? The Indians?"

"No, Parks," Dajwal said, looking hurt. "The Americans. Marcharelli Space has made an offer."

"Bob wants to buy the LEF?"

"He and his partner, Winnimer Jenks."

"That guy. Who the hell is that guy, anyway," Parks steamed.

"They have a pretty interesting approach, Parks," Dajwal explained carefully. "They avoid the isssue of 'ballistic transport'. And for power beaming, they are using what they called Sub-Solar Transmission, or SST. The selling point is that they raster the energy across a constellation of relay sats with a switched duty cycle that never gets more intense than the microwave energy coming naturally from the sun. They claim it is too diffuse to make into a weapon. They claim they can serve almost every customer on the globe with a single energy-generating location."

"Sounds like a piece of cake," Parks said, and got up to go get a glass of wine.

Afren followed him into the kitchen, placing a restraining hand on the liquor cabinet. Consulting Mindy's list, he prepared Parks a plate-full of pills, and filled him a glass of water from the RO tap. Dajwal also came into the kitchen. They were now closer together, looking at each other as they hadn't in years. Parks placed his hands flat on the kitchen island, resigning himself that he would get no alcohol.

"So, how far along are they?" he asked Dajwal.

"So far," Dajwal said," they have gotten pretty far overall. However, their negotiations over the LEF are holding them back. It's central to their business case."

"*My* business case," Parks exclaimed, suddenly angry again.

"Parks, please," Afren reminded.

"Are these negotiations going slow, because of you?" Parks guessed.

"Oh, I would never manipulate things like that," Dajwal answered casually. "But there may be times when I withhold help that I could have provided, if I did not believe something to be in the best interest of the organization."

Parks raised questioning eyebrows, wondering what that meant. "What does that mean?" he found himself asking.

"What it means is that there is a window of opportunity here before anything is solidified."

"Oh, I see," Parks said. "You want me to form a competing company, bid for the LEF, and hire Afren's graduates. Is that it?"

"No, that's not it at all," Dajwal said.

"Good, because my wife has left me because I have such bad business sense. My kids are gone because they either believe their mom, or hate my guts. I'm a month from Chapter 7 bankruptcy, and I have a congenital heart condition that means I could die over any given high-stress event that comes my way. So, I hope it's not your plan, Dodge, to involve me in any of your schemes, because either way, I end up either dead, divorced, or both."

"Parks, it's a good thing you do not want to be involved, because there is so much excitement," Dajwal said, his voice a mixture of sincerity and mocking enticement. "You see, the Russians at Energia are developing a manned lunar transport based on the Soyuz. When they strap on solid rocket boosters they pick up at LEO from an earlier launch, they can get to the moon for half the cost we can do in the US."

"So what," Parks said. "The Chinese have manned launch costs down to almost peanuts."

"Yes, Parks, but I also happen to know from my counterpart at ESA that they are very keen – Very Keen – to help fund a permanently-manned base on the Far Side, and operate a deep space radar telescope and a very large optical array to find earth-like planets. They want to extract lunar oxygen to fuel a future Mars trip. And they have talked about bringing you on-board."

"Right. No, Dodge. That's complete hogwash. The ESA knows as well as anyone that I'm a pariah. Any partnership with me would hurt the whole enterprise from the beginning."

"In the US, maybe," Dajwal countered. "But they want you as a sort-of executive producer. Your inventions stand on their own outside this country. In China, you are still something of a hero. Even in Europe, people see you in a more balanced way than you are seen here at home. They like that one of your astronauts was using slag bricks to build an igloo. They want to use this technology to build a base. Parks, you can bring this all to them."

"That's ridiculous," Park said, but his body language indicated otherwise.

"Can they use my laborers?" Afren interjected.

"Exactly," Dajwal said. "The ESA is more interested in performance than nationalism, by definition. So, here's what I've been thinking."

Parks and Afren looked at each other. Both found agreeable attitudes. From independent perspectives, they had both learned to listen to Dajwal Chellapilla with a healthy measure of scepticism. Yet, they had also learned that from his hype often came a useful jingoism.

Reading them, Dajwal spoke on. "So, here's what I'm thinking. Jenks is already helping us with this SST ideal. Next, we need to work public opinion on ballistic transport. Call it something else. I made up this name, see if you like it: Inertial Orbital Transfer Action, or IOTA. Sounds like a small thing, no? And if this time, instead of shooting canisters from the Near Side, always hanging over the Earth, we start from the Far Side – safe and out of the way. We give a gentle luna launch to the L2 point where a recoil-less mass driver sends it spiraling down to GEO. We capture the canisters, using your Payload Catcher invention, but this time the orbit is slow, the angle would not make it a weapon at all. IOTA –what do you think?"

"How can you manipulate public opinion?" Parks asked.

"Not manipulate," Dajwal stressed. "But to re-cast the language. It is often the only way that public opinion can be changed. We use new words with different connotations. With enough help from enough places, and given enough time, we can make some modest movement. If other things go well, then change is possible. We must never give up hope."

He looked straight at Parks with this last, challenging him.

"Forget it Dodge," Parks said. "Any kind of management role is out of the question. But, I do need money. So, if they wished to hire me as a consultant, and I could work from home…"

The phone rang and Parks pointed a finger at it. "Hello."

The face of Yoshi Hanazawa showed on the screen. His looked at Parks, and then his narrow eyes cut around the room. He saw Afren and his eyes opened to their full extent, the whites showing a a thin, broken line. He looked at Dajwal, and gave a head nod, then turned back to Parks.

"Hello, Parks. Mindy told me you would all be there."

'Yeah, and she told me you were the only hold-out. What gives?"

"So sorry. Actually, something important has come up. I will need to come see you after all."

"Great. When will you be here. We're about to order pizza."

"Ginger tofu veggie wraps," Afren corrected.

Yoshi smiled, showing chicklet teeth, widely spaced. "Excellent. I will bring some non-alcoholic sake for you, Parks."

"Ugh. That stuff's bad enough when you just want to get drunk. I'll have beer."

"Quinine water," Afren chided. "With fresh lime."

"I will need a short time alone with you, Parks," Yoshi said, his face dour.

"Don't look so serious, Yoshi. There's nothing in the universe you can't discuss in front of my best friends. Just get your butt over here so we can eat."

Yoshi stared immobile, head-bowed quickly, and signed off.

*   *   *

Yoshi Hanazawa was ambushed at the door by an elderly but playful Afren Kwapili. The two aged judo masters grappled in the hallway, banged into the wall once, tilting a blue-hued Vambolent family portrait; then scooted the settee across the hardwood, until they broke, laughing and smiling. The two men gripped forearms, testing each others' strength at the same time they renewed an old friendship.

"If you two boys are done playing, we can welcome our guest properly," Parks said. He shook hands with Yoshi, and then handed him a paper plate grease-stained with a wedge of cheese pizza. "How are you, Yoshi?"

Yoshi went still again, spooking Parks. Then he smiled, and shoved the point of the pizza slice deep into his mouth and ripped through the bread like a tiger at feed.

"Good, Parks," he mumbled around his food. "You?" Yoshi slap-hit Afren's shoulder, and the latter stumbled into the living room ahead of him. Parks was amazed at the easy comaradarie of the two men, and how very different his Kenyan friend became.

Dajwal stood, looking awkardly straight, and greeted the new guest formally. "Hello, Yoshi. Welcome. How are you?"

Yoshi head bobbed as he swallowed. "Very good, Dajwal. It is very nice to see you again." He placed his hands beside his thighs and bowed. Then, he stuck out a hand to shake.

"Oh, I don't know if I can wrestle like Afren, there. Maybe I should just give you a nice, Japanese bow instead," Dajwal said, and imitated the bow.

Yoshi, for the first time, made a fake lunge towards the smaller Indian man, who flinched mightily. Grabbing his hand, and the opposite shoulder, Yoshi held him, looked him in the eye, and smiled broadly, his eyes narrowing to the finest of black lines. "Good to see you, Dajwal."

"Oh, my, it is very good to see you, Yoshi. Very, very good. When Mindy said you were not coming, I was disappointed. Now, I am very glad you have come."

They all sat. Yoshi picked up his plate and finished the slice in another two bites. The rest of them waited for their host to start the conversation.

"So, you said you have some important news?"

Yoshi grew sober at once, chewing his remaining mouthful with his eyes openly staring at Parks. His Adam's apple bounced and he spoke.

"Yes, Parks. I have a message to deliver to you. Unfortunately, as I explained, I am to deliver this to you in confidence – no disrespect intended to you Afren, or to you, Dajwal."

Parks waved a hand dismissively. "I'm past all that, Yoshi. You can speak freely here. If there was ever a circle of my most trusted friends, it is right here. Thanks, all of you, by the way. This is the most fun I've had in a while."

Yoshi cut his eyes at the other two men. "Parks, I must insist. I am under strict instructions, and following a solemn oath. So sorry."

"You sound much too serious for tonight's festivities," Parks said. "Please, Yoshi. There's not a thing in the world I would keep from these two."

Yoshi frowned, clearly uncomfortable. Several long seconds passed as he decided how to proceed next. "Then, perhaps I will come at another time."

The brows of Parks Vambolent turned inward and down. "Look, Yoshi. We're all having a great time here. Don't spoil it by being a killjoy."

After another pause, Yoshi spoke more softly, his body language stiff and formal now. "I must remind you of a mutual friend. Mifune-san, if he were here, would understand."

Parks grew livid. "That's not fair. Whatever you have to say, just say it. Stop with all the cloak-and-dagger, drop all the hush-hush stuff. Because if this has anything to do with anything we've done before, I'm not interested. Like I told these two, if it involves space, I face either divorce or death. My wife and son are in Africa. My daughter fled to Indiana. I have a few more years to watch the world go to hell, then it's not my problem anymore. So, whatever it is you have to say, say it, or get out of here."

Yoshi waited during the dead silence in the Vambolent living room, waited for enough heartbeats that Parks did not recant his position. He drew a deep breath, and then stood, slowly. "Then, I must leave. I am very sorry to have disturbed your *wa*."

"I will see you out," Afren said, rising.

Dajwal looked crushed. His natural curiosity burned to know more, but in his wisdom he did not interfere.

Parks pulled the handle of his recliner, and forced his way back, effecting a dis-interested demeanor. Afren and Yoshi were just leaving the room when the front door slammed open. He heard the swish of a slide, and watched as his son's backpack slid down the hallway, coming to rest with a thud in the corner of the stairway.

"Eric?"

"Uncle Yoshi!" Eric's voice cracked in his excitement. Yoshi and Afren disappeared into the foyer. "Uncle Afren!" the adolescent voice said, this time with a forced low voice.

Parks struggled getting his recliner to come forward, and heard the exchange from another room.

"Uncle Yoshi, what have you got for me?"

"I didn't know you were here."

Afren's voice: "Eric, where is your mother?"

"She dropped me off, and went to a friend's house. She'll be back in a little bit. Is my dad here?"

"Hi, Eric," Parks said, coming around the corner.

"Dad!" Eric hugged his father around the waist. The boy was still short. Parks felt a patriarchal warmth at being so recognized.

Then, Eric was back on his own feet, and turned back to Yoshi, who he had not seen in many years.

"So, did you bring me a present? Mindy called my cell. She said you got her something."

"I am so, so sorry. I have nothing for you. I did not know to expect you," Yoshi said. Then, he quickly shifted tactics. "I brought something for your father. But he does not want it."

"I'll take it!"

"Oh, really. You don't even know what it is yet," Yoshi said. "Your father also did not know what it is, and he rejected it out-of-hand."

Eric looked at his dad. So did everyone else. Eric stood taller and spoke in a suddenly deep and mature voice. "So tell me, Yoshi. What's the deal?"

The three men who had not grown up in the United States laughed spontaneously. Parks was nonplussed. Eric led them back into the living room, and took his mother's chair, sitting as regally as a sultan prince. Dajwal sat on the ottoman.

"So, Uncle Yoshi, tell me what you have brought for me?" Eric's voice was low, and remarkably charismatic. Afren clamped his lips onto a very big smile.

Yoshi settled into his chair, and took a slow breath. He looked several times at Parks, several times at Dajwal and Afren, and then, appearing to make up his mind, focused his entire attention on Eric.

"What I have brought you is half the resources for you to retrieve a comet. I was instructed to make this offer to your father, but your father has lost his will. He does not want this boon. So, it is now up to me to decide how to dispense with this. Since you are the rightful heir to the Vambolent family, I have decided to make the offer to you. Should you learn and develop the skills to use this fortune wisely, I will bequeath it to you."

Eric treated this as the game he supposed it was, still in his imperious personna. "So, Master Yoshi, you have brought me a fortune to either go halfway out to a comet and back, or enough to go all the way out there, but not come back. Either way, I am not interested."

Dajwal was laughing behind his hand. Yoshi suppressed mirth, but answered in perfect seriousness.

"Maybe you can be more clever. Maybe you can find a way to accomplish this for half the cost. Or, perhaps you will invest this boon so that when you are finally ready, it will have grown to an amount sufficient to this purpose."

"Okay." Parks said. "Okay. Eric is still a minor. I will decide if he receives such a boon." He watched as Eric folded petulant arms across his chest. "If you are to make such an offer to my son, you can make it to me."

Yoshi's eyes momentarily showed mischievious triumph, but just as rapidly became earnestly serious. "Parks-san, my mentor, Hiyate Kindoh, was a great supporter of your work. He spent many hours anticipating your next moves. Before he was so aggreviously murdered, he set aside funds commensurate with a 50 per-cent share of what he expected your next need would be. During this time, he knew that his life was in jeopardy. He took pains to ensure that his desires would survive him. In this, he made me executor of his estate – in more ways than one."

"What does that mean?" Parks asked.

"I am not at liberty to discuss," Yoshi answered. "But I can tell you that Kindoh-san was very precise in his work. Of me, he extracted a solemn promise. So, I must ask you, Parks Vambolent, did Hiyate Kindoh extract a promise from you?"

In the silent moments that followed that bombshell question, his son and best friends watched his face as emotion after thought after emotion played out on that tableau. When the brain behind came to a resolution, they all leaned forward.

"Yes." Parks said. He sat back, glancing at Afren Kwapili, glancing at Dajwal Chellapilla, and then looking hard at his own flesh-and-blood, Eric Vambolent. "Yes, he did. He asked me to build a solar power satellite installation over Africa. The facility would supply power to that continent, but also, through submarine cables, to south-east Europe."

The verbal silence in the room was overwhelmed by the mental machinations of every male present. Men, boy, glanced at each other in a rapid-fire exchange of significance. A different atmosphere formed in the air of the Vambolent living room.

"The Russians can get you to the moon very cheaply," Dajwal said.

"We have trained workers who can operate installations there," Afren said.

"My satellite company can provide an L2 relay station," Yoshi said.

"And you have the patents to make it all happen," Eric added.

In a flurry of talk, exciting hyperbole, and down-to-earth practicality, the four men and a boy fell into a free-fall of ideas. Decades of development came to a head in the plans of a small cadre of individuals whose life-paths brought them to a point where they could create something new and enduring. Yoshi's boon was the critical mass which made it all possible. Eric watched the ideas bounce from man to man like a spectator at a tennis match, but he was in the circle as well. "Can we do this?" Parks said at last.

"Yes," said Eric, and they all agreed.

In the stillness of the decision, the front door opened. Three seconds later, Elise Vambolent stood at the entryway to the living room. Every male became as a stone. She stared every one of them in the eye.

"What. Is. Going On. Here?"

Chapter 16

Switzerland

Black sedans swept past the last parking lot on the mountain. They bypassed the train station feeding tourists to the village of Zermatt, where combustion engines were not allowed. A large sign guarded a gravel road, warning of the alpine environment in three languages. Each car rounded the turn and passed through a settlement 700 years old, where each residence was built over the top of a stall housing the family ass.

Hard, wiry, ancient farmers looked up and scowled as the progression wended between the beam and plaster homes tucked within the arms of a primordial mountain, holding their valley safe. Stunted trees dotted the grazing pastures where horned sheep moved like cumulus clouds.

Waiting at the far end of the settlement, behind a stony outcrop, and out of sight for the tourists, an electric bus awaited them. From the second limousine stood Winnimer Jenks, a long black coat blowing open in the sun-warmed alpine air. Stepping from the lead car came Noelle Wehrlichs, a tall woman with full brown hair waving around a face of chiseled classic beauty. She flashed a smile at Jenks, then reached in to offer a hand to her mentor, Gentian Banderberg.

Banderberg served as host tonight, the newly-promoted leader of their group.

The vehicle listed as Banderberg shifted his weight closer to the door. His thick legs swung out first. A broad hand with sausage-like fingers gripped the frame as he heaved his bulk vertical. Once on his feet, he was steady and nimble. Banderberg shrugged heavy shoulders, stretching after the long ride up the mountain, and rested a meaty arm on the roof of the car. His driver emerged from the other side, and walked over to the bus.

Jenks pulled Valerie Schzecherman from his own vehicle. She stood up, pulled an overnight bag to her shoulder, and froze at attention, facing away from the vehicle. Jenks did likewise. Their driver aided the aged Count Hilbert to his feet. Hilbert, board chairman for DMBG, the largest industrial conglomerate in the EU, lurched upright, aided by power-assist mechanisms hidden under his well-tailored garments. Once walking, his stance and gait were indistinguishable from normal. Yet, his struggle up against gravity betrayed evidence of his prosthesis, and he was loathe for anyone to see it.

Hilbert walked around the car, head held proud, drawing in the thin, clear air through his narrow nose. He choked suddenly as rock dust irritated his sinus membranes. He gave out a hacking cough, then bent his head and spat a small wad of brown goo into the rocks.

"Pardon me," he said, covering his mouth as if to hide a cough, and then coughing politely. "Auspicious weather!" he said, sweeping a hand across the striking panorama of raw nature all around them. Departing clouds had dropped dew everywhere, and left the atmosphere charged with energy. The sun was painfully bright, sparkling through the dewdrops and turning them into wisps of steam that would form clouds in the night, lower down the mountain.

"Indeed," Banderberg called over. "Powerful forces come together at the equinoxes."

The two men converged in the tiny road, shook hands briefly, and led the way to the bus. The rest of the party followed the two great men at a discrete distance.

Jenks stepped between the two women and offered each an elbow. Both hesitated for the same amount of time, but for different reasons, then slipped a hand in the crook. They watched the last few clouds slide down the valley. From their underside, shimmering rainfalls of water fell in banners of lighter grey.

"I'll bet someone can see a rainbow over us, if they were standing up on that mountain," said Schzecherman, pointing.

"That must mean we're at the pot of gold right now," Jenks quipped.

Werhlichs laughed a throaty laugh that pierced Jenks to the coccyx. She pulled her breasts tight against his arm, and he felt a tingle of heat race up it. It electrified his heart. Then, she let go of him altogether, and he felt suddenly disoriented, aching for her to touch him again. Jenks made a mental note to guard against her sexuality in the future.

They climbed aboard the bus, a low-ceilinged khaki-colored six-wheeler with nearly a meter of clearance for the undercarriage. Electric lifts brought each passenger up to a comfortable level so they could step in, head bowed, and take a seat.

Banderberg's driver clicked a key to ON. Without noise, the bus began to move. They turned uphill onto a rock-strewn path. At times, the road was barely wider than their ATV. A hands-breadth to the right, and their bus would have rolled off. The driver never seemed to slow for these dangers, coming to a stop only for a goat to get off the road, or further along, to get out and move a large rock that had tumbled onto the roadway.

Around one last bend and up an even steeper grade at the very end, they came to Zermatt. They were on a broad avenue of packed cobblestones, in what appeared to be the back edge of the village. The buildings faced away from them, in towards the town center. The only other traffic was comprised of utility trucks: small, green, electric totes, carrying cargos of stones and refuse in opposite directions.

A sudden downpour cascaded from the sky, pelting against the roof of their bus. They trundled on, followed the road down slightly, where it disappeared inside a sublevel garage cut from the rock of the mountain. They entered, and took the ramp to an even lower level. The driver stopped before an elevator door marked Hotel Alex, and his passengers disembarked.

Banderberg pulled from his pocket a remote control, which he aimed at the elevator door. It opened immediately, revealing a spacious cabin with tinted mirrors and gilt trim. A smooth ride brought them to the lobby level, where giant blocks of native stone held up the floors above, and massive timbers separated each room. Each wall was painted with mountain scenes, and the floor was an intricate inlaid stone path that wound around into a bar, and back towards the pool and spa, smelling of chlorine and Italian salt.

"Welcome to the Hotel Alex," Banderberg told them. "Your keys are being held at the front desk under your names. Would anyone like to freshen up before we go?"

All shook their heads except Schzecherman, who looked to her mentor.

"I will be taking my leave now, Count." She gave him a forlorn look, flashed a glance at Wehrlichs, and walked around to the front desk. Most of the plutocrat member had lieutenants or protégés, but only one could attend each solstice or equinox meeting. Tonight, Wehrlichs attended as Reciter. She made a passable imitation of not appearing smug.

Jenks was there as an acolyte. He knew there was another, but not the man's name. It was fairly certain the other acolyte was Japanese, to take the place of Hiyate Kindoh.

"Well, then. Everyone ready?" Banderberg asked. He aimed his remote at the elevator again, and it opened. They stepped inside. This time, Banderberg entered a series of numbers on his remote, finishing with a flourish. The elevator started back down, and continued for a long time. It came to a gentle stop, and the doors opened.

They were inside a stone cavern. The air was cool and slightly musty. A single bare bulb lit the landing. When the doors clicked fully open, the sound echoed.

"Come in, come in," Banderberg boomed. "Enter the hall of the mountain king!"

Count Hilbert walked directly to a four-seater open rail car parked directly across from the elevator, and sat down in the front. Jenks and Wehrlichs sat in the back. Banderberg aimed his remote at the elevator, and the doors closed, leaving them alone, deep underground.

Leaning heavily on the handrails, but not stumbling, Banderberg lowered his great mass into the open front seat, behind a windscreen. The leaf-springs creaked under his weight, but worked perfectly – one of mankind's more practical inventions. He tilted side to side to settle himself to his liking.

"I'd let you drive, Count," he said, "but this is the only excitement I am likely to get this evening.

After checking his passengers, he grabbed a lever and pushed the stick forward. They started down the track. Left behind was the glare of the lone light, and they disappeared into darkness. A cold wind flowed around them as they glided along the smooth rails. Even the clack of expansion joints was missing in this environment where the temperature never changed and the rail segments could be welded together.

"What do you think, Winnimer?" asked Banderberg. In the darkness, his big voice seemed to originate inside Jenks' ear.

"Quite a ride, Mr. Banderberg."

"Yes, quite. I always leave the lights off for first-timers. What does it feel like?"

Jenks thought a minute, keenly aware that the formality of his admission to their brotherhood tonight depended very much on everyone's impression of him.

"I can't tell if we're really moving," he said. "Maybe the car is now at a stop, and you just have a fan blowing wind in our faces."

"Excellent!" thundered Banderberg, laughing so hard the railcar shook. "Excellent. It would be hard to fool someone so astute."

Suddenly the lights turned on. Jenks' eyes had been wide open, and he squinted against the glare. The line of sconces converged forward into the indistinguishable distance, where far away, they disappeared around a slight bend. The bend approached them rapidly, and Jenks suddenly realized how fast they were really moving. He felt the bitter bile of stark terror rise high into his throat.

Jenks half-listened to Banderberg ramble on about his Zermatt holdings, and how he acquired the tunnel. It was started almost a hundred years before as a secret passageway from Switzerland to Italy. Jenks was keenly aware of Wehrlichs beside him, because the toe of her crossed leg kept tapping his calf randomly. Jenks gave himself a pep talk inside his head, calling upon his usual confidence and bonhomie. It would not do for these men to see him uncomfortable and out of his element.

He watched the two men in the front seat, trying to absorb some of the calm, disdainful ease they portrayed. It would be simple without the woman distracting him. He wondered if this was part of the test, and resolved to find a way of disarming her.

"Ah, here we are!" Banderberg announced.

They had entered a tall space, and coasted smoothly to a stop. It was more a massive fissure than a cave, extending upward into the darkness far, far above them. The walls were natural rock, unscored by dynamite or drill bits. The two long sides tilted slightly, not parallel, but converging towards a line that would be a thousand meters up in the darkness. The skew to the room conveyed the gut sense that the whole scene could fall over, despite the logical inconceivability. A landing ran along the rails and opened onto a wide semi-circular area of flat rock that formed a small amphitheater. There was a simple stage of plywood on aluminum legs, and stacks of folding chairs rested against one wall. A couple of darkened spotlights that looked like WWII era equipment rested on stands next to two side steel doors that probably enclosed a storage locker.

Banderberg opened his throat and warbled a few basso notes into the chasm. They returned, magnified, feeding on themselves and filling the space completely with sound.

"I love doing this," he told them, and tried a few more notes. He seemed to be experimenting with the resonance.

Jenks was just about to thrum a tenor harmony when Count Hilbert broke the veil of vibration which surrounded them. "Come, Gentian," he said, scolding. Jenks looked at Noelle, and saw that her neck was stretched as if to sing, and he felt something catch in the tendons of his groin.

Banderberg, unfazed but complying, pressed an inconspicuous button on the near wall. From high above them, there was a metallic click and the slight whine of a winch unrolling. They all looked up, and for a half minute nothing happened. Then, they saw a warm yellow glow that crept down the high black walls, illuminating both sides. A red elevator car then descended into the lights of the floor level. Braided steel cable ran through a large pulley on the elevator's roof. The car came to a precise stop just off the gritty floor. It swayed very slowly.

"Gotta get my damn engineers to eliminate this swing," Banderberg grumbled. "Sorry folks, but only two at a time. Count Hilbert? Would you and Miss Wehrlichs like to go first?"

"If it's going to break under your weight, Gentian, I'd rather be trapped down here with Noelle. I insist. You go first."

"Of course, Count. We shall see you upstairs," he motioned to go. "Jenks?"

Jenks walked over, opened the latch, and entered as he had been taught to enter a canoe. Banderberg swung the door open wider, and when he stepped in, the cable stretched. The bottom of the elevator car screeched and crackled against the floor. After adjusting himself, he closed the door. A round black button protruded from the door frame. A gargantuan thumb mashed it.

They reeled upward, and soon the landing was far below. Banderberg's tone changed completely.

"I hate that old bastard, Hilbert. He's hardly even a Count. His family lineage is noble, that is true, but he's more like a Baron. A 'landed baron', if you will. It is said that he bought the title by funding a private little war for the secret government."

"Isn't that the only way?" Jenks asked.

"Why yes. I suppose that's true," Banderberg laughed. "Maybe *I* could be a Viceroy. Ha! How about that? And you, Jenks. What title would you prefer?

"I'd be a Governor, say, of an island."

"Oh, that's rich," Banderberg chortled. "Very rich. Ha, I've punned again!" He looked over at Jenks. Sitting, his eyes looked like little pig eyes. Standing, his face was much more open and easier to trust. "I like you, Jenks. You've got class. Frankly, we need more of that in our group. Things are getting stale. Old men with old ideas need to step aside. Don't you agree?"

Jenks looked at Banderberg, wondering cautiously how much of the banter was truly spontaneous, and how much was carefully calculated. What could he be insinuating? Think?

Banderberg caught the hesitation, and changed his demeanor again. "Oh, I forget myself. You lost your mentor since our last meeting. Bruno was a good man. I'm so sorry."

"Thank you," Jenks said.

"But his departure has left a space for you. And then, with your skillful editorial work when our presence was suspected after the assassination, you've certainly earned the spot. Hilbert is grateful. You saved his reputation. His vote will be with you for sure. And as Z, he carries the most weight."

"He would ascribe that quality to you, I think."

Banderberg narrowed his eyes at Jenks, then decided suddenly it was a joke, and that the joke was really against Hilbert, without being. Banderberg laughed heartily, and clapped a thick hand on Jenks' shoulder.

"You'll do just fine here, Jenks. Just fine. Now watch your step, it's a long way down."

They reached the top of the lift, and Banderberg slapped open the door latch. He maneuvered out, and the car lurched as he exited. Jenks was keenly aware of the great open depth beneath him, and climbed out quickly. Banderberg touched another button on the wall, and the car began to descend.

In the foyer, Jenks noticed artwork by grand masters, and never doubted they were authentic. The rock walls were textured, with patterns of grooves cut into a polished surface of gneiss. An enormous Persian rug covered the floor, its colors and design complementing the Impressionist paintings on the walls. From the next room, they heard animated voices and the clink of glasses.

"They can find their own way in," Banderberg said into the chasm, and crossed the room with Jenks at his side.

Inside the reception room, Jenks greeted the principals of the organization, whom he had met during his turn as Reciter in France just three months ago. They were all men, all tall, and most quite old. There were ten true members. The two spots left by the deaths of D'Argeste and Kindoh were to be voted upon this night, at the Equinox Meeting.

Count Hilbert and Noelle Wehrlichs joined the group a few minutes later, and they were complete. As host, Banderberg allowed time for them all to fawn over Hilbert and seek his favor for the projects to be reviewed. Others talked with Jenks and the second new candidate, Yoshi Hanazawa.

The two knew well of each other, Jenks and Yoshi, but had never met. Each eyed the other across the room as they were chatted up. Jenks had a certain feeling that the Japanese man would not be admitted as a replacement for Hiyate Kindoh. Something in the body language of the plutocrats he spoke with made Jenks guess that Yoshi was being considered as a politeness, out of respect for a dead man's wishes.

The maitre d' made a discrete signal to the host.

Banderberg cleared his throat loudly. Flesh under his chin quavered. "Gentlemen, Lady, your dinner is prepared. Please join me in the banquet hall."

The groups of people filed in slowly, and took their places at the one huge, square table. Hilbert sat in the center of the North side.

Banderberg pointed to Jenks' seat, reminding him in a stage whisper that the seat was not truly his yet.

Jenks stood behind his chair, feeling chided unfairly. Banderberg sat down, took one look at Jenks, and waved him down into it. Then he laughed out loud.

As the guests settled in, waiters filled water goblets and offered choices of wine: Merlot, Chardonnay, Riesling or sake. A thin soup with stuffed dumplings was brought next, in shallow bowls of bone china. Hilbert nodded to Wehrlichs, who sat, not eating, at a small table by the side of the room. As Reciter, her role was to relay news items of interest to the group.

She gave a small cough, and the table conversations quieted. All dozen pairs of eyes were on Wehrlichs. She smiled demurely, but with a mischievous look in her eye.

"Our first piece of news regards the longevity drug, Extensar, which was denied approval for distribution in the US. As you know, Extensar has been available in the EU for years. What makes this of interest tonight is the hostile takeover in progress. Blaxer, which holds the patent on Extensar's telomere preservation properties, is the target of Kelso, the US pharmaconglom. Our analysts suspect a collusion between Kelso and certain elements within the FDA. However, our sources have not been able to confirm or deny."

"Does anyone have comments which bear upon this matter?" asked Hilbert, sounding officious in his first role as Z. All of their roles had reshuffled, peaceably – by lots, but weighted by seniority. Elder members got three chances to deny, while newer ones got two, or one for neophytes. Hilbert had been there for nearly fifty years, inheriting the family's concerns at the young death of his hideously wealthy father.

"This is clearly economic protectionism on the part of the US of A," said an aristocratic-looking man of Mediterranean lineage. "The denial is a stalling technique. Of course, Blaxer will contest the decision. But during the delay, you can be assured that the US firms will be working overtime to beat the patent. And the proposed takeover is their way to hedge their bets. What of the Justice Department? Do they remember their own Taft-Hartley act? Kelso plus Blaxer would clearly be a monopoly."

An elderly gentleman, seated at the ZX spot on the table spoke next.

“I take Extensar. There are no side effects. I would not even know it was working, except that my doctors tell me vascular degeneration has stopped, and mental acuity is as sharp as ever.” He looked about suspiciously. “Does anyone else take Extensar?”

“I do,” said Banderberg. “But I’m more concerned with knowing who is calling the shots with the FDA. If there is a hidden source of control, we must discover it at once.”

“What you *should* worry about,” said Hilbert, sitting next to him. “Is control of your weight.”

Banderberg laughed, and it almost sounded genuine. “My dear Count Hilbert, my philosophy is to eat, drink, and be merry – but to invest heavily in life extension.”

“Are you taking stem cell injections?” asked one of the younger members.

“Certainly!” boasted Banderberg. “Plus cholesterol-lowering pills, ginkgo biloba, shark cartilage, an antioxidant complex, and a handful of other goodies I can scarcely pronounce. Why, I probably use every method known to man!”

“Except one,” piped an innocent-sounding voice from Noelle Wehrlichs. She had been routinely coached not to offer any comments except for news at the command of Z. Several members narrowed their eyes at her, but most were curious to hear if her interruption were worth the breech of protocol.

“And what might that be?” asked Hilbert, for Banderberg.

“Caloric restriction,” she said simply, and shrugged her shoulders. Her hair responded in rippling waves of shimmering auburn. “Scores of studies have shown longer lifespans in laboratory animals fed only a subsistence diet.”

“Ah, but what miserable little rats they were.” Jenks interjected, completely poised for his moment, when the moment finally arose.

Banderberg nearly choked, laughing. Gobs of laughter rolled off him and his fat palm slapped the wooden table. Wehrlichs managed to shoot eye-sparks back at Jenks while smiling beatifically, apparently taking it in stride as a good sport. Laughter around the table dissolved the tension which had started to escalate.

Waiters cleared the soup bowls, and refilled wine and water glasses. Hilbert signaled again to Wehrlichs as they began to bring out the vegetable dish.

"Item two," she announced, "regards the environmental conflicts arising in the Asia theater. There are three airborne pollutants which cross national boundaries in the prevailing winds. First is smoke, also called 'haze', from cooking fires and forest clearing. Second, illegal logging and legal over-grazing in China has fueled desertification, so desert now consumes 30% of their land area. Wind storms have carried yellow sand and dust throughout Eastern Russia, shutting down airports, seaports, and filling their meager hospitals with respiratory distress victims. And finally, coal burning in China, despite assurances of proper scrubbing, is still pumping huge amounts of sulphates and nitrous oxides into the atmosphere. These precipitate as acids on the islands of Japan, where they have poisoned rice paddies, and denuded cherry trees.

"The Imperial Japanese Navy is believed to have deployed missile subs arrayed along the Chinese territorial boundary. Their objective seems to be surgical strikes at coal-fired power plants. It is another example of the martial forces in Japan taking authority over traditional government roles. There is talk of a feudal system re-arising, based on legal rights enforced with lethal force."

"Does the Chinese intelligence community have this information also?" asked a member.

"We do know that their recon sats can detect sub wakes. However, several of the Japanese submersibles use a lateral fluke, which makes no wake. The Chinese are almost certainly aware of *some* submarines, they may not know the full extent of the buildup."

"Hanazawa-san," said Hilbert, "did you know of these developments?"

"My position as JAXA Science Director does not allow me to learn of military operations. Yes, we helped design the fluke subs, but the ship building and operations are left to the Navy. I do also know that the Chinese have stealth subs of their own" Yoshi said. "Kilo class. Russian-built diesel electrics. They are old now, but no less stealthy for their age."

"Tell us how this news affects us now," asked Hilbert, then took a bite of vegetable.

"Our analysts believe," Wehrlichs said to his obvious prompting, "that the assassinations of Messrs. Kindoh and D'Argeste were conducted by a Japanese extremist group – former Special Forces units. Kindoh had publicly offered engineering talent to help re-build the Three Gorges Dam. Because of the state of war, the offer was refused, but it might have changed the course of history. Anyway, these anti-China groups have been making death threats to him for years."

"What about Monsieur D'Argeste?" asked one man.

"We believe his death was incidental," Wehrlichs said, and looked to her mentor for confirmation.

"And the coincidence with Parks Vambolent? We all know he was far too pro-space. So was D'Argeste," said a sandy-haired man with a woolen jacket.

"Just coincidence," Wehrlichs said. "Somebody had to be nearby."

"And you, Miss," said the Mediterranean gentleman, "are presumptuous to assume that anything about these events is coincidental. I suggest your 'analysts' study the matter in more detail."

There were nods of agreement around the large, square table. Not one of their ranks had attained this auspicious level of influence and control without creating enemies – armies of enemies – along the way. The seemingly simple rail ride and elevator trip to the redoubt were thoroughly sensored, heavily booby-trapped and could repel almost any attack conceivable. The rooms themselves were high in the solid rock of the Matterhorn, unassailable to nearly any weapon known to man. Still, the plutocrats had discussed for several years the option of holding virtual conferences. For all of them to meet in one place jeopardized the continuity of the millennium-old organization they ran.

"As you wish, sir," she said, head down slightly.

Jenks smiled. He did not smirk, but was all smirks inside. She watched him, very nearly stared at him, the whole time he ate. He had rarely felt so good.

The main dish was salted beef in a creamy wine sauce, served over egg noodles, and garnished with slivered almonds. The men devoted themselves to the food, and after the first precious mouthfuls, several conversations broke out around the table. Jenks watched each person in turn, guessing at what motivated them, what each feared most. Now that he was seated at the table, although not yet a full member, he began to see that these men he had sought so long were no different from other captains of industry he had met. Of course, these men were far wealthier than most, and many of them had long lineages, either in royalty or in the field of commerce. That continuity, which gave them staying power, could also be their undoing. The world is changing faster every decade, Jenks thought. If this group doesn't learn to adapt, it's power will fade away. They need a leader with vision.

"Miss Wehrlichs," Hilbert called as the plates were being finished up. "Another news item, if you please."

"Of course, Count," she said, beaming for him. "Ahem. We have several new developments in the Russian proposals for a union with Canada."

"Those are foolish rumors," said a wide man with mixed Mongol and Slav features.

"The US would never allow it," said Jenks, who had been in a position to know.

"Yes," said Wehrlichs smoothly, "that is undoubtedly true, and that is why Russia has been very careful not to antagonize the US. However, there are a number of seemingly unrelated developments that form a pattern we just recently discovered. We believe Russia is planning a surprise attack."

"What?" shouted several people together.

Hilbert hushed them, but it took several seconds. All heads turned squarely on Wehrlichs.

"As you no doubt realize, the Arctic polar ice cap is excellent cover against satellite surveillance. However, we have substantiated reports that the Russian Navy has launched a moveable undersea base from Murmansk. This submersible base, which has four levels, contains over 20,000 square meters of floorspace. What has helped the Russians elude detection of this project has been the modular nature of the work. Multiple shipyards along the coast of the Barents Sea have been assembling special cargo ships. From overhead, these ships appear to be normal freighters, but once out at sea, they can be docked together. The lower portions of the mated ships can then submerge, while the shallow upper decks remain afloat. These decoy freighters can further help by hiding the powerful wake of the base until it can move under the ice cap."

"Do they have nukes aboard?" Banderberg asked her.

"No sir, there is no radioactive signature. We believe the base carries troops, rigid hull inflatables, CnC gear and sufficient ammunition to commandeer islands and oil rigs in far northern Canada. At this time, we do not know where the moveable base is located, nor how many others like it may already be assembled or built. Yet, with a sufficient number of these bases, we believe that Russia could occupy most of the lands north of the Arctic Circle, about the 67th parallel. This land mass, while a tiny fraction of Canada's total area, constitute nearly 40% of the newly discovered oil fields."

"Did you know of this, Dimetr?" asked Hilbert.

"The Northern barbarians have their own agenda," Dimetr said. "There has not been one of their kind at this table in 115 years. Who knows their desperation. It is very cold in Siberia. Oil is expensive like never before. How could they not?" The waitstaff had cleared the main course, and brought out Brunsli and fresh fruit for desert.

"What is the expected response? US response? UN response?" asked a member named Toloisk, wearing round thick glasses.

"Well, dismissing Canada's military, and given the slow workings of the UN Security Council," Wehrlichs said, "the US would be first respondent. If the Russians execute, the US could have bomber squadrons to Ellesmere Island in 6 or 7 hours. However, the Russians are sure to expect them. Their most likely strategy, according to our analysts," here she looked at the man who had maligned them earlier, "is to strike several points simultaneously, hold the locals hostage against a bombing attack, and then petition the UN for diplomatic negotiations."

"They'd never dare!" said a man with shiny, steel-colored hair. "There would be sanctions at the very least!"

"They'll have oil," Hilbert said. "The most valuable commodity of this entire century. It won't be cheap, but it will be theirs. That mobile, undersea base is almost certainly laying down the return pipeline as we speak."

The room was a stunned silence.

"Thank you, Miss Wehrlichs. Now, if you'll excuse us…" he said. She left the way she came in, the giant metal portal pounding closed behind her. The waitstaff shut their access door one last time, slowly, but with finality. The latch caught and clacked. The evening was about to begin.

* * *

Hilbert and Banderberg led the group into the next room, a library eight meters tall, lit by a true chandelier, and surrounded everywhere with dark wood and leather bound books. It smelled of age and permanence. The bulk of the room consisted of 10 overstuffed chairs arrayed around two others. An overhead spotlight was centered on the two chairs. The air was warm, but there was a cool draft underfoot.

The men gathered around the bar where Banderberg poured brandy from squat bottles into balloon snifters, and handed them out. He had opened his spacious humidor, and several men gathered around to select a cigar.

"Gentlemen," Hilbert announced when they all had brandy, "we have a solemn duty tonight. Two of our brethren have fallen. Hiyate Kindoh-san and Monsieur Bruno D'Argeste. To them, we toast."

Twelve glasses were raised in unison, held up against the light of tallow candles. The amber liquid shared by each glass shimmered as if part of the same overhead river. They swirled and sipped, then Hilbert bowed his head. For a full minute, the dozen held reverently silent.

"Would anyone like to say a few words for our lost comrades?"

Banderberg cleared his throat, jostling his wattles, and delivered a long speech on Kindoh. He paused, then gave a similarly long speech on D'Argeste. Jenks learned much from the recital, despite his own intense study of their lives and accomplishments. After another respectful pause, Banderberg turned to Jenks.

"Winnimer, you were mentored to Monsieur D'Argeste. Could you say a few words in his honor?"

"Of course," Jenks said, and cleared his throat as if it were on the threshold of being choked by tears. He was sharply aware of them all staring closely at him. This group would not be swayed by most of his usual mannerisms. He had to step up to the big time and do it right now. Think hard, think through, and think beyond.

"Bruno, er, Monsieur D'Argeste, was like a father to me," Jenks began, and there was thickness in his voice. "He was more than a father. He was a friend. Bruno had so much energy, so much enthusiasm; such a drive to succeed, to make the world a better place. He brought out these same qualities in others. It was infectious." Several were nodding agreement. Jenks took heart. "He wanted to live life to the fullest. To accomplish each day some new dream or idea. He was constantly thinking up new business plans."

Banderberg was smiling, his eyes lightly closed as if remembering just such a scene. He inhaled through cavernous nostrils and put a large hand on Jenks' shoulder. "And what were his last words to you, Winnimer?"

"Our last conversation of substance was on economic growth models," Jenks said, the perfect acolyte. "But his last words to me were: 'Watch this!', and he proceeded to proposition a young lady at the hotel bar. He left with her, and that's the last time I saw him."

Most of the plutocrats chuckled knowingly. Jenks calculated that a majority vote would already put him over the top. But in any group like this, there is the alpha male. Was it really Hilbert? Or did Banderberg carry the most weight when push came to shove?

"And you, Hanazawa-san. You were mentored to Kindoh-san, a relationship that lasted many years, I hear," Banderberg said. "What can you tell us of his last days."

Yoshi set his brandy glass on a bookshelf and stood at parade rest. "My master was a great man. He left his mark on many, many initiatives both inside Japan, and around the globe. Most of his work will never be known by the world at large. Yet, despite his good intentions, he knew for a long time that there are those abroad who prefer him dead to alive. Why, I do not know. But my final promise to him, made long before his death, but anticipating that event, is to avenge his murder, and bring his killers the justice they deserve." Yoshi stared pointedly at Jenks.

"A sentiment we all share," Hilbert said, nodding. "Gentlemen, I needn't remind you that their killers are still unknown, and therefore still at large." He paused to look around at them, reading their expressions. "We are all in danger until we do the needful. Do you all have the spine for it?"

A thin man with tightly frizzed black hair stepped forward. "I suspect the Islamic fundamentalists. This man Hababek has whipped his followers into a frenzy. They have declared *jihad* against all Western values and Occidental meddling, as have so many before them. He has the rank-and-file mobilized to carry out his directive, but their weapons are limited. All they have accomplished is to frighten off the tourists who are so important to their economy. But using the principle of *zakat*, the payment of taxes outlined by the Quran, he uses the guilt of sin on wealthy Muslim businessmen. Thus he exacts this holy tax for the work against the 'Agents of Satan'. While gathering wealth itself is not considered evil, any proceeds extracted from an un-balanced business deal would weigh down their souls on Judgment Day. Hababek calls it his 'sacred duty' to relieve them of this terrible burden. As a result, he controls billions in US dollars and gold reserves."

"What does he do with all this money," asked another.

"My analysis group believes he has been paying scientists from the Ukraine to develop launch vehicles, large ones. There does not seem to be any nuclear developments within Egypt. Perhaps they are content to send conventional explosives to highly-specific targets anywhere on the globe?"

Jenks spoke up. "Won't the US missile shield be quite effective against a single missile?"

"Of course," said the thin man. "But US interests are spread throughout the world, yes? We suspect he will hold hostage a city, somewhere, perhaps to leverage firepower, perhaps for religious goals. We do not yet know his plans."

"He is a small player," Hilbert said, "but we should keep abreast of any developments. Golen, will you please provide us periodic updates?"

"Of course, Count."

A gaunt man with hooded eyes stepped forward. It was Toloisk, but his demeanor had changed. He seemed taller, darker. A ceremonial hat covered his pate, and he clasped his hands in front of his stomach and addressed them all in a formal tone.

"Gentlemen, I would like to announce a question and answer period for our two new candidates. I shall serve as moderator. Mr. Jenks, Mr. Hanazawa, will you please be seated?"

He indicated the two separate chairs, and they sat. Hanazawa settled comfortably, but Jenks sat ramrod straight, hands on his thighs. It was a pose he learned watching US presidents.

Most of the early questions went to Hanazawa, who responded quickly and intelligently, obviously very well-informed. Their curiosity over political developments in the land of the rising sun were inexhaustible. His answers seemed to spontaneously generate new questions, and the bulk of the questions went to him. Yoshi told them there were even rumors of a military coup, with a senior commander or minister to take the ancient title of Shogun, undisputed ruler of all Japan.

There was a pregnant pause. It occurred to Jenks that they were waiting for him to relax. He sat back in his chair, and was at once fired a question.

"Mr. Jenks, what is your position on the development of a space-based economy?"

The question did not sprung from the thread of the interview, but Jenks had been expecting it.

"I see this issue very simply," he began. "Economic growth is always good. Development of space is an eventuality. Therefore, I am fully in favor of it." He paused, searching the audience. A few heads nodded assent and there were some murmurs of agreement. "However, the pace of that growth must be managed. Growth too fast risks internal collapse. How can we forget the Tech Bust of '99? The repercussions are still felt today! We must have steady, slow growth. This is the only sure way to build a stable base for production as well as consumerism. The space market is still poorly-defined, otherwise we'd already be out there in force. It makes sense, but let's not get ahead of ourselves."

"Aye-aye," said one.

"And you, Mr. Hanazawa?" directed Mr. Toloisk. This was his first interjection, and it seemed to Jenks that they were now at the topic that would decide their induction.

"Mr. Jenks makes a compelling argument," Yoshi said, and looked genuinely impressed. "He is a good student of history, and I believe it was George Santayana who said if we do not learn from the mistakes of history then we are bound to repeat them. Moving into the future well means not forgetting your past, and keeping your present actions aligned with a suitably favorable future state. In my opinion, the time is right, right now, to make a full-scale effort for space development."

Several of the plutocrats coughed on brandy fumes in alarm. Yoshi continued.

"Our time is running out. The last man who didn't believe in global warming, honorable Monsieur D'Argeste, is now regrettably gone," Yoshi continued, on a roll. "But with his passing, I hope you all realize that the entire fate of the world hangs in the balance. Not politically, not military, but from our own waste. Humankind has soiled this planet, and we are overdue for a mighty correction by the ecosystem. If we do not act now, it may be many generations before we can try again."

"That is certainly a gloomy account of things," said Count Hilbert. "Jenks, what do you think, and also, given your voice on the Business Review, how do you assess public opinion about these issues."

"Mr. Hanazawa's views represents much of the popular opinion, fueled by that teacher-scientist from Florida and his impending deadline," Jenks said. "Yet, he neglected to state that another large segment of the global population is very much anti-space. These negative feelings are the lingering after-effects of so many failed promises. We need a step-wise approach, with each plateau self-sustaining. We must build the markets over time, finding niches which can be exploited by the unique environment and location of orbital factories and stations. Public opinion must be brought behind these projects to support them. That will take time. The technology will take time. Time and money. Neither are inexhaustible sources. Keep the long-term focus, but don't neglect the short-term. We've all still got to eat."

"Bravo!" said Banderberg, and raised his snifter in a toasting gesture to Jenks.

"Mr. Hanazawa?" prompted Toloisk.

"The common people do not know what is in their best interest," Yoshi said, and looked from face to face around the room. "Individual people are greedy, selfish and short-sighted, on average. Collections of individual people share these same properties. My mentor Kindoh-san once said it is the greatest blessing of the human race that a group of men, as he hoped this group exemplifies, can operate above these traits and work for the common good. Only in this way can the individual people be saved and developed into consumers. Consumers who can help to grow markets, once they are shown the way. It is we who must open up space. Grab the infrastructure, and everything that follows will depend on us. It is a once-in-a-civilization opportunity."

Several stunned faces betrayed agreement with his premise.

"Mr. Jenks?" Toloisk arbitrated.

"How many civilizations have been destroyed by new technologies? If people are truly as horrible as you make them out to be, then consider this: if we open up space to every sociopath with an axe to grind, what do you think we'll get? Anarchy and piracy, that's what. You can't turn people loose without controls. With a carefully developed plan, implemented at an 8 to 12 percent compounded annual growth rate, full economic realization is delayed at most a few years over a more aggressive growth, but the likelihood of sustainability is much, much greater. As the greatest good for the greatest number, it is in our interests, as a group, to nurture these developments at a sustainable pace. Sometimes that means we have to go faster, and sometimes it means we have to go slower."

Count Hilbert broke in: "As demonstrated again and again by that American inventor, Parks Vambolent."

"Exactly," said Jenks.

All eyes looked at Hanazawa. He knew he was out. He was making a point. And, he was making a very big gamble. The answer he gave was practiced, and sounded so.

"Power generation systems present the greatest economic opportunity for space-based development. Yes, the dangers have been demonstrated, and we have already learned from that. Failsafes and multiply-redundant designs managed by a military-style organization with stringent security measures at all levels will be *de rigueur*. Orbital installations will be under the protection of a single governing body, UNOOSA, which will have authorization to build a Lunar Polar Observatory and have a fleet of space-planes and passenger rockets available to oversee all cislunar orbits. With adequate power, an entire cislunar economy suddenly becomes much cheaper. Entrepreneurs will find opportunities to profit from our products. This organization can help lift humanity from the planet, and give us a foothold to the stars."

Hilbert asked the last question: "What about Vambolent's ideas, Master Jenks?"

"There's a fundamental flaw," Jenks said at once. "Large-scale power beaming will always be politically unacceptable. It's fundamentally prone to abuse. What we need is a system that is fool-proof, idiot-proof, and damn-fool-proof. Robert Marcharelli Jr. has reactivated an idea from the University of Houston, Clear Lake where we send small packets of energy to relay satellites that bounce it down to individual peoples' houses. No packet is ever large enough to harm even a sparrow. It's inherently safe. It takes a little longer. But it will work."

"Very convincing," Hilbert said. Banderberg beamed at Jenks.

"Mister Hanazawa?" prompted Toloisk.

"Parks Vambolent is a great man," Yoshi Hanazawa said, sitting up straighter and becoming formal, expressionless. His heavy face relaxed and his eyes were smoky points. "Others try to tear him down." Here, Yoshi let his eyes slide over to Jenks. "When any body of people, or any individual will stop at nothing to undermine a rival who is further ahead, the market is spoiled for everyone. The pie is smaller. If instead, we cooperate, the pie is larger, and everyone profits."

"Vambolent got what was coming to him," Jenks said, hammering the last nail into Yoshi's coffin. "Everything happened as it should. Nothing can be imbalanced for long. Every effort of his reached too far, too fast. He gets far, I will give him that, but then it all snaps back. It's Vambolent's Achilles' Heel."

Toloisk stood, and the two acolytes rose with him. "Mr. Jenks, Mr. Hanazawa, you will please come with me while the vote is taken." He walked to the bookshelves, where a section of deeply-carved wood panels separated large sections of the library. He removed a thick book from the edge of the shelf and reached his hand into the space left behind. With a slight whoosh, the wood panel swung out into the room. A cold mass of air fell into the room from behind it. Jenks caught his breath, which didn't come in as full as he wanted.

Toloisk led them into a stone passageway, lit from above and only wide enough for one. The door sealed behind them, and the exchange of air stopped. The space inside felt still – like a tomb.

Following Toloisk further into the tunnel, the air grew colder and they could see their breath. After many steps they stopped at a niche carved into the native rock, where fur-lined parkas were hung on pegs. Each man selected a coat and zipped it up gratefully. They pulled their hoods to, and jammed hands in their pockets, as Toloisk demonstrated. He took them around a final few bends, where the wind could at last reach in to grab their body heat. When they arrived at the end, there was an irregular hole cut out directly into the elements, with a rough sill where they could rest their elbows.

In front of them, thrashing clouds whipped by, torn into ribbons by the sharp edges of the Matterhorn. Through breaks in the scattered clouds, a sweeping panorama opened out in all directions. Above them, a full moon shone extra-bright in a field of stars crisp and large with altitude. To their left and down was the ski bowl, and beyond that, the yellow glow of Zermatt was visible just barely.

Jenks marveled that he had traveled so far in the rail car and the elevator.

"We are still far from the top," shouted Toloisk over the roar of the wind. "But we are a very long way from the bottom."

He pointed down, and Jenks and Hanazawa leaned over to look. The sheer drop below them vanished into clouds thousands of feet below their perch. Jenks stepped back inside. Yoshi reached over the lip with a closed hand. A blast of frigid air raised the fur of their parkas. Yoshi's hand came back empty.

"You killed my master," Yoshi said, turning on Jenks suddenly.

For just a moment, Jenks was caught aback. He recovered before the moment had even passed. "You forget my mentor was killed as well," he said icily. "How dare you make such an accusation."

"Gentlemen," Toloisk said, calming them. "Please."

"Your fellows will invite this man to join them tonight. But he is viper in your midst."

A buzzer sounded. Toloisk leaned down to lift a shroud. Under it was a mouthpiece. "Send in the first candidate," said a voice, probably Banderberg.

"Yes, sir." Toloisk escorted Hanazawa back down the hallway. Jenks heard the sound of them removing their overgarments, and their footsteps down the hallway. He turned back to the vista below, reviewing his plans and plots in his mind. All was well.

Was that a scratching sound? He heard something over the sill, and looked out. Nothing. Must have been falling rocks. He felt sure Yoshi would be rejected. That would leave an absence – a place he would try to fill with a candidate of his own choosing. Being D'Argeste's official protégé granted him many rights and privileges once he was accepted as a member. In time, he would lead, be Z, and then control of the group would be his. With an unshakable foundation in space-based power generation, developed with the Marcharelli family, he would preside over the phase shift he had predicted back in his schooldays. There were many steps, but he had already progressed faster than hoped.

Toloisk returned, smiled at Jenks and settled himself shoulder-to-shoulder on the rock window sill. The freezing, thin air was cleansing and remarkably pure.

"What is your letter, Mr. Toloisk?" Jenks asked at length.

Toloisk regarded him for a moment, then his face softened. "Oh, I suppose you already have it figured out. Your genius is legendary, Jenks. And I suppose there's no doubt you'll be voted in among us. So, there's no harm in it."

He paused, almost reconsidering. Then, he explained quickly, teaching the neophyte.

"There's a Matrix, 3 by 3, and a Triumvirate. The matrix is like a map, using letters X, Y, and Z, as addresses on two adjacent sides. Z is highest, so ZZ is in the top left corner. XX is bottom right to the head of the table as viewed from above. Across the top is the Triumvirate, with only single letters X, Y, and Z. That's about all I can tell you before the Initiation."

"What is the significance of the adjacent sides?" asked Jenks.

"It has meant varying things in our history. Today, the top side is enablers. The left side is actors. The Matrix keeps us well-rounded as an organization, but the distinctions are flexible, and can change to meet new situations."

"How do you advance in rank," Jenks asked. "How do you become a Triumvirate?"

Toloisk looked over at him, and in the thick, round glasses, Jenks saw a dancing red laser spot. Toloisk's eyes grew wide. Jenks realized the spot was on his own face and ducked. The first shot missed him, but parted his hair. The second shot ripped a hole in the side of Toloisk's head, tearing off his parka hood. Jenks was spattered in the face; hot blood stinging into his eye. His ears roared with deafness.

Toloisk fell in a slump, making a heap of dead on the floor. Jenks crawled over his legs, and left him behind, scampering down the passageway on all fours. The crack of rifle fire came staccato now, and he was sprayed with chips of rock that pelted off the back of his parka. He scrambled around the corner, the only time he was exposed to the current line of fire. His feet were slick on the floor and he scrabbled. A rifle bullet caught the sole of his shoe, ripping it off his trailing foot, and twisting him halfway onto his back. He fell awkwardly, but clawed his way upright and began to run further into the passageway.

The doorway into the library was closed when he finally reached it, and he ran up to it, calling for help. He threw his body weight against it, but it held firm. He banged a fist hard, several times. Banderberg called through the wood.

"Who is it?" his muffled yell came.

"Jenks!"

"Who else?"

"Nobody. Quick. Before they come," he shouted between gasps of the rarified air.

The door opened, and Jenks staggered into the warm room. As he shrugged off his parka, four of the plutocrats were pointing pistols at him. They looked ridiculous. Old men wielding single shot dueling pistols, probably from the 18th century. They relaxed as Banderberg shut the door behind Jenks.

"What's going on out there?" Banderberg demanded after swinging the door shut quickly. He opened a panel next to the door seam.

Jenks was doubled over with his hands on his knees. His words came out in short phrases as he caught his breath. More men came back into the library doorway. They noticed his one bare foot, but it was the blood on his face that riveted their attention.

"Snipers," he gasped. "Toloisk was hit. He may be alive, but he's down. There were several shooters, at least. They may be coming in. I just ran. I'm sorry…" he finished, shaking his head.

Banderberg stood beside the door, watching a video monitor he had revealed behind a panel. Four quadrants showed the tunnel at intervals. In the bottom right corner, they saw two diminutive black-clad assailants moving in a crouch towards the camera. One raised a slim rifle, and the feed went to snow. Banderberg pulled a keypad from below the monitor. It formed an angle to the wall. He twisted an arming switch, and poised his finger over a red firing button.

"Let's wait to see how many more of those buggers come into the tunnel," said Banderberg. The two figures had moved into the bottom left quadrant, and again shot out the camera. Banderberg waited until he saw them appear in the top right quadrant, and pressed the red button.

A strong concussion, like a constrained thunderclap, came from directly behind the door. A deeper, subsonic rumble echoed on through their feet. All four quadrants of the monitor went to static and white noise.

Banderberg pulled out several books, dumping them onto the floor. At the back of the bookshelf was a panel which he pushed to a detent, and which responded by popping itself out towards him. He swung the front open, revealing an infrared headset, which he placed on his head in a single fluid motion. Also inside was a .445 SuperMag handgun with an IR/visible laser sight. He hefted the gun, checked that the chambers were loaded, and spun them into the revolver. A row of extra ammunition lay on foam, and he scooped them up, dumping them into his jacket pocket.

"Time to do a little hunting," he announced. "Who's with me?"

Hilbert pushed his way past the pistol-wielders and stepped forward.

"Gentian, this is no time to play hero. We are under attack. Our first priority is the safety of our members."

"That's exactly what I'm about to go do."

"You're off to get yourself killed, is what you are about to go do. There are more snipers waiting. There are more of them coming. No one would attack us unless they were sure of victory. We have to get out at once."

Banderberg hesitated, struggling against his bloodlust. He looked down at the dead monitors. Hilbert was right. There could be attackers as far as the third camera location already.

"Okay. I'll cover the rear," he said. "The rest of you take the elevator. That thing can hold me, so it can take any three of you. Hell, you can probably go down four at a time if you hold your breath."

The men began dispersing without further prodding. Jenks stayed behind, standing erect now. Hilbert let the others flow around him.

"Gentian," Hilbert said softly, when the room was otherwise cleared, "I received a message from Valerie. She says that armed commandos have stormed the Hotel Alex. Whoever sent these *assassins*," he spat the word, "is surely waiting for our people in the hotel elevator."

The two men stared at each other, then looked toward the other room, where the sounds of panicked men were clearly heard.

"You said the tunnel was originally intended to reach Italy," Jenks said, remembering. "Was it ever finished?"

"Ah, yes, but it is a long, long walk. And neither of us are in any shape for such a trek," he said, indicating Hilbert and himself.

"But we must try," Jenks said, looking down at his one bare foot.

The two leaders looked again at each other and nodded. Banderberg shrugged. "We may as well try," he said. "But I have just one more thing to do. You two go to the foyer. I'll be right there."

He turned to the monitor and control pad. Jenks and Hilbert did not move, but watched him, curious.

"Go on, you two," Banderberg said, waving his fingers back. "It is about to become very unsafe in this room."

He keyed in a series of numbers, and hit an Enter key. Jenks and Hilbert left.

The monitor changed to a graphical interface, and he typed in 00:30, pressing Enter again. His finger hovered over the red button, shaking. His heart ached over destroying the passageway, and he hesitated.

In the silence of the room, he distinctly heard the scuff of a boot, just on the other side of the panel door. He activated the red button, and backed away from the door, cocking the .445 as he moved away.

Banderberg had just turned the corner when the wall panel was blasted into the room, missing him narrowly, and banging clattering into the dining room. Red laser lines lanced through the grey smoke of the explosives. Banderberg continued backpedaling, and started to fire shots from the massive handgun into the library.

He reached the heavy metal blast door, and maneuvered behind it. A tremendous shock wave slammed it closed behind him as the stone corridor to the rock ledge was blasted into rubble. Banderberg was knocked off balance. Jenks steadied him as dust from the rock ceiling fell into their hair.

For a few seconds there was no sound. Then a rifle shot pinged against the door.

"There's no way to jam this door from the outside!" Banderberg told them. "We won't reach the bottom before they get through. All they have to do is cut the cable." The fear was evident in Banderberg's face.

"Have we no choices at all?" asked Hilbert, strangely calm, perhaps resigned to his death.

Banderberg looked over at Jenks, fixing him with a serious stare. "There's only one," he said, still holding Jenks' eyes as he answered Hilbert's question. "There's an emergency 'chute. It's developed specifically for this ledge. There's no time to explain how it works, but just remember to pull the rip cord *as soon* as you jump. Got it? *As soon* as you jump. Here," he said, handing the .445 to Jenks, "you'll need this."

"Where?" Jenks asked in a rather small voice. They could still hear rifle shots against the library door. There was a distinct sound of splintering.

"Behind the Manet." He fished in his pockets, and removed a handful of bullets. "These too. Good luck, old man."

"Thanks," said Jenks quietly, accepting the ammo.

Banderberg joined Hilbert who had just stepped into the elevator car. He slammed the door, and slapped the Down button with a meaty palm. "And by the way, Jenks, you were voted in." The car dropped quickly out of sight.

Jenks gave himself just one second for a winning smile. Then, he pulled back the frame of the indicated painting, and found behind it in a small niche, a parachute pack. He donned the pack, attached the buckles, and then began exchanging spent cartridges for live rounds in the .445, walking over to the precipice as he worked.

For a moment, he thought about firing the revolver into the cables paying out before him, thus advancing his position by two. But there was a explosion from behind him. The metal door bounced inward slightly before the latch yanked it back. Jenks spun up the heavy revolver, cocked it, and held it out with one hand. With the other hand, he gripped the red rip cord. His knuckles were deathly white.

Several seconds passed, he was breathing heavily, and then there was the sound of many footsteps. The latch on the heavy metal door lifted, and opened slightly. He fired.

His first shot struck the metal door and pushed it half-closed. Jenks looked down at the weapon with respect. He thumbed the laser sight, switching it to visible, and knelt down. Leaning close to the ground, he aimed under the door. There was a slight gap, and he could see the shadows of feet. He pulled the trigger twice, quickly, setting the aim of the second shot with the recoil from the first.

One of the feet was suddenly gone, blown away in perfect revenge for his own missing shoe. A split second later, a body crashed heavily to the floor. He saw a flurry of movement. Jenks fired again. But in another second there were no shadows under the door.

He stood and turned to look behind him, down into the darkness where the elevator car left. The cable had stopped moving. He uncocked the weapon and tucked it into his waistband. He jumped.

Cimarron Hills

Elise and Dajwal sat across the kitchen table from Parks.

"Now, explain to me again where I fit in?" he asked them, arms crossed over his chest.

"If you agree, you'll be the president, Parks," Elise said. "President of LunarSource."

"But I don't really have any authority, right? You and Dodge are – what did you call it – co-CEOs."

"Dodge is on loan from NASA-lite as part of their technology transfer program to jump-start the orbital space industry in the US. With me as the other senior manager, we qualify for favorable status on government contracts. And because I'll own the majority of stock, we can get SBIRs and STTRs to build our business."

"But solar power satellites are not a part of the business plan."

Dajwal opened his hands placatingly. "The people are not yet ready, Parks. And if they were, it would be Marcharelli Space who would be doing this."

"Why them? Why not us? You both know that is my only reason to do all this! Afren wants me to build arrays for Africa. I promised Kindoh I would do that with his money. I have no interest in building some kind of niche market while Jenks and Marcharelli grab the market I've worked for these twenty years!"

"Calm down, Parks," Elise warned. He stayed silent, but his face began to redden.

"Parks, my good friend. I must tell you something that we had hoped not to tell you."

"Oh, great. Another secret."

"The technology jump-start program originally offered my services to Marcharelli Space's new spin-out StarSys."

"What!" Parks came up out of his seat.

"Please, Parks. Please calm down. You must know that I refused right away."

"How can I know that," he said, sitting slowly. "Most of what goes on I don't know about. Yoshi gets promoted to Director General of JAXA, and I hear through NASA Watch. The FCC opens up the 2.45 gigahertz window for space solar power at great expense and effort, and I learn about this through an information bulletin from my phone service. Elise opens up a space training school in Ecuador, and how do I find out? From a solicitation for fund-raising! Now you tell me you almost went to work for my competitor!"

"No, Parks, I did not almost work for them. It was simply an idea of the new Administrator. Of course I told him that I could not serve in that role with a clear conscience."

"But I bet you thought about it."

"Parks, you're being unfair to Dodge. He's here, isn't he? We want you involved. We *have* to have you involved. You alone control the money from Kindoh that we need to start-up. You are the focal point of our entire business plan. It's a good plan. It will open up lunar facilities, orbital transport, materials technology, and maybe even lunar tourism. And Parks – we will make a *lot* of money."

"You two make lots of money while I give tours to schoolchildren and greet visitors. And when I'm not playing figurehead and corporate icon, I sit in my corner office dreaming big dreams. Does that about sum it up?"

They were both quiet.

"I see. So, I'm neutered and put out to pasture."

"No, Parks, it's just that…"

"And any business plans I might have, like going public to raise money to retrieve a comet, are off limits. I can invent gadgets. I can lay down technical roadmaps. I can work special projects when our astronauts are not busy with operations. But if I start espousing policy in public, or if I make business arrangements with customers, or if I meddle with the government in any way, then you two pull back on my leash."

He stared at them.

They stared back.

Elise's mouth worked, wanting to soften his words. But he was close enough to the mark that it would only provoke a fight. It would be a fight she'd win, but her husband's spirit would be broken. She resolved that if their marriage were not to survive this transition, it would not be due to her immaturity. She swallowed her explanation, and sat back in her chair.

"Yes, Parks. That's about it."

Dajwal started to speak, but Elise hit the side of his leg with the toe of her shoe. He, too, sat back.

Parks steamed and stewed, looking from one to the other and back again. At length, his shoulders softened, and his blood pressure eased off its dangerous high. He heaved a sigh.

"So. President, huh?"

"With a corner office on the top floor," Dajwal said brightly.

"Staff?"

"A dedicated administrative assistant, and during the summer, college interns to work special projects," Elise said.

"Conferences?"

"That will actually be one of your most important roles, Parks," Elise said, folding her hands on the table. "You'll work with our PR department. We'll get you invited to forums, conferences, workshops and seminars on space and space business. Your face will be the face of our corporation."

"As long as I toe the party line."

"Yes, Parks. As long as you toe the line."

"And if I don't?" he challenged.

Elise looked straight at him, into him, all business now. She didn't have to say a word.

Parks could guess what would happen.

He finally sat back in his chair, acquiescing. In that moment, LunarSource became a going concern. With the money he controlled, Elise and Dajwal would create a profitable space company. But in his mind, he knew that somewhere along the line, he would somehow find a way to get his comet.

PART IV

Chapter 17

Lunar Far Side

Unbroken blackness filled the panoramic windows of the Far Side Equatorial Base. With no Sun and no Earth in their sky, the meager starlight could not compete with the full spectrum plasma lamps inside. Their power was coming full strength from the solar array at the edge of Mare Smythii, named after a British Astronomer, on the slopes of the mountains surrounding crater Schubert, named after the Viennese composer. The array lay on the Near Side's easternmost mare, and was originally dropped by the third manned mission to build the once-defunct Lunar Energy Facility.

Connecting it to Far Side were rails laid by the East gleaner. That machine was once built by a contractor to MIT's designs, commissioned by Space Manufacture, LLC, funded by the US Congress, and paid for with tax dollars. Now it had been reconfigured by Vambolent's new company, LunarSource. He leased the machine for a monthly fee reconfigured it with a dropsat and sent it around the backside of the moon. Its twin rails carried high tension power which was currently in plentiful supply. That source supplemented the microwave beam sent down to them from the Japanese transfer station at L2, hovering due 'up' from the Far Side.

The gleaner had been retrofit again. This time the old chutes were replaced, and new bits welded onto the digging surfaces. The gleaner would travel North, towards the planned Lunar Polar Observatory from UNOOSA. This time, as it made a double-wide swath out of the regolith, it would lay a light rail. They were calling it the Aunty Lise Line, and it would connect Far Side to the LPO in another couple years. A couple more years later, the other rail, and supports, and they might drive a train over it.

With a transport line, LunarSource planned to ring the lunar pole with a heavy gauge twin rail system that would allow an isotope separation factory to follow the sun. It was an idea originally envisioned by Elise Vambolent in Kenya many years before.

Inside Far Side Base, the walls vibrated with the sound of loud drumming. Four small-statured Africans manned the drums. They were canisters temporarily unloaded to make resonant cavities. Another man played a thumb piano. In the adjacent crew quarters, Dallas Barstow covered his head with a pillow and tried to sleep.

For the last 13 days, he and the men had it easy. But the sun was coming. Now, with just 18 hours left of night until dawn, Dallas was trying to get his seven before the pre-dawn work shift started. That meant he had to go to sleep this very minute, but the other men gave no sign of slowing. How these young people could still function with so little sleep amazed him, and he felt old.

For the ninth time, he reviewed the work ahead. Memorization by rote repetition made his brain regurgitate the data. If he didn't flush it all out now, he'd dream about it. And then, he would awake exhausted, and be loathe to don a pressure suit. Just one more time through, and he could get to sleep!

His pressure suit was no longer his friend. Early signs of arthritis had accelerated in the two fortnights since their crews' arrival. Low-g life leaches calcium from the bones into the blood where it can clog the joints. He was already looking forward to a VIOXX II with his first coffee of the next day, the day of the sunrise.

He noticed that his finger was tapping to the beat. He recognized the tune. It was his favorite. Restive, he kicked off the silver blanket and bounced gently to his feet. Coming around the bend into the common lab, the drummers noticed him and smiled. Good guys, he thought to himself. Damn hard workers!

He sat down, picked up a stylus and tapboard, and pressed a hotkey he had made a few days ago, but never used. With a magician's flourish, the top of his tapard became a bongo drum, with drumhead tones mapped expertly to the picture. He tapped just a few times with his thumbs, and the five went wild with excitement. They surrounded him in a heartbeat of time.

"How did you do that?" one asked.

"May I try next?" asked another.

"Get your own," Dallas growled good-naturedly. They broke off and retrieved tapards from around the station, and asked him how to configure their own. Soon, he had them all set up, and he sat back to enjoy listening to them play and experiment and learn. It made him feel warm and relaxed and fatherly. He yawned after a bit, and showed them how to adjust the volume. Then, he went back to bed.

"Last time," he said aloud. Lying on his back with his hands behind his head in his bunk on the Moon, Dallas Barstow heaved a contended sigh. He went once more through the protocol. The ore they gathered in the first few days of the dark would last them this whole lunar day of 700 hours. Their isotope separator was set up for platinum and iron. Every kilotroy would be electromagnetically launched gently up to L2, and caught there by reverse process. The iron-rich canister holding the precious metal would then be used as reaction ballast to send the naked platinum to Middle Earth Orbit.

Interceptor satellites were in place to capture the bundles and prepare them for descent. Each bundle was sized within the recently-adopted Canada Rules for ITO. Featherdrops released from the Interceptors would then deliver the kilotroys to ocean sites off the coast of Japan, far away from population centers, but close enough to defend. The featherdrop technology began with the Cold War spy sats which dropped film from orbit, and was further developed by Old NASA briefly, but well. The project went underfunded, and was picked up later by the Japanese under Yoshi Hanazawa, now Director-General of JAXA.

So, they get their platinum, put it into fuel cells, and voila, the Hydrogen Economy, courtesy of LunarSource catalyst and their new fleet of underground pebble bed reactors. In just two more days, they were planned to make their first shipment. By the end of this coming lunar day, Far Side will have shipped enough tonnage of pure platinum to keep Japan happy for another 14 days. Then, he would jump from Far Side to the tug at L2, and be home for Christmas.

The rote routed, the worries rested, he heard the other men diving softly into their own bunks. He slept, and his sleep was as blank and unbroken as the panoramic windows.

New York City

Senator Mason Barnes rode in a black limousine with a lead All-Purpose Vehicle, and another one as chaser. The small motorcade stopped outside the hotel where L. Marshall Washington was staying. Washington had just been named director of the recently-expanded UNOOSA, the UN Office of Outer Space Affairs. Senator Barnes had just been re-elected, retaining his Chairmanship of the Subcommittee on Science, Technology, and Space. Both men were to be honored tonight by the most influential space lobby group in the country. There would be fancy food and lots of speeches.

Washington stepped from between his two bodyguards, and entered the limo. Seated, the two shook hands, and each smiled with genuine warmth. The two long-time politicos first became acquainted in the aftermath of the Limited Nuclear Exchange in Kashmir. Over recent years, they had served together on space law committees, as well as architecture roundtables. The UN-to-USA connection on the Lunar Polar Observatory, which would draw heavily on American technology, was managed by these two men. The LPO would serve as a central traffic control center for all of cislunar space. It would also enforce the newly-upgraded and ratified Moon Treaty, if necessary.

"Gloomy weather," commented Barnes, not wanting to jump right into the real reason for their celebration.

"Every time I visit, it seems to rain," Washington said. "Vienna can be that way, too. Maybe I carry it around with me?"

"No, it's not you, Marshall. Damn hurricanes haven't let up in two months. What we get here is less severe than further south, but very nearly as wet. I hear the sewers are already at overcapacity."

"Well, let's hope this banquet hall has an upstairs," he replied, and they both laughed.

Barnes grew serious. It was important that they talk. "We both know they've invited us here with a specific agenda, right, Marshall?"

"Yes. I assume they want you to lease them the Lunar Energy Facility. And I would not be surprised if they ask you to take back the power you are now leasing to Parks Vambolent's company."

Barnes gave a plosive sound of agreement and dismissal. "And for you, Marshall, they clearly want you to build the LPO from lunar materials. Popular sentiment favors that too. So, they want a lock on your bid. Guaranteed profits, and no more worries about patent infringement. They could be billionaires if they pull this off."

Washington nodded, looking seriously concerned. "There is subtle shift in the public opinion, Mason. Do you catch wind of a sea-change?"

"I hear many things," the Senator said, wagging his head sadly. "I see editorials from Winnimer Jenks syndicated now. He has a lot of pull, and he's helping people see space as ready."

"He will be here tonight?"

"Yes. Robert Junior and he are partners on the lunar energy thing. Do you hear how they market this now? They plan to send the energy in little bursts. Kind of bounce tiny packets around to thousands of low-orbit satellites, which bounce the energy down to the roof of your house."

"Amazing, if they can pull it off," Washington said, then asked. "Why didn't Vambolent go after the LEF?"

"His first company – actually, it was his second by that time – Space Manufacturing, went into receivership. They lost the equity in the LEF. Ownership transferred to a congressionally-controlled company that reports directly to me. He's got some kind of angel investor backing him, I think, but some really good customers: ESA and MITI. But I think he's gun-shy about space solar power."

Washington tilted his head, growing thoughtful. "I have heard that his old friend Afren Kwapili wants SPS technology. Would he buy it from the Marcharellis?"

"They seem focused on the US market."

"Then I wonder what his plans are. He has even stated publicly a further use for the power beams. The source for many of your hurricanes is just off the Barbary Coast. The power beams, applied correctly, can weaken them into mere tropical storms."

"Astounding concept. But will it really work?"

"That remains to be seen. But with all the displaced persons in the US alone, I can easily imagine people around the world supporting his idea."

"How are people responding in Africa?"

"Egypt is in an uproar. They have not joined the Greater African Union. The people are violently opposed, largely because of foment by Hamid Hababek and many other mullahs. They refer to their country as the first in history to be attacked by beam weapons. They have promised to prevent their use in Africa, regardless of consequence."

"What is your new UNOOSA position on power beaming, Director?" Barnes said, smiling to ease the tension that had gripped them.

"Well, Mason, as I see it, any form of energy in large amounts is a security risk. But we are a species that thrives on energy. And we are growing that need rapidly. So, we must learn to harness it safely. When you consider all the risks, it is far easier to police an energy installation in space than it is underground."

"You've obviously heard about the Canada Rules?" Barnes asked.

"Another of Winnimer Jenks' contributions, right?"

"Yes, I think so, but he claims he pulled it together from several sources at McGill and in Montreal and Edmonton. It's a simple set of rules on what we used to call ballistic transport and power beaming. Now, it's called SST and IOTA, and with the limits imposed, it's impossible to concentrate kinetic or electrical energy in sufficient concentration to be a threat to someone on the ground or in the air."

"So, do we give the Marcharellis what they want?" Washington asked.

Barnes nodded a non-verbal assent, but he said "Let's see what they have to say."

Their vehicle pulled up to a cordon of security officers. With the nation at Red for so long, Security continued to be the fastest growing career in the country. Twin rows of large men formed a corridor to the front door. Outer rows of even more guards searched the facades across the street. The sidewalk was remarkably clear of pedestrians.

Barnes and Washington exited the limo and walked to the building.

Inside was a long, immense foyer. A red carpet ran from their feet all the way to the far end of the room where a grand stairway led up to the banquet room. On the far side of red velvet ropes on brass stands, plus one step further back, stood the assembled delegation of the space lobby group. They erupted into enthusiastic applause. There was no whistling or shouting, just the rich, full chorus of many people clapping together for the two guests. Elder statesmen of the space movement were further down the row, and the recent growth in space interest was a lifetime dream of theirs. Internationally, these two men were seen as the key government officials to unlocking the vast new markets of space, enabling the manned exploitation of our universe.

Barnes knew many faces, and shook each of their hands. Further down, he knew almost everybody, and had to issue blanket hallos. But a few hands just had to be shaken.

Washington was looking at the base of the staircase, where Robert Marcharelli Sr. and his son stood waiting on the first step. As they approached, a narrator's amplified voice introduced them, along with their new and continuing roles. The two guests were welcomed by the hosts, who stepped down to greet them at the base of the staircase.

Marcharelli quieted the applause still scattered across the room, and accepted a microphone from Winnimer Jenks, who stood almost unnoticed just off the main crowd. "As you can see Senator, Director, you are very welcome here…" The audience interrupted him to applaud again. This time there were a few cheers of 'here, here'. "We invite you to sup with us, and then talk about the future. We stand at a turning point in history, and you men are helping us to make that tack."

Barnes accepted the microphone that Marcharelli passed him, trapping him into an impromptu speech.

"It's been your faith and hard work throughout the years that have brought us to this point," Barnes said. "With leadership like Robert Marcharelli provides, great things are possible." He passed the microphone, amid generous applause, to Washington.

"I am honored to be here," he said, bowing his head to the people on the left and then to the right of the red carpet. "I thank our hosts, the Marcharellis, and I say thank you to all of you for acknowledging my new role. I hope to learn much tonight from what each of you have to say." He passed the microphone back to Marcharelli, Sr.

"Now, we go to the Egyptian room for a delicious repast. Don't crowd our guests on the stairs, folks. The bar's open, so refill your drinks and then join us." He handed the microphone to Jenks. Jenks switched it off, and passed it to a young female assistant. She accepted it from him with an arm so slender the skin wrapped around the bones. She also fumbled with the switch, not having seen Jenks turn it off.

"At least we are going to be upstairs," Washington said as an inside joke to Barnes.

"Yes, but so many of them," Barnes said, staring upward. His trick knee clicked and popped on stairs, and he knew he would be sore at the top of these 30 or so steps. But in his secret aspiration to be President one day, he told himself that as long as he could climb stairs, he could run for President. "You up for this, Bob?" he asked the senior guest.

"We'll just take our time," Marcharelli said. "We can talk on the way up. Robert?"

His son came over, and was introduced. They also called over Winnimer Jenks and introduced him. Both Barnes and Washington looked him over carefully. He was completely handsome in a totally nondescript way. But when finally noticed, it made him endlessly fascinating to look at. Jenks winced playfully at their scrutiny, smiling crookedly. They all three laughed.

Robert Marcharelli held the elbow of his father as they took the first stair. Jenks moved in beside Washington, ready to steady him if needed.

Barnes and the elder Marcharelli discussed the LEF arrangements that were also being pursued through official channels. The actual substance of the leasing would be decided in dozens of meetings over many months of lobbying, investigation, panel hearings and committee reviews.

On the other side of the staircase, rising slowly, one step at a time in sync with the other three, Washington quizzed Jenks lightly about his space architecture.

"Explain to me the Canada rules, please?"

"They were started by a young student there, sir," Jenks began. "Her Senior Thesis at U Alberta was a study of energy packet distribution in space."

"I don't quite get that, Jenks."

"Basically, her premise was that any form of energy was safe if distributed in time or space. Consequently, the higher the concentration of energy, the greater the risk to security and to safety."

"That makes sense," Washington said, getting slightly out of breath.

"We've applied her concept to wireless power transfer," Jenks explained. "From the lunar surface, we send tiny packets of energy to relay satellites in LEO. We can send thousands per second, each too small to harm. But with our electrical receiving power systems, no bigger than a refrigerator, we can power your home 24 hours a day."

"You'd need a … polar location for that, wouldn't you?" Washington asked.

"Yes, sir. Once we achieve breakeven at the LEF, we plan to build rails to the Moon's South Pole, if the UNOOSA approves, of course."

"That would be the one spot in cislunar space that our LPO could not see you," Washington said.

Jenks smiled assuredly. "The L2 relay point currently occupied by the Japanese would be an ideal place for a bounce between the poles. We would be linked by rails, and linked by radio."

On the other side of the stairs, halfway up, Barnes looked behind them to see who might be listening. Only a few old coots stood in loose groups around the base of the grand stairway, subtly holding back the eager youngsters.

"So, how are you funding this?" They had just been talking about the incredible expense needed to bring the abandoned base back up to safe operational status.

"I've been around the globe," Robert said proudly. "You'd be surprised how many people are willing to invest in Lunar Solar Power."

"Why would I be surprised?" Barnes asked.

Bob Marcharelli hiccoughed.

"My partner, Winnimer Jenks, brings some funding also," Robert said defensively, looking concernedly at his father who had turned pale after the Senator's remark.

"Yes, I wanted to learn more about that," Barnes said. He took another step, and heard his kneecap click, and knew it was not good. He looked over at Jenks, and beckoned him with his face.

At the next step, Jenks moved one over to stand between Barnes and Washington. Robert Marcharelli had taken his father's elbow, and they were negotiating the stairs one by one.

"Yes, Senator?"

"Robert was talking funding," Barnes said, fighting not to grimace at the pain. "And he mentioned that you were contributing substantially."

Jenks' expression did not flinch from his genial attentiveness.

"I believe the Senator knows I was mentored to Monsieur Bruno D'Argeste. After his death, I was given control of one of his pet projects, space development. But the finances are simply something I manage. My real role is as Systems Architect."

"Yes, too bad about Monsieur D'Argeste. Almost killed Vambolent, too, that blast. Where were you, exactly, at the time of the unfortunate event?"

Jenks showed shock at the accusation. Bob Marcharelli, hearing all this as he labored up the stairs, hiccoughed again, and there was a rattle in his throat. Suddenly, he began choking, and his body stiffened up. As he straightened, he leaned back onto Robert, whose hold on his other arm caused the old man to twist backward. He flailed a hand, which caught Senator Barnes in the middle of the chest, reeling him. Quick as a flash, Jenks had stepped down behind Barnes to steady him. Barnes used the support to recover.

Marcharelli doubled over now, and a gasp went through the huge room, echoing on itself. Robert was pushed back further, and his foot went to the next stair. But the foot didn't quite make it off the higher stair, held down by his father's dropping weight. He began to topple backwards.

"Jenks!' he cried on an in-breath of total panic. His arm began to pinwheel, fingers splayed.

Jenks gauged himself against the Senator. In that one hesitating fraction of a second, Robert crossed the very narrow gulf between just in time and too damn late. Their eyes made contact as Robert's backward fall began to accelerate. Whites showed huge at the top of Robert's eyes. Jenks was calm for only a cat's moment. Then, giving the Senator a perfectly light forward shove, Jenks dove for Robert, yelling his name in desperation.

He was too late. Robert fell back onto his neck and his feet bounced up and over him, folding him into a tumbling ball of loose limbs and lolling head. Old men at the bottom at first held up hands to stop, to help. But as Robert tumbled faster, more violently, they stood aside. Hands slapped him as he flew past, but he was hardly slowed. Marcharelli Jr. twisted out in a last attempt to stop himself, and careened over to the corner of the grand stairway.

The final time his head cracked down onto the brass base at the end of the velvet rope, the microphone held by the female assistant recorded the sickening sound, and amplified it through the whole hall. She screamed, bringing the turned-on microphone right into her maw. Her amplified voice crescendoed into a squeal of feedback, and the sound system overloaded, sending drinks falling, glass cracking, and eardrums splitting.

* * *

After the facility ambulance had taken both Marcharellis away, Jenks stepped in to salvage the meeting. Having rushed down the steps to retrieve the microphone, Jenks had directed everything. Paramedics were already on call as a request by Washington's assistant, and had been available to deport the father and son within an incredibly short time.

"Is there anything more we can do?" Jenks asked them over the public address system. Nobody answered. He waited. They were all looking to him to take charge, as he had been so capably doing. "Then, let us continue the party. The Senator and the Director are taking the elevator. You can all come this way."

He led them up the staircase, and their exodus upwards was metaphoric. The red carpet belonged to all of them, and they were climbing up higher in their gravity well. Jenks orchestrated the caterer, and served as master of ceremonies to the many speeches that were delivered in the presence of Barnes and Washington. The dinner was excellent, the service superb, and the speakers were by and large engaging and informative. It was a successful event, and Jenks knew it.

Barnes had been watching Jenks surreptitiously during the evening, looking for signs of inconsistencies. His mind was beginning to put together a few connections, and he had grown a bit suspicious of Winnimer Jenks.

At the end of the ceremony, the assembled crowds had given a standing ovation to both men that lasted minutes. A few people came up afterwards and chatted at the head table. At length, Jenks had to motion everyone away. He offered to take the men back down via the elevator, and they both agreed.

When they were alone backstage but for a few discrete security officers, Barnes stopped and turned to Jenks. "Wasn't your D'Argeste an associate of Banderberg, in Zurich?"

"They were acquainted over many years," Jenks said, handling himself well despite being hemmed in by the two men who held his future within their power. He decided to tell more rather than less. "Their countries adjoined, and they both acquired mass holdings. Overlap was inevitable. So, yes, they were business associates."

Washington stepped closer to Barnes, so they both faced Jenks square on. "Wasn't Banderberg associated with the UN Secretary-General getting shot down in Central America? A great quantity of small-arms fire. Banderberg had brokered them."

"Freedom fighters sometime cause collateral damage," Jenks said, his face now tight.

Barnes stood even straighter, looking Jenks in the eye. "We know Banderberg and Count Hilbert are good friends. Hilbert runs one of those secret societies. What is it, Marshall, the Illuminati? The Majestic? What do they call themselves these days?"

"I haven't heard a name," Washington said, "But I've met Hilbert downtown in Vienna. He's a very powerful man. A very influential man."

"Are you suspicious of a conspiracy?" Jenks asked.

"I don't subscribe to conspiracy theories," Barnes said. "But I'm sure they do exist. So, we have all these tycoons getting together in supposed secret to do what? They aren't all in the same industry, so wholesale collusion seems out. There's no evidence of monopoly, regardless of how deeply one investigates their business practices. Which leaves me wondering. What do they talk about? Why are they all present every time?"

Jenks began to feel faint, wondering that if Barnes knew about their meetings, he may know about the ninja attack on the Matterhorn redoubt. Maybe he was part of the assassination attempt? Was there a connection between Barnes and Kindoh? Jenks did not know of any. He elected not to respond to the rhetorical questions.

"Are you a member of their group?" Washington asked.

"I'm simply an apprentice," Jenks said, and both men could detect the half-truth. "My relation with Bruno gave him the confidence to will to me control of a fund with space-oriented requirements. My partnership with the Marcharellis is based on Monsieur D'Argeste's dying wishes."

"Weren't you also active in the op-ed pieces surrounding power beaming?" Barnes asked, now that the nagging in his brain had gelled into a line of questioning. Jenks nodded his affirmation, and Barnes continued. "There was one other, or maybe two other voices that became dominant in that debate, as I recall. I remember wondering if maybe there was one person behind all these people. Only you, Jenks, of the three of them, ever had a public face."

"But the whole outcry on power beaming happened when that Chinese student was attacked," Washington said to Barnes.

"Yes, but the perp was traced back to a man who was a known fanatic of that Henry L. Williams, who opposed SPS and the whole China Energy Fiasco. In fact, that one attack, deliberately planned to discredit power beaming, became the turning point in popular opinion, and led to the abandonment of the whole LEF." Barnes stared at Jenks with deeply-narrowed eyes.

Jenks had no answer for them. Washington was visibly exhausted. He thanked Jenks formally for the dinner, and stepped into the elevator. "Mason?"

"You go out to the car first, Marshall. I'll be right down. I want to talk to Jenks for just a moment longer."

Washington nodded, pushed the button, and left. Barnes turned to Jenks. Then, he shooed away the nearby guards and stepped right up into Jenks' face.

"I don't know everything you're up to, young man, but I'm onto you," he said.

Jenks did not appear intimidated. He waited for the elder statesman to speak again. He did not, but just waited. Jenks took the chance he had been leading up to the whole evening – his whole adult life.

"Sir, my only ambition is to bring energy independence to America, and become rich doing it. You can help me, Mason. Think of it, America growing again, and all because you had the foresight to promote it. A man who could do that would be President. Guaranteed."

Jenks noticed the reaction in Barnes' face. Though Barnes recovered his poker face quickly, his own secret ambition had been wrested from him. He looked closer at Jenks again, somehow always unable to quite focus on his facial features.

"And besides, Senator, I just saved your life."

"What?"

"That's right, sir. Bob's arm knocked you backward. It was me who held you in place. I kept you from also falling."

Barnes gave pause, remembering the hideous fall and the terrible sounds at the bottom. His ears still rang from the PA system feedback. "I guess I had forgotten that fact during all the excitement."

Jenks bowed his head respectfully as he shot for the moon. "So, I suppose you owe me one, sir," he said, voice low.

Barnes looked at him with respect and deadly caution.

"Yes, I suppose I do. But only one, Jenks. I'll give you the LEF. After that, you're on your own. Debt paid."

"And if you want to run for President, sir?"

"Then, we'll talk again, Mr. Jenks."

Atlantic Ocean

December hurricanes stretched from the Barbary Coast to the Eastern Seaboard, separated by slots of nice weather. From space, the hurricanes looked like whirling bullets, fired in machine-gun automatic. And so they were. The energy balance in the atmosphere had changed. While some still called it global warming, most realized that term was now outmoded. There was simply more energy in the biosphere, introduced by the collective and accelerating activities of the human race.

According to Le Chatelier's Principle, energy added to an equilibrium state will disturb that state only temporarily, until it corrects back to a new stable point. As the average temperature gradually increases, some species will die, others will thrive, and new ones will develop as the equilibrium point shifts. That is a gradual change, occuring over generations. But the short term fluctuations through the transition are not so gentle or steady. Long before the planet becomes too hot to be habitable, terrible storms of greater and greater energy will rip some of mankind's greatest creations to rubble, as they had already done to the Washington Monument. Once, it was the tallest freestanding masonry structure on the planet. Now, the marble blocks were sold as souveniers. As atmospheric energy continues to increase, humans will be forced underground, undersea, or out into space.

Kokomo, Indiana

Afren Kwapili's caravan turned into the entry way of StarSys Corporation like they were entering another world. Leaving behind the lush greenery of north-central Indiana, the landscaping looked gray and stark. Passing through gates of artificial rock mountains, the vehicles drove onto a dust and talcum field simulating moon dust. Rising before them loomed a seven-story building suggestive of a rocket. Three-story buildings spread out from the central donjon with facades that mimicked Habitats A, B and C of the Lunar Energy Facility. A larger-than-life statue of a suited astronaut, hand raised in greeting, completed the scene.

Security guards from the caravan took up standard positions. When the captain was satisfied, he opened the door of the central vehicle. Kwapili stepped out, and was followed by his aide Mwahaha. Guards formed a circle around their charge, and they entered the lobby through an entryway shaped like an over-sized airlock.

Standing inside, Winnimer Jenks stood flanked by two young women. Kwapili remembered the first time he had driven up to a space facility, in Malindi years ago, and how different was this arrival from that one. He and Jenks greeted each other and introduced their assistants.

"Robert apologizes he is not here to greet you, Afren," Jenks said. "He will join us in the conference room upstairs as soon as his conference call with the Senator is complete."

Kwapili knew from the preliminary work Mwahaha prepared that the StarSys spin-out headquarters had been located in Kokomo as a concession to Mason Barnes. Barnes had guided passage of the LEF contract with the government legal corporation to Marcharelli Space. The Marcharellis formed a new corporation for the ground-based receiving units called StarSys, and named Winnimer Jenks as the chief executive officer. Their agreement with the Senator was to draw upon the local talent for lithium-polymer battery expertise, antenna technology, lean manufacturing, and systems engineering.

"That's fine," Afren said. "Is there somewhere I could wash my hands?" he asked, brushing fine dust from his sleeves.

Jenks escorted Kwapili to the washroom, talking conversationally, but clearly loading facts into his guest's head.

"Our new company is called StarSys, Afren, because we are truly building systems from the stars. Instead of selling to centralized utilities, we sell power directly to the point of use. No middleman distributors to raise prices, no power lines to clutter the landscape, and of course, pollution-free and perpetually self-sustaining."

Kwapili farted, leaning over the washbasin. "And your relay satellites? What does their ground path look like?"

"Oh, if you want a technical explanation, I can have one of our technical fellows bring you some charts and take you through the orbital parameters. But for a quick answer, we use a Molniya orbit that lingers over North America, then swoops quickly across the rest of the planet, and back up over the States again."

Kwapili activated the hand drier in the middle of Jenks' monologue, causing his host to speak louder. "That was an orbit used by spy satellites in the Cold War, I believe."

"You are well-informed," Jenks said as they moved into the foyer and towards the open elevator. Mwahaha and the two ladies entered with them, and Jenks pressed the button for the top floor. The button was obvious, but surrounded by a densely-packed mock instrument panel. A rumble and muted roar like the sound of a rocket engine rose as the elevator ascended.

The doors opened on the seventh floor, and Jenks escorted them to a large conference room. They arrayed themselves around the cherry wood table, and one of the girls served them refreshments from a tray at one end of the room.

Marcharelli drove in on a wheelchair. "Oh, don't stand up on my account," he called to Afren. "Please, sit, sit. I was just on the phone with our Senator. He wanted me to extend his greetings to you, Afren."

"Thank you, Robert. And how is your father doing?"

"He's fully retired, living on partial life support, but he's as cantankerous as ever. Got another good 10 years in him, with proper life-extension."

Kwapili noted that Marcharelli spoke well. Apparently his mental functioning was not much affected by his fall down the grand stairway. He knew the man was a quadriplegic, but still served as CEO of Marcharelli Space.

"And you? How are you doing?"

"Oh, just fine, I suppose. Riding the elevator is about as much as I get it up. But there are ways around that. Just ask my wife!"

Kwapili grimaced slightly.

"Robert here is proud of his prostheses," Jenks said. "I'm surprised he hasn't shown off to you yet."

"I was saddened to hear of your fall. You seem to be handling it quite well," Kwapili said.

"It's all mind over matter," Marcharelli said. "You set your mind on something, and never, never give up. There's always a way to get wherever you want to go. It's just a matter of being patient, persistent, and clever."

"You make it sound as if nothing is impossible," Kwapili offered.

"If you can dream it, you can do it," Marcharelli said. His hand made an expansive gesture just a shade later than the words.

"I have a dream."

"Yes, of course. That is your reason for the pleasure of your company. Am I right?"

"Indeed," said Kwapili. "I have set my mind on a world in which the African continent bypasses fossil fuel growth and directly implements a solar-powered hydrogen economy. In my dream, our collective of nations advances to first-world status, greatly expanding world markets, and greatly improving the quality of life on this Earth. Already, we have an education system to support this reality. Already nanotechnology is wiping the scourge of AIDS from our people. And with one or two exceptions, we have complete agreement with all of our nations. This is something never before achieved. And in my dream, the world's greatest success story rests on the foundation of energy supplied by Marcharelli Space and the StarSys Corporation."

"You make it sound as if it is already a reality," Marcharelli said.

"In my dreams, it is. Now, I will practice patience, and persistence, as you say, until the day that all men can see it, with their own eyes."

"You could not be more correct," Marcharelli said. "And we will be glad to help you, provided your patience and persistence are strong enough. Right, Jenks?"

Jenks leaned forward, an earnest expression on his face. "We are very interested in expanding our markets, Afren. Right now, we are growing as fast as we can. Hundreds of thousands of our earth-based rectenna and power storage systems have been shipped to waiting customers. The solar cell production and the transmitting antenna at the LEF are proceeding apace. Relay satellites will be launched within the month, and our first deliveries will follow immediately. Yet, even with all this, our growth rate is finite. It may be some time before we can serve Africa."

Kwapili looked thoughtful. "I can understand that. We are a large market, but ready to expand quickly. The scarcity of power in Africa provides a unique opportunity to you. Rather than displacing existing energy technology, you could easily capture the lion's share of this fast-growing demand."

"That sounds very attractive," said Jenks. "Over what time frame do you expect this development to occur?"

Kwapili shrugged slowly. "We could begin next month. When your relay satellites are launched. They will pass over our continent, will they not?"

"Yes, they will, but briefly," Jenks answered.

"During that time, you cannot relay power to your North American customers. So, it should be a simple matter for you to beam power down to Africa."

"That sounds simple," Marcharelli said, "but in reality it's just not that easy."

"How so?"

Jenks answered. "Until we have a polar facility, we can beam power only 13 days in every month. During those 13 days, we have to pump as much power as we can down to our US customers. The energy must be stored in their batteries to cover the other 15."

"So you see," Marcharelli finished, "we really have no excess power to share. Not yet at least. In a few more years, yes. But today, no. I'm sorry."

Kwapili sat quietly a moment, thinking. "And if you had more relay sats. Would that not solve the problem?"

"The power limitation is on Luna," Jenks explained. "In a few more years, when our transmission capacity grows, we can grow to other continents. We would be very interested to include Africans in our customer list."

"And at that time, after I have been patient for 'a few more years', will you have the same business model? Where individuals or buildings must buy a receiving system from StarSys before power can be delivered?"

"That's right," Marcharelli said, pleased.

"Yet, by that time, relatively few of our people will be able to afford such a system. Our needs will be met much more effectively with a central receiving facility which then distributes power to rural users, much like the United States did in the 1950s."

Jenks frowned. "Unfortunately, Afren, our adherence to the Canada Rules would seriously restrict the size of a centralized facility."

"Hmm. It seems to me the 'Canada Rules' do not benefit Canada so much as they do Marcharelli Space and StarSys," Afren said.

"What can you do?" Marcharelli asked rhetorically. "We don't make the rules, we just follow them."

"I was led to believe that you did make the rules. Isn't that right, Mr. Jenks?"

"Just Jenks please, Afren. And you're right, in a sense. I did gather the rules and helped to codify them. A woman did the preliminary work, and I merely adapted it to our situation. Yet it is adherence to these rules which makes our system intrinsically safe, and therefore universally appealing."

"I see," Afren said, standing up. "However, I find it less than appealing in my own situation."

Jenks stood as well, as did the three assistants. "I hope we have not offended you, Afren. We mean well. It's just the timing. Once Africans begin to buy our systems, we will be more than happy to supply them with power."

Kwapili did not bother to argue the chicken-and-egg improbability. "Then, I shall be patient," he said. "But I will also be persistent. Good day, Mr. Jenks."

"It was good to see you, Afren. You are always welcome here. Please feel welcome to call me anytime."

They shook hands and Afren moved around the table to Marcharelli. As he walked over, Robert's wheelchair rose him to near-standing, and a mechanical assist, operated by scalp electrodes, lifted his flesh and blood hand to shake. Afren shook it, holding lightly, but the hand did not grip back. It just jiggled up and down a little, from the assist.

Afren fought down a visceral nausea, and bade farewell to Marcharelli. Followed by Mwahaha, he located the green exit sign, and walked down the seven flights of stairs, careful to hold the hand-rail. In the lobby, adorned with space photos, animated displays and space hardware in glass cases, he was re-absorbed by his phalanx of guards and left the building. They entered their cars and drove away in a hurry.

Indianapolis

Winnimer Jenks was enjoying a moment of well-deserved self-congratulations when the bad news arrived. The Rotary Club International had just named him Businessman of the Year. Handing him the plaque was none other than Bryce Vambolent, Sr., Parks' uncle, the President Emeritus of the Indianapolis chapter of RCI.

The elder Vambolent extolled Jenks and his accomplishments on the eve of their first launch of a relay satellite. The words still echoed in Jenks' mind.

"Winnimer is the President Pro Tem of the country's largest space lobbying group. He is the CEO of StarSys, the company that is putting Kokomo back on the map. Now he is prepared, in just a few days time, to begin freeing the United States from foreign oil. He has positioned himself to become the Titan of Cislunar Space."

Jenks let that phrase reverberate again and again. What a lovely choice of words. How apt. How perfectly poetic. To be thus compared to greats such as Rockefeller and Gates. It was heady wine, and he enjoyed it most thoroughly.

What they didn't know was his association with the remaining plutocrats under Count Hilbert. After their ranks were thinned by the ninja attack, Jenks had found it almost trivially easy to tap into the group's vast wealth and considerable influence. The older men were cast adrift as markets changed and shifted. Established mores and practices had become much more volatile in today's continual state of global crisis. The singleness of purpose that Jenks provided gave them a rallying point, and he was bending them all to his will. Once he became Titan of Cislunar Space – gotta love that title – he could make his play for the Z position.

The continuously running video wall of his high-rise penthouse suite flashed, indicating a news item of specific concern to him. He acknowledged the wall with the cast of a finger, and the message time-shifted to the start of the relevant segment.

A face was speaking, but the translation was another voice. The face was haggard, ugly, old. Dark fleshy lips speaking from a grizzled beard. A wide nose with forward-facing nostrils flared with anger, and were matched by blazing eyes under tangled eyebrows.

"I declare personal *jihad* against the C. E. O. of the following companies. First, StarSys. I declare *jihad* on you, Winnimer Jenks. Second…"

Jenks stared in disbelief. The dissonance between what he was just thinking and what he now heard jarred his mind into a rare paroxysm.

"What?" he yelled. Then, hearing himself, he quickly recovered. With hand signals, he told the monitor AI to gather relevant threads and detailed analysis, summarize and digest for later. He waved off the sound, so he could think.

What did this mean? Why a holy war on him, personally. Why not the US government? Why not Western civilization as a whole? He had never heard of a Muslim fundamentalist targeting a specific individual.

Was Hababek a threat? Could zealots break through the incredible security he had erected around himself since the attack inside the Matterhorn? Could he hurt the satellites?

No. Not unless he had lots of help. That was worth investigating. Activating a light-keyboard, he configured and launched a query agent to gather intel on possible orbital capability in Egypt, or by any extremist groups in the area. Using skills from his early days at RAND, plus contacts gathered through association with the plutocrats, the search would fill out quickly. He still had a few minutes to think hard, think through, and think beyond.

The first launch of a Marcharelli Space rocket with a relay sat aboard was scheduled four days away. Jenks would be flying to Oklahoma for the event. Activating a micro-mike embedded in the skin of his neck, he issued a quasi subvocal missive for his chief of security.

"Tighten and increase security at the launch site. Randomize my travel schedule. Triple sensitivity on all anomaly monitors."

Burns Flat, Oklahoma

The thundering roar of a Marcharelli rocket carrying a StarSys payload ripped through the air above and pounded into the ground below the Oklahoma Spaceport. Tens of thousands of spectators, well clear of the launch pad, watched through binocs, monocs, telescopes, and personal magnifying viewers. The slender rocket lifted quickly, belching huge gouts of flame and leaving behind roiling, tumultuous clouds of vapor. The craft arced gracefully upward into a LEO injection burn. When it rose above the point where the vapor trail was visible, the first stage was cast off. Its trajectory would carry it out over the Atlantic Ocean, but it would burn up into smoke and dust long before falling all the way to the water.

The second delivery stage was a small booster, and gently lifted the payload into the exact, looping orbit that Jenks described to Kwapili. Although the relay sat would circle the Earth in a simple elliptical orbit, the ground track of that orbit heavily favored North America. Without much ado, the maneuver was completed, and the payload separated from the booster. Using its remaining fuel, the booster did a kamikaze dive towards the Pacific Ocean. But like the first stage, its useless hunk would only reach the ground in pieces, days or weeks later, as part of the gentle rain of space dust continually blanketing the planet.

Unfurling in space, photographed and monitored by parasitic microsats, the relay satellite assumed its final form. Its overall shape was a very large octagon, a huge stop sign in the sky. The inside skin was metal, but coated with a nanoengineered material that could be electrically adjusted for thickness. Using exceedingly tiny adjustments at tens of millions of points along the inner surface, a bounced microwave pulse could be guided with great accuracy to a receiving antenna on the ground.

After considerable ceremony for the press corps, Winnimer Jenks stood on the podium at the spaceport waiting for the satellite to come back up over the continental US, ready to bounce its first packets of solar power from the moon. To demonstrate the safety of the process, he held the receiving antenna in his outstretched hand. There was absolutely nothing metal in his clothes or on his person, although the crowd was not informed of this precaution.

Connected by a wire to the hand-held rectenna was an old-fashioned Edison-style light bulb – a green one – which Jenks held in his other hand. Hidden inside the base of the bulb was a tiny battery, should anything go wrong with the demonstration.

Marcharelli, wisely sitting away from the locus of the coming microwave beam, narrated as Jenks waited confidently. Using a microphone strapped to his inert hand, he explained the whole setup, described the path the microwave beam would follow, and what to expect up on the stage, where Jenks stood alone. He counted down the seconds, and the expectant crowd counted with him.

"Five, four, three, two, one…"

There was another second of delay, and then the green bulb winked on. It winked off. It winked on, brighter this time, then winked off. One more time it lit, but not as brightly. That was it. Jenks bowed. The crowd cheered. The relay sat was a success.

Groundpath

Along the ground path of the relay sat, people gathered in a nearly continuous line across the planet's nightside landmasses to see it. The reactions were recorded real-time on thousands of streaming channels, and broadcast to the rest of the world.

"It's HUGE," said one boy, with eyes like saucers.

"It looks very close," an elderly woman said. "Could it fall down on us?" Her shoulders shuddered involuntarily.

"I shot it with my thirty aught six," announced a billed-cap man with leathery skin. "But it kept right on flyin'."

"So, this is like just number 1, right?" asked a teenage girl. "So, what happens when there are, like, *thousands* of these in the sky?" She scrunched her eyes like she could not even imagine what that would be like.

"My building is one of the first customers of StarSys," said a proud man in a business suit. He stood outside his company headquarters and pointed to the rooftop. "Up there, our rectenna catches lots of small bursts of power that we use to run our HVAC system. It's going to save us a lot of money!"

"What I want to know," demanded a mother with hands on hips, "is how *dangerous* are those power bursts. Say my son is flying a kite. What if the kite gets hit by one, and it lights him up like Ben Franklin. What then?"

An owlish man with a scowl on his face spoke vehemently, and spittle formed on his lips. "These will ruin my NEO studies. Amateur astronomers everywhere are banding together to stop this outrage. My time-lapse photographs will all be *ruined*!"

As the large satellite passed overhead across the northeast edge of Africa, there was a great furor. Many more people shot at the sky with rifles and sidearms. Millions of people stood under the stars and shouted curses and invective at the glowing, glittering octagon that seemed to pass right over their heads.

Further south, a small sounding rocket stood ready on a launch pad.

Kokomo, Indiana

Marcharelli and Jenks met with their senior technical staff to discuss the poor performance of the relay satellites. Customers were calling from all over the country, complaining that the power levels they were receiving fell well short of that promised. The second relay sat was scheduled to launch at the next half moon, doubling their output. But even then, their cost per kilowatt would be shy of the design levels, and uncompetitive even with ground-based solar power.

"Now, explain in terms I can explain to our SHAREHOLDERS," Marcharelli shouted, fear and anger showing on his face, "why our power levels are so low."

Frowning deeply, teeth gritted, the StarSys Director of Engineering explained using his hands – one an open fist, the other with pinky finger extended. His arms were stretched to their maximum reach apart from one another. "This is Earth. This is Luna. At the scales in space, my arms would need to reach from *that* wall to *that* wall across the room. Okay? It's a long-assed way, Robert."

"Fine. So what?"

"In between are the van Allen radiation belts. These are symmetric around Earth, but are aligned with the Sun. So they are oriented along the line-of-sight between the Sun and the Earth. With me so far."

"Yes. Go on."

"Our microwave energy packets – can I call them MEPs?"

"Yes."

"The MEPs travel on a straight line from Luna to Earth."

"Obviously."

"But Luna circles the Earth. That means every day, every hour, every minute, every second, the orientation between the MEP line-of-sight and the Earth-Sun line-of-sight is changing. Does that still make sense?"

"We'll have to show charts when we go public with this," Marcharelli said, "but I think people will follow."

"Okay, so the forces interfering with each MEP are constantly changing. To account for this, we reconfigure the phased-array antenna for each MEP, once a millisecond. With me still."

"*I* am, of course. But we're working on shareholders. I think you're still okay. Proceed."

"But also, each millisecond, we direct the antenna output to a different point in space. Once the second bird goes up, and then more and more, our distribution of MEPs will be scattered across an even wider angle."

"Makes sense."

"So, when we compute the antenna reconfiguration for each new MEP, to guide it accurately to the relay sat, our best estimate is based on the last time we shot in the same direction. Okay?"

"It's getting a little deep, but barely okay."

"We're almost there. Now, the updates for each target will be spaced further and further apart the more targets there are, right? So, as we scale up our operations, the updates will be more and more out-of-date."

Marcharelli couldn't control himself. "So that means as we scale up, our accuracy decreases, and therefore our transmission efficiency falls off."

"Exactly," the director said, and flopped back in his chair.

Jenks folded his hands on the tabletop. "Your explanation is good, but it doesn't explain why we've got poor efficiency even with just a single relay sat."

The director waved a hand as if shooing away a fly. "Part of that's just tuning the antenna. We've got neural networks training on every shot, tons and tons of good data. Our efficiency will come up as the adaptive electromagnetics get calibrated. Of course, we need a full year's worth of data to train the nets completely, but we should be 90% of the way there after next lunar month."

"At 90%," Jenks asked, "how do we compare with our cost targets."

The finance director took the question. "Based on our small scale testing, we came up with a range of values for the full-scale version. We're in that range, but barely."

"But does that level make us cost-effective?" Jenks pressed.

The finance director's eyes shifted nervously. There was a twitch at the corner of his mouth. "Um, that depends upon global energy prices. If they continue to rise at the historical rate since 2001, we should be competitive in, um, a few years – less maybe."

"This is unacceptable!" Marcharelli said. His hand came down on the table, late by a half-second, and without the satisfying thud that he wanted for emphasis.

"Robert is right," Jenks told them all. "We must find an engineering solution before we go public. It's all right if we're not cost effective just yet. We're still uncontested in this market, and that gives us some breathing room. But we cannot release our metrics until we have a more positive story to follow up with. Now, let's hear some ideas."

The chief engineer for relay sats, Larry, raised a hand and began speaking. "The simplest answer is to increase the diameter of the relay sats. With a larger target, the aiming accuracy is less critical."

"Oh, that's just great," Marcharelli said sarcastically. "That's all we need: more concerned citizen griping about light pollution and saying 'the sky is falling'."

"Is there a way to make the relay sats so they reflect microwaves, but not sunlight?" Jenks asked. "If they are less bright, we might forestall public criticism."

The relay chief jotted down a note. "We'll have to look into that, sir. I know we can make *some* improvement, but there are fundamental physics we can't get around."

"Can we modify our orbit so it's higher, or only flies over sparsely-populated areas?" Marcharelli asked.

The flight ops chief engineer answered. "There are a class of orbits we can move into, but each one has a higher altitude. The farther from the surface we orbit the relay sats, the larger they have to be. The angle subtended to the naked eye will be unchanged."

From a video presence, the mission commander currently on Luna addressed them, cutting into Jenks' next question. Jenks stopped short to let the man speak. "The other answer to what Larry said is to make our transmitting antenna larger. We can boost accuracy a factor of two if we quadruple the size of the array here. Would that help?"

Jenks answered his face floating above a place at the conference table. "We've already got some legal battles with UNOOSA on what people are calling 'large scale habitat destruction', which is prohibited in the Moon Treaty, both the original and the updated one. Apparently, our size is already bordering on what can be seen with the naked eye. A four-x increase in the transmitting antenna would be clearly visible from Earth. I don't think that's an option. But thanks for the idea, Fred."

No one spoke for a few moments. "Well?" Marcharelli asked. "Is that it? With all the money we're paying you people, that's all you can come up with?"

The engineering director frowned again, making deep parentheses in the skin connecting his mouth to his chin. "There is another answer, Robert. But we've all been careful to avoid it."

"Spit it out, man!"

The director narrowed his eyes at his boss – but briefly. "Continuous beams. With a continuous beam, our feedback is real-time, as fast as physics allows. We can keep the relay sats the same size. We can keep the transmitting antenna the same size. Each relay sat gets serviced for several tens of seconds, or a few minutes, and we can get nearly theoretical transmission efficiency."

Marcharelli almost erupted with a command, but caught himself. "But we can't, right?"

"That's right," said the director. "Canada Rules."

They all looked at Jenks.

"You're all wondering if we can change the rules," he said. "Believe me. Not a day goes by that I don't wonder that myself. The conclusion I've reached is that we can. But not now. Not yet. The issues are still too sensitive. The technology is too new. The Oasis Massacre is still in the racial memory of the generations alive today. Right now, the Canada Rules enable us. Protect us from criticism. Provide us legitimacy. To abandon them wholesale invites back every obstacle we've already overcome."

"Except the technical ones," challenged the director.

"You didn't hear me say this," Marcharelli began, "but have we considered simply switching to a continuous beam, but not telling anyone?"

Larry shook his head. "We'd be caught within minutes. Hours at the most. Anyone checking their operation logs on their StarSys ground units would notice at once. Any government agency sampling our beams with airborne monitors would know instantly. We'd never get away with it."

"You knew that, of course, Robert," Jenks said.

"Of course. I'm just frustrated, that's all."

"Okay, people, what else?" Jenks asked.

No answers were forthcoming.

"Okay, then, let's start thinking outside the box. What alternative technologies can we exploit to boost transmission efficiency?"

"Lasers." Fred said, from the Moon.

"Explain, please."

"Coherent radiation doesn't disperse much," he said. "Compared to this huge beam we've got, with its lossy side lobes, a laser would dramatically simplify the transmission problem we've got. At the right time of day, we can even deliver direct from Luna to our customers on Earth."

“We can’t use power lasers in space!” cried the corporation’s chief council, who had been silent until this point. “That’s a totally ridiculous idea, Fred. For the same reasons Larry explained, but a hundred times worse. Once we were caught – and we *would* be caught, gentlemen – we’d be shut down before you can say ‘Bob’s your uncle’.”

Marcharelli raised himself to his full seated height. “I want an answer, and I want it tonight. Each of you go call in your entire staff. Right now. Get everybody here, and find me an answer!” He moved to leave the room, and banged his knee into the table leg. It would have hurt terribly if there were any nerve connections to his brain, but he felt nothing. The wheelchair reversed sharply, pitching his torso forward, then he wheeled about and lurched forward, snapping his head back against the seat. They heard him give a cry of pain as he left the room.

Burns Flat, Oklahoma

The second Marcharelli rocket climbed into the sky to a much smaller crowd. News of low power output helped allay fears of the technology, but seriously hurt stock prices for Marcharelli Space and StarSys Corporation. Engineering teams worked around the clock to boost efficiencies, and both Jenks and Marcharelli issued positive, confident assurances that design efficiencies would be met. The first stage dropped away over the Atlantic, and the second stage booster oriented itself for the next burn.

Shortly before it passed over Egypt, a slender white sounding rocket leapt off a launch pad, rising on an exactly vertical line. Too small to achieve orbital velocity, the vehicle nevertheless could achieve altitudes of several hundred kilometers before running out of momentum.

At the apogee of its flight, the rocket exploded, shattering into thousands of sharp pieces. They spread throughout the black sky of space, expanding in outward arcs, and leaving a cloud of shrapnel that would soon start the acceleration back into the atmosphere.

But before that could happen, the second relay sat and its booster engine passed into the edges of the debris cloud at 25 thousand kilometers per hour, and fired up. The shards of the sounding rocket shredded the relay sat and booster, punctured the fuel tanks, and triggered a powerful explosion. The streaking fireball was clearly visible from the ground. The relay sat was totally destroyed, and left a trail of dangerous space junk that would litter LEO for years.

Honolulu, Oahu

Elise welcomed the LunarSource board members to the Vambolent's new home with a sour expression. Worry lines creased her brow, and age finally began to pull her body down and out. Mindy, dressed in a black maid's uniform with a lace apron, was taking a break from homework at UH to serve their guests. Eric, who was in his last year of high school, drove his father's platinum-colored Cadillac hybrid fuel cell vehicle back and forth to the airport as people arrived on-island.

Ned Austin had come the day prior, and was now bright red from sunburn. He sat in the cool, dark living room, sipping chilled rum and coconut milk, trying to move as little as possible. Yoshi Hanazawa arrived earlier in the day, and was now busying himself around the grounds, tending to the landscaping. Dajwal Chellapilla stood on the far side of the pool deck, talking out over the hillside, his voice picked up by a throat mike and relayed to Colorado via satellite.

Parks and Afren returned from a walk through the gated neighborhood as Eric drove up with the last carload. Franklin Juneau climbed out quickly and rushed to the front door where he helped a very pregnant Sellenya Laurencino to her feet. Elise trotted outside and embraced her, enjoying the contact. Parks and Afren strolled behind, watching the scene with amusement. The board was now complete.

Shortly after, they sat for dinner. Mindy brought each of them a bowl of vegetarian saimin, garnished with baby bok choy and green onions. Accompanying the noodle dish were pot stickers filled with spiced, crumbled soy protein. Homemade nori wafers were fanned around julienned carrots and beets at each setting.

"Saimin is the state dish," Elise explained. "It evolved historically from the confluence of many national cuisines. Plantation workers in Hawai'i coming from the Phillipines, from China, Japan, Portugal, as well as island natives would bring ingredients to share, and make a communal dish. Over time, this meal has developed into a staple of the islands, and can be found everywhere, from fast food joints to the finest restaurants."

They sampled their food and complimented the chef.

"In a similar way," she told them, "all of us here tonight have come from many cultures." She raised her wine glass, and the other raised theirs as well. "Perhaps tonight we can find a way to combine all of our ingredients, and make something that we can all enjoy."

Parks frowned at her obvious manipulation, clearly aware that she and he were on opposite sides of the debate they had all come to resolve. Unaware of the hidden tension, Franklin asked to be caught up on the news.

"Parks, I hear you have had some surgery. Did it go well?"

Elise dropped her spoon, which clattered loudly on the china.

"Yes, Frank. I got a new heart."

"That's what I heard."

"I was tired of taking it easy. Tired of being tired. I was growing soft, and getting ready to die. So, I flew to Seoul for the procedure…"

"Without telling me!" Elise said, glaring at her husband.

"…and everything went fine. Now, I'm stronger, fitter, and I can eat just about whatever I like."

"Which is disgusting," Mindy chimed in from the sideboard. "Dead ham in the fridge – yuck! Do you know what they put in it to make it look red?"

Yoshi smiled at her.

Parks chuckled and sat up straighter. "I feel great, Frank – like a million bucks!"

"You sound like a new man," Dajwal observed. "I have not heard you sound this strong in many years."

"I am, Dodge. I've had it, letting others take the lead in my race. No more bleeding-heart apologies – no, sir! No more Mr. Nice Guy. I'm out for blood. I'm out for money. I've got a new lease on life, and this time, there's nothing gonna stop me."

"Even if it tears our company apart," Elise commented bitterly.

"I think, people," Afren said. "That we should reserve this discussion for our meeting. Personally, I am very interested to hear more about you, Ms. Laurencino. Do you have a name selected for the baby?"

Sellenya told them about her and John's plans for the baby, chatting brightly in a way none of them would have ever imagined coming from her. "If it's a boy, his name will be Artemis. A girl – we plan to call her Diana."

"Moon names," Austin observed. Sweat beaded on his forehead from the spicy food, exacerbated by heat from his sun-damaged skin.

"You know, Ned, that's where we first fell in love."

"I didn't know that," he said, wiping his brow. "Only later did we figure out that's why you stayed the extra two weeks. Can you tell us the story?"

She did, and the dinner passed pleasantly.

Mindy cleared their places, and set small bowls of cut fruit and rice pudding before them. She poured tea and coffee, then left to let the adults digest their meal.

* * *

Around the Vambolent living room, beside each seating place was a folio of materials for their board meeting. Elise sat next to Sellenya on the sofa. Dajwal settled next to Elise, with Afren on his other side. Franklin and Ned flanked Parks, who sat opposite Elise. By their apparently casual seating choices, the battle lines were drawn.

As co-CEO, and being the more neutral of the two, Dajwal introduced the meeting's topics.

"Okay. We are all here, so let us begin." He wrung his hands, looking quickly from face to face. "A summary of our financial standings can be found on the third page of the packets beside each of you. The next several pages outline our bookings, our debt position, a capital summary, and projections for the next four quarters. Narrative summaries follow of our lunar operations, written by Mr. Dallas Barstow, and a briefing on the Japanese portions of the transportation system, provided to us by Yoshi-san himself."

Most of them leafed through the pages, scanning the material. Dajwal cleared his throat uncomfortably.

"Ahem. And then, at the back of your packet, behind the tab, you will find the new business for today's meeting. Our president – Parks, that's you – has a proposal for a new direction in the strategy for our business implementation. I believe, Parks, that you are best suited to explain your proposal," he finished quickly.

Parks sat comfortably straight in his chair, and looked from eye to eye around the room. His face was tanned and deeply lined. Silver hair graced his temples, framing the steely-blue color of his eyes. His cheekbones seemed higher, and his chin more pronounced since his visit to South Korea. An air of authority seemed to surround him, an utter confidence evident from the set of his jaw. Even his mouth seemed more firm, smiling perhaps in optimism, yet still traced with the hint of a smirk.

In that moment, Sellenya decided to herself that she would vote for him, no matter what happened. She even began to reconsider names for her baby – if it was a boy.

"I am very proud of our company, of LunarSource," he began, looking straight at his wife. "Elise, you and Dajwal have done an outstanding job building LunarSource into a profitable enterprise. Yoshi, with your cooperation, we are building a powerful new economy in Japan, and have begun expanding our platinum group metal delivery to other markets. We have the premier on-going enterprise on Luna, and the best operations in cislunar space. These are incredible achievements."

Parks stood, and began pacing against the picture windows. The hazy sky outside silhouetted his figure, black against the medley of colored flowers and palm trees lining the pool deck.

"We've benefitted greatly from the fiasco by NASA in crashing that asteroid into the Far Side. Our profits were buoyed by the stratospheric price of platinum. Now, as you can see in your reports, the market has levelled out. Our supply of PGMs, together with expanded mining operations in South Africa, the Pacific Islands and Southeast Asia has taken some of the luster off platinum. And with gold prices so high, nations have been selling off their reserves. Russia, which has an enormous hoarde, has been increasing their sales, flooding the market with their bullion. Palladium, used in hydrogen storage, has also softened. And even more recently, advanced catalysts are being commercialized for fuel cells that could cause even platinum itself to roll over. Then, we'd be dead."

He came back to his chair, standing before it, hands curled and fingers tight.

"And then we have the recent events over Northeast Africa. Winnimer Jenks and Robert Marcharelli, their StarSys enterprise, and the lunar energy facility, are on indefinite hold until stability can be found in Egypt. Even so, technical problems abound. Franklin tells me they still don't have answers to huge problems, right Frank?"

Juneau nodded rapidly, transfixed by Parks' oratory.

"Solar wind, and the electromagnetic flux are factors you can't ignore between Luna and the Earth. As Luna orbits us, those patterns change constantly. Sometimes the sun burps, it farts, it bubbles and boils. Yet, their control loop has a feedback delay measured in seconds. Correcting for this is hundreds of times harder than the adaptive optics used on Earth-based telescopes like at the Keck observatory, just down the highway from us.

"And then, they have a business model which requires huge, fast, expensive satellites to fill up low-earth orbit. Orbital debris is an ever-worsening problem. Atmospheric drag means they need to boost up every couple of weeks. All of these problems scale up as they sign on more customers. As Yoshi explained years ago, the whole enterprise will eventually collapse of its own complexity."

Parks sat down, almost finished. His hands gripped the arms of the easy chair, denting the fabric, and paling his knuckles.

"Don't you all see? Jenks has created the market. He has people wanting space solar power. But he can't deliver. StarSys is doomed to failure. And what will fill that need? What other answer does the world have to the pressing problems of finite petroleum, atmospheric warming, and nuclear proliferation? That is the question we are here to discuss."

He sat back, relaxing, and looked at them all individually. They waited for him to finish.

"Us. Those of us here, my friends. We have the answer. We had the answer years ago, but the time was not yet ripe. Now, the time is ripe. We have the know-how. We have the infrastructure. We are ideally positioned to capitalize on this incredible opportunity. This is our chance to take our company to a whole new level. My proposal is simple. It's all in your packets. We double our staff at Far Side Base. We build solar panels and launch them to geostationary orbit. We capture them with payload catchers. We assemble them with waldoed spiderbots and deliver low-density microwaves to rectennas beside major energy markets. We sell to the utilities and use the profits to expand our operations. We build habitats on the moon and in orbit using the igloo technology that the Eskimo-man started at LEF. People, this is it. This is the time we've all been waiting for. The time is now. It is our time at long, long last. Thank you for your attention."

Sellenya felt her hands start to clap each other, but stopped them. Ned's face stretched in admiration, but feeling the burn, pulled it back to a neutral visage. Franklin beamed at the image Parks created, seeing the installations grow, the power delivered, people populating habitats, beating back Chaos from the doorstep of the Earth.

"Parks, I thought your dream was to retrieve a comet?" Afren asked politely.

"Comet-shcmomet," Parks spat. "We're going to build an array over Africa. Businesses around the world will crowd into the continent, ready to use talented, low-cost labor and cheap power."

"How about climate change, Parks?" Elise said. "The past three hurricane seasons have been pretty mild in the US, your father's prognostications notwithstanding. People have been moving back into Washington, D.C."

"You're right dear. But people have short memories, and that is just a fact. We all know that climate is chaotic. A few years of respite is merely the calm before the storm. My father saw this years before it happened. When the bad weather swings back to the Western hemisphere again, it will be worse than ever before. The time to invest is now. If we wait, it will be too late. World economies will be crippled even further, until the point where a space-based solution becomes too expensive for anyone."

"Your plans are ambitious, Parks," Dajwal said. "Ambitious plans require considerable capital investment. I do not know if LunarSource can withstand the drain on its resources. I can very easily envision a scenario where both enterprises collapse, should the market fail to materialize exactly as you envision."

"A valid point, Dodge. Which is why I am proposing a spin-out company. LunarSource will be a minority investor, but I'll be free to find other funding sources."

"Such as…" asked Yoshi.

"As you know, Yoshi, I have some investment funds still available…"

"Parks, you promised me," Elise protested, stopping herself before revealing that he was intending to spend their nest egg, their safety net, the assurance that their children could finish college.

Parks ignored her.

"…which will be supplemented by advance sales from arrays over Asia. Monsoons and typhoons have not slackened, but grow worse. As China and India struggle to bring their populations up towards developed nation lifestyles, their hunger for electrical power will never be sated. Japan, China and India have all done the math. They all want to do this, but don't have the foothold we've already established. There are trillions in untapped potential worldwide."

"And how do you address the developments in the Middle East," Afren asked quietly. "This man Hababek has many adherents. There is talk of making him the Caliphate of all Islam. As much as I loathe to admit this, these ideas are in direct response to the popular call for me to become …. Osagyefo." He spoke the last word with a sour look on his aging face.

Parks spread his hands.

"Look, I can't solve all the world's problems. But energy is something that affects everyone. If we create this spin-out, and we can deliver carbon-neutral, uranium-free, unlimited energy to the earth at costs that actually decline with increases in scale, that takes away the pressure which brings nations into conflict. What more can we do? But if we don't do this, won't things be that much worse? Without power from space, terrestrial energy is at best a zero-sum game. Worse, it's a diminishing resource. The human race has no choice but to go solar or die!"

"What will you call the new company?" Ned asked in the quiet which followed.

"Vambolent Interspace Industries," Parks enthused. "The acronym is the Roman numeral for the number seven. Unless I've missed my count, this is the seventh initiative I will have started on the way to this ultimate goal. We will have divisions for space electrical power, for raw materials, for lunar transportation, for cislunar delivery, and for habitation. We will open up space for settlement. It won't be just energy paying the bills then. No, sir, it will be tourists, entrepreneurs, sports extremists, and even colonists. We can even build ships from lunar materials which can take us out to the asteroids, to Mars, and eventually, eventually, to capture a comet."

"I'm in." Franklin said. He stared at Parks with complete conviction.

"Me, too," Sellenya said. Elise looked at her blankly.

"You can count on me, Parks," Ned announced, "but I'll need a bit more sunblock if I'm going to be a space tourist."

"Yoshi?" Parks asked.

Yoshi pushed his lower lip very high on his face, blinking rapidly. "I am willing to help you build the payload catchers at GEO," he said slowly, "provided you can show me the funding you propose. It would not do well to harm the LunarSource business model if you do not come up with the money you need for this."

"I guess that leaves it up to our co-CEOs," Parks said, and folded his arms across his chest. The rest of the group turned to look at them. Mindy, who had been listening from the kitchen, peered around the corner to watch her mother's response.

Elise's face was furrowed in concentration. She assessed the resolve in each person's face, and then turned her eyes to a blank spot to think. She turned to Dajwal, who looked back at her silently. He knew the real struggle here was between the two Vambolents, despite his position in the company.

Elise drew a breath in through her nose, and set her hands on her thighs.

"Dajwal and I will each have a seat on your board, and the total number of board members cannot exceed seven. LunarSource will pledge a small staff to get you started, and in return, LunarSource ownership of V.I.I. cannot be diluted below 19 percent. Any intellectual property you generate will be jointly owned by the two companies. Anytime your operations conflict with the planned business strategy of LunarSource, the needs of LunarSource will be given priority. And finally, you personally, Parks, will be limited to no more than 60 hours of work in any given week. I mean it!" She looked at Dajwal for confirmation. He nodded, ever so slightly. "If you agree to these conditions, then you have our vote."

Parks looked at her with awe and respect. Speechless, he nodded his assent. In his self-absorption as president of their company, he had hardly noticed what a fine businesswoman his wife had become. It almost brought tears to his eyes.

But because he had no human heart, his eyes remained dry.

Cairo

Two days later, Winnimer Jenks was in Egypt.

They met in Mohamed's Mosque after the building and grounds had been cleared by Jenks' security people. Hamid Hababek arrived with only a young boy to help him up the stairs. He passed through the perimeter of guards, scowling at them contemptuously, and removed his sandals. He stepped onto the rugs which covered the floor of the main room, which rose high above them and was filled with colored sunlight. Hababek waved the boy away, and approached Jenks alone.

Jenks waited with an interpreter, a slender woman dressed courteously in a full length skirt, and a high-necked sweater. She stood behind Jenks, and whispered to him the appropriate words of greeting.

The two men settled down on the clean but worn rugs. Jenks was careful not to show the soles of his feet towards Hababek, as he had been coached. The interpreter sat to the side, and performed a near-instantaneous translation between Arabic and English. After just a few moments, both men could ignore her, and speak directly, face to face.

"So, you are Winnimer Jenks. I expected you to be older."

"I feel older, since you declared *jihad* upon me."

"You are the agent of Satan. You bring fire from the sky. Your people have rent the Moon and cause water to spring from the Earth. It is the end of times, as the Quran correctly prophesied. It is my holy duty to remove you from the Earth, and send you to Allah for judgment, as he surely will."

"Your rocket destroyed my satellite."

"Rocket? I have no rocket. I am a simple cleric. Poor, humble in the face of Allah. I possess no rockets."

"Your people then. Their actions could be construed as an act of war. My government is considering taking action."

Hababek smiled smugly, his dark lips stretching and cracking. "What do you know of war, young man? If my people are testing rockets, that is their right, surely. You launch rockets all the time. It was unfortunate that their rocket exploded. For you, it was unfortunate that your rocket was also destroyed. But this is the will of Allah. If He did not wish it, it would not have happened. He is very wise. You should learn from his Acts."

Jenks tried a different tack. "Our rockets are designed to help people. To help reduce suffering on this Earth. To allow more people to live in peace."

"Your rockets are designed to make you money, and for your greed you shall be punished. And what do you know of suffering. I see your soft hands, your smooth face. A little suffering would serve you well. Yes, I think it would."

"You were very upset when the Mossad attacked Egypt," Jenks said, trying yet another approach. "That was a long time ago, and we have learned much since then. With the Canada Rules, it is no longer possible to use this technology as a weapon. It is perfectly safe."

"Safe, you say? Safe? I do not believe you. My advisors tell me that your equipment is perfectly capable of firing a continuous blast of power even more destructive than the one used on Egypt. How can you look into my eyes and tell me this is safe?"

"It's the way we operate the equipment that prevents mis-use. Unless someone completely reconfigured the equipment, it is impossible."

"Do not insult me, Jenks. The eye of Allah is upon you in this holy place. If you are so foolish to believe that your power beams cannot be used for evil, then go away and tell your puppet master to come talk to me." Hababek turned his face away from Jenks.

Jenks bristled at this remark, and fell back to try yet another angle. "Your people have rockets. This one exploded. Perhaps my people could come to help you. In return for a truce."

"We already have people helping us with our rockets," Hababek said blithely. "Very smart people from the Ukraine. They are very good with rockets. Sorry, but we do not need your help."

Jenks did not respond to this revelation. His search agent had already informed him of the high probability of this development. He knew that now was the time to play his trump card.

"Mr. Hababek, I understand your concern. I have traveled here to tell you that my company, StarSys, has decided on a different approach."

The old man's eyebrows stairstepped up his forehead, curiosity showing in his rheumy eyes. "I am listening."

Jenks leaned forward, pouring sincerity through his facial expressions. "The man you want is not me, but another. Parks Vambolent."

Hababek reacted, and Jenks knew he had taken the bait. Now, it was time to set the hook. "Parks Vambolent, the man whose solar power satellites were responsible for the Oasis Massacre, is planning to do the same thing again, but on a much larger scale. He is planning with Afren Kwapili, even now, to put an enormous solar array in orbit over Africa."

Jenks watched the hatred return to Hababek's face. The names 'Vambolent' and 'Kwapili' were powerful levers, and Jenks worked them masterfully. "In just a few years, maybe less, Vambolent's installation, hidden on the backside of the Moon, will start shooting canisters down towards Earth. They will be collecting them in orbit over Africa, and assembling the largest solar power array ever constructed. This installation will beam continuous power straight down. Hundreds of megawatts, and eventually hundreds of gigawatts, right down over your head. Even a tiny fraction of that power is enough to burn the entire city of Cairo into ashes in a matter of minutes."

Hababek's breathing became ragged and his cheeks flushed. Jenks pushed harder. "Kwapili seeks to unify Africa. It is his ambition to become Osagyefo. With all his power from space, he can force Egypt to join the AU. You will have no choice but to surrender, or be destroyed."

Hababek clambered painfully to his feet. His fists were clenched and shaking wildly. His teeth gnashed and his wide nostrils flared even wider. Jenks felt like he could see directly into the man's brain.

"Please, Hamid, please sit down," Jenks soothed. "There is an easy answer. An easy answer, my friend."

Hababek allowed himself to be calmed, then regained his self-control. He lowered himself to the floor, and looked at Jenks under half-lidded eyes.

"Now, I am listening again. I am ready to hear your 'easy answer'."

Jenks explained his proposal.

Both men left the mosque smiling.

Chapter 18

After decades of escalating tensions worldwide, several modest developments helped a cooling of conflicts, concerns and crises. Weather patterns were mild for a few seasons in a row, and hurricanes in the Atlantic returned to their earlier intensity, despite the doomsday predictions of the previous years. Parasite-born tropical diseases had spread, true, but medical advances on several fronts were providing cures and vaccines. Droughts and warming climes shifted the fertile bands for many staple crops, but farmers adjusted, moving their operations farther from the equator. This eased the incredible pressure on land prices that had been caused by the escalating storms on coastal regions.

Energy prices, which had been gobbling a greater and greater share of expenses for homeowners and businesses had leveled out. With oil prices still rising, and nuclear energy still costly because of the IAEC regulations, alternative fuel sources were having a renaissance they had not seen since the 1970s. Wind, solar, hydroelectric, geothermal, tidal and biomass power plants were being built at a rapid rate. Hydrogen synthesis from coal, tar sands and frozen hydrates became even more attractive as energy prices offset the sequestration costs for the carbon.

People everywhere, except for the most die-hard status quo believers, began to realize that this was a temporary reprieve before things turned even worse than before. Governments and businesses scrambled to build infrastructure during the respite. Emergency management agencies developed disaster plans while their economies were enjoying a welcome year of growth and expansion.

Orbital factories, hotels and retirement habitats began to fill low-Earth orbit. Booster-assisted space planes made travel to orbit accessible to the moderately-rich. Zero-g hotels were a popular honeymoon location. For the less wealthy, sub-orbital parabolic flights were now routine, and no more costly than a transcontinental flight. UNOOSA had at last authorized the building of the Lunar Polar Observatory, and granted the iron and aluminum feedstock contract to Vambolent Interspace Industries. With two going concerns on the Moon, there was buzz about building commercial habitats there, a place for the enfeebled, the eccentric, and the elite.

StarSys, still led by Winnimer Jenks, was caught retooling during the energy price lull. After de-orbiting the only relay sat that remained, StarSys began the changeover to a completely different architecture. Instead of bouncing lunar solar power from LEO, they now planned to capture 19.6 GHz microwave energy at GEO, heterodyne it down to 2.45 GHz and transmit to rectennas in the US Southwest. Yet, investors in StarSys and Marcharelli Space were distraught with the ensuing collapse of their stock values. Financial analysts kept a sharp eye on the power race between StarSys and VII.

Senator Mason Barnes, halfway through his announced last term, was furious with Jenks for pulling StarSys operations out of Kokomo. Nationwide, a class action suit had been filed against StarSys to recover money for all the receiving systems they had sold, but which were still useless. Barnes ordered an investigation into the business practices of StarSys in general and Winnimer Jenks in particular.

Parks Vambolent worked around the globe, avoiding only the US. Early adopters of his space solar power installations began placing orders at discount rates as an incentive to bootstrap the operations. The contract with UNOOSA for LPO came at a good time. Russian rockets were busy ferrying African workers to Luna, and the growing stock of raw materials was amassing ever faster as they used part of their production to build new manufactories.

Despite the good news, soft energy prices held his business plan from closing in on a positive cash flow. Parks knew the only way to cover the gap would be to sell his company in an initial public offering. A moderately succesful IPO would enable him to complete the African cluster, but there were still not enough direct investors in African infrastructure, despite loan guarantees offered by the World Bank. In this chicken-and-egg scenario, Parks knew he needed a breakthrough, or a miracle, to go over the top. So, he bullied on, refusing to surrender his utter confidence in success.

Still, waking next to Elise in the middle of the night, he wondered. What would be the outcome of this, his final gamble to make space solar power economically viable.

Geostationary Earth Orbit

John Lomish was running out of the vast patience that was his hallmark. Three months on-station with Benza Loray, and they still had not fired up the magnetron. In fact, the whole array was turned 90 degrees to the Sun, so their hundreds of megawatts generating capacity was sitting idle.

Just like he was.

Like clockwork, the payload catchers did their dance and jerk with each new canister from Far Side Equatorial Base. Dallas Barstow was back there now, on his amazing fourth trip to Luna – the world record holder. Lomish spun his workpod to look at the Moon.

The waxing crescent meant it was almost high noon over Far Side. *They* would be busy. Ah, to have the Sun overhead, all that equipment glowing, gleaming and cranking out product, all that machinery breaking and getting fixed, gumming up and getting overhauled – now *that* was work!

At Far Side, they never saw the Earth. But that meant Earth couldn't see them, either. From Luna, he recalled, the Earth was the size of a coin. Beautiful, yes, but small and remote. There, you were safe. Here at GEO, the view of Earth was much more imposing. You definitely saw it as *planet*, a ball, a sphere, a big round rock. It loomed large in almost every field of view. To Lomish, it looked menacing. Just the thought of all those people down in Africa below him, hating him, wishing him dead, made the hair rise along his spine.

On their way up, while parked in LEO, he saw the enormous rectenna which should even now be soaking up their power. Built on the arid Bateke Plateau in the Gabonese Republic, it was as far from Egypt as Kwapili and Vambolent could place the equatorial receiver. The Gabonese were delighted, as their oil revenues were exhausted, down from being 80% of their exports just a few decades ago. But they had yet to taste the first erg of power from the Africa Cluster.

Lomish knew that Vambolent and Kwapili were waiting for things to settle. The death threats against Vambolent were considered totally serious. Crazy man Hamid Hababek had updated his *jihad* hit list. Topping the charts this time was Parks Vambolent, CEO and namesake of Vambolent Interspace Industries. Given the previous year's attack on Jenks' StarSys satellite, everyone was being extra careful.

Kwapili was doing the diplomatic thing, but to no avail. Secretly, Parks had told Lomish and Loray that they would see action soon, since Kwapili was about ready to just pick a date and start beaming. Probably a few days after Ramadan, before the Islamic community became grumpy from fasting. Until then, they were to keep building onto the array. The work took only a few hours each day. The rest of the time, they were in a holding pattern – orbiting without seeming to move. It was maddening!

If they had asked *him*, they should fly the whole array around to the US and beat Jenks to the punch. The StarSys heterodyne facility was a distant pinpoint in his telescope. They were already beaming modest amounts of power to a rectenna in New Mexico. But even they had been stymied by Greenies wailing about the fate of endangered species like the sonoran green toad and the astragalus hypoxylus. That plant, the Greenies said, had never been domesticated, yet showed immune system boosting unmatched by man-made drugs. In this age of designer viruses and the constant threat of plague, they did not want the US government to jeopardize such a potentially valuable resource.

Meanwhile, in John's opinion, *mankind itself* was about to go extinct.

Yeah, if they asked him, Vambolent should just snatch the LEF back from Jenks – everyone else seemed to be after him anyway – so they could build clusters 'round the clock. Then, they wouldn't even need the rails around the poles, like Parks had planned. Trains take a lot of work, a lot of maintenance.

But then again, that wouldn't be too bad. Trains had been John's third love as a kid, behind dinosaurs and rocketships.

All the automation was getting to him. Years ago, he had watched Barstow sit back as AI piloting routines landed their spacecraft. Today, human backups to autopilots were a luxury not always afforded. But now, they were automating *his* tasks! Back when Washmick and Loray were assembling the first cluster, the waldos had learned their tasks. Refined and streamlined in the years since, the waldos were clearly superior to hand work. They never tired or got distracted. With their nano-midges, they self-repaired everything but the occasional meteorite strike. Even those repairs would be automated soon.

Sometimes, John wished he had more to do. Daydreaming about his beloved Sellenya and fawning over movies of nine month old Arty Lomish only filled so much of his days of endless night. On Luna the days were impossibly slow, but noticeable. Here, the Sun and Luna whirled daily, but without a frame of reference, they were just *there*. Earth always looked the same except for clouds – random but somehow always the same. Boring.

He wanted something to work on. He wanted to start sending power to Gabon. He wanted to rotate out of this ancient tin can and get away from the always busy, always chipper Benza Loray. He wanted to get back to his wife and baby boy. He tried to refocus on the payload catcher schematics he was supposed to be reviewing.

He wished for sleep.

Orange County, CA

Jenks left the country the same day a federal grand jury decided to indict him on charges of conspiracy, racketeering, and obstruction of justice. Also working its way closer to him was a re-activated FBI investigation into the microwave assassination attempt at Klamath Falls. The lease held by StarSys on the LEF was not renewed by the legal corporation holding the assets, and Jenks had been delivered a letter telling him to vacate the premises. Environmentalist groups, powered by the ever-stronger Green Party, initiated lawsuits to save species and habitats, sponsored boycotts of lunar solar energy, and staged protests at the gates to the rectenna facility.

The bad press from all that brouhaha, coupled with the suspicions surrounding the CEO, was compounded by internationally-coerced FCC take-backs on the 2.45 GHz bandwidth used by the heterodynes. StarSys stock tanked. Marcharelli Space stock sank as well. Layoffs and closures at their fabrication and assembly plants turned workers against the company. Their plight galvanized the nation against the whole enterprise. So elected officials, from His Honor, the Mayor of Kokomo, to Indiana Senator Mason Barnes, to the President of the United States were about ready to decide the US had had enough of space solar power.

Insiders had warned Jenks to leave for days now. People avoided him around their office building south of L.A. He wrapped up his affairs, and left with only what he could carry.

Including hundreds of millions in bearer bonds.

He landed in Honolulu, dropped a letter into a local mailbox, and then changed carriers for a flight to Narita. In Japan, he changed terminals to the spaceport, and bought a ticket on a commercial suborbital spaceplane to Athens.

Debarking in Greece, he took several taxis, leaving one and passing through buildings to catch another, changing hats once or twice, and shedding his overcoat. He even bought new shoes and left the old ones in the store. Then, thoroughly cleansed and free of tails electronic, visual and orbital, he finally arrived at the marina.

Waiting there was a sleek cutter, his retirement vessel, prepared in secret over many years. The captain recognized him through the monocular, and signaled his men. They pulled in aside Jenks, who strode up the gangway and disappeared inside a hatch.

In just minutes, the ship backed out of its slip, wheeled about, and steamed for open water. Once past the last marker buoy, the twin diesel, 12 cylinder engines revved up the forward-facing propellers, and the small ship dashed across the surface of the Med. They sailed south to a point at the far end of line-of-sight with a receiver on the top of the remaining pillar from the Library at Alexandria, whose books had been burned in 642 to fuel the public baths. There, they waited.

Addis Abbaba, Ethiopia

Afren Kwapili made a public announcement, broadcast all over the continent, that the SPS cluster would begin beaming experiments 3 days hence. Hamid Hababek, anticipating this, had a videotape prepared. He protested the fire in the sky, and vowed that before any beams could be shot from space Allah would strike their evil machines from the heavens, and they would all be destroyed.

That was it. A sound byte the perfect size to pass through the media channels like a slippery fish. Within hours, just less than a day of Earth's revolution, all news-savvy people on the planet knew of the face-off over Africa's Solar Hydrogen Economy.

Olympia Mountains

Parks got a call, patched through a secure line from Colorado Springs to the safe house where the FBI had sequestered Elise, Eric, Mindy and him. The bust and voice of Mason Barnes filled the pseudo-3D display. He looked careworn, past his prime, but still soldering on. Gravity pulled at his aged body mercilessly, and the effects showed, despite an obvious effort not to.

"Parks, Elise, good to see you both." The voice was slower now, strained and edgy. Firmness of purpose and tacit authority still came clearly through the old voice.

"It's a pleasure to see you too, Senator," Parks said.

Elise stood behind Parks, resting her hand on her husband's shoulder. "Hi, Mason," she said, smiling broadly.

"Hello, Elise. You know I'm still a vegetarian because of you." He winked.

"You've been saying that for years, Mason. These days, almost everybody is."

"Ah, you're probably right. I'll have to find a new thing to remember you for."

"How about protecting my husband?" she asked.

Barnes looked struck. "Yes, well, there's something to be done about that. Yes, there is."

"What, exactly? Can you protect John Lomish? He's got a baby down here."

"Yes, well, I've just dismissed my advisory team. They're prepared to do everything they can. Everything they can do, will be done. You have my word on it. Meanwhile, how are the children?"

"They're fine," Elise said, and kneed the back of Parks' chair when he looked up at her with a frown. "But what exactly *can* they do? I mean, assuming the wacko has another rocket over *there*, our defenses are over *here*, right?"

"That's right, Elise. Yes, that's right. Our missile defenses can protect us over here. And to a certain extent, we can protect John and the other man – what's his name?"

"Loray," Parks said. "Benza Loray."

"Yes, Mr. Loray as well."

"But Mason," Parks said, "can our defenses really protect them?"

"We can train our defenses to the space around your Africa Cluster. But I must point out, Parks, that this is your first space installation not done under the auspices of the United States. You were aware of this, I know. But by space law, protection of that facility must come from the sponsoring nation."

"We don't really have a sponsoring nation," Parks said. "The AU has not been officially recognized by the UN as a sovereign entity. Plus, they have no military. So, I guess we are defenseless after all."

"Now, don't get so grim, Parks. Not so grim just yet. I've got the personal assurance from the President himself, as your Commander in Chief, that those two Americans on-station will be protected to the full extent possible."

"You've said that several times, Mason," Elise broke in. "What do you mean? Can they stop an attack, or not?"

"Well, seeing as you put it so directly, Elise, there are certain attacks we can stop, and there are other attacks we cannot. My advisors seem to believe that Hababek's followers have a GEO-capable rocket loaded with nothing more dangerous-sounding than rocket fuel. But that rocket fuel can certainly be used as a bomb. Apparently, rockets are nothing but a controlled explosion anyway, according to my top scientists, so making one blow up is about the easiest thing you *can* do."

"So, what does that mean for Benza and John?" Elise asked.

"Well, Elise," he said, and then fell into a pause. With a start, he began talking again. "If a *spaceship* were to approach their station, like a boarding party, with soldiers and all, we have an escalating series of measures that would surely deter them. If they would not be deterred, they could easily be disabled. If necessary, destroyed. But my analysts believe they have something else in mind. From their attack on the StarSys relay satellite, they know full-well it doesn't take much of an explosion to do a lot of damage to something in orbit. My people believe they are *counting* on us to destroy their rocket, since the debris will do their job for them, but they can blame us. If we do not attack, my people believe they will simply cause their rocket to explode – calling it an act of God, no doubt – and this will amount to the same thing. Either way, the facility is ruined. That's what my best people believe anyway. They could be mistaken." But the look on his face indicated that such hopes would be foolish.

Parks stood up at the revelation that he was trapped. "Can't you just send in the Marines, and destroy their rocket?"

"We can't do that, Parks. It's a different world today. You have to trust me, there are wheels within wheels. We just can't do that."

Parks' eyes flashed at that phrase. He bristled with anger. "So, you're going to do nothing?"

"Parks, I understand your frustration, but you are being unfair," Barnes said, straightening himself. "We're doing all we can. You know me. Trust me when I tell you we-are-doing-all-we-can."

Parks sat down heavily. Elise had nothing to say. Barnes had nothing more to tell them. They said muted goodbyes and well wishes, they ended the call.

"What now, Parks?" Elise said. "What about Arty Lomish? That's his father up there."

The tough guy that Parks had become since his heart replacement suddenly snapped. He felt as if the ground beneath him had dropped away. He felt heavy and falling at the same time. He felt empty and inert as a stone, all at once. A spasm of a sob gripped his chest and rattled the steel and plastic inside.

"I just kill myself," he said. "I've failed for the last time, Elise. Everything I've ever worked for is shot. I'm too old for this. Somehow, I thought it would be different this time. But it's not. Some things never change. The world has no further need of me."

"Parks, don't say that. You've got me, and I need you…"

"No, Elise, you don't. I've realized that ever since my retirement party. You're staying with me out of obligation to the kids and our marriage vows. In fact this whole thing with the new heart, the new personna, Elise, it was all for you. I thought if I finally became the kind of man my cousin Bryce once was, the first man you ever fell in love with, that you would finally truly love me."

Elise's hand went to her mouth, and a lake of tears filled her eyes. "Oh, Parks! What are you *saying*?"

"Didn't I cause all this grief? Of course I did. Think of all the trouble that would have been avoided if I'd just stayed in Indiana and worked for my uncle. Cloy Morgan would still be alive and riding bicycles across continents. The perfume factory workers, or whatever they were, would still be alive. Sugeto Mifune would not have killed himself. He at least was spared the escalating horror that I've caused. That Chinese boy in Oregon wouldn't have been fried just to stop my microwaves. All of those people in Washington, at Purdue, at MIT, and everywhere else could have been working on some other technology to save the planet. Maybe they would have already been successful. And now, little Arty Lomish waits to see if his father will be blown to bits in front of everyone. Elise, the whole world would be much, much better off without me!"

Elise was deeply shocked, and didn't respond fast enough when he lurched upright and strode from the room. "Parks!" she cried after him. "PARKS!"

The door of his temporary office in the safe house slammed, and she heard the lock engage as she ran to it. She crashed into the door with her whole body, arms up defensively.

"Parks, you're scaring me. Open this door right now!"

"Go away. I need to think."

"You're going to do something stupid, Parks."

"I'm not!" he yelled through the door. "Just leave me alone."

Elise opened her mouth to speak, then had a thought, and changed her mind.

"Okay. I'll leave you alone, but just for a little while. You promise not to do anything stupid, right? Promise me!"

"I promise," came a small voice.

She left, and went out to the security captain, who had probably already been informed of the scene. She quickly explained to him, and asked that he post someone to monitor the surveillance camera in the office, and if he took any actions which may threaten his life, to come bust down the door at once. He agreed. She took a stool from the kitchen, and set it a few yards away from the office door and sat there, praying for her husband.

*   *   *

The room was dark and Parks sat slumped in the captain's chair. His brain felt tired, too tired even to be angry. He wished for oblivion, and wished for it hard. Behind his eyes came a penetrating pain as he tried to force his body to shut off, to release him.

He thought of Sugeto Mifune, forgotten for years, and his friend Horse, also forgotten. Both dead because of him. What if they were waiting on the other side? The thought gave him pause, and he stopped trying so hard.

A deep depression enveloped his entire being, and he felt drawn down into the mire, more and more negative, pulled into the bowels of hell. He had never wanted to die more pointedly before, even after the Oasis Massacre. Back then, he had his MMID to turn off the thoughts which haunted him. That machine was long gone, and he had not meditated in several years.

The thought of meditating grabbed him. He wheeled over to the easy chair, and dropped from the captain's chair to the floor, leaning against the soft upholstery. There, he sat cross-legged, and began the breathing exercises. In his mind, he cast about for a focus object.

Emptiness, he thought. Death, darkness, void. No Sun. The Moon. The New Moon, which was a stupid name, he thought. It should be Dead Moon, because it was the absence of all light. The Empty Moon represents the ultimate yin, the complete receptivity of the feminine nature. And, he remembered, today was the start of the New Moon, and it would be completely dark for a little longer than a day. He chose a Black Luna as his focus object.

Right away the thought popped into his head that had once occurred years ago while he used the MMID. He tried to force the thought away, but could not shake it. It nagged him like a song fragment that plays over and over in your head, holding you captive to its catchy rhythms. '… when the New Moon is Full' echoed and echoed in his head.

"What!" he said out loud in frustration, breaking his fragile concentration. "When the New Moon is Full: What?" He asked of the empty room. It was the same question he had asked many years before.

Then, the answer came.

"Oh," he said quietly, humbled.

Rising, he moved up to the desk chair and wheeled back to the workstation. On the wide screen before him, augmeneted by interactive holographics around the perimeter, he pulled up a multimedia design tool. Using network connections he had set up for his transglobal enterprise, Parks set up to flood them all, and send cascades everywhere. The words he would use were already forming in his head, in an odd dual-path sequence of thought.

He turned the camera on himself, and looked as his visage on the monitor. The man there looked haunted and maniacal. He straightened his hair, summoned his normal, professional demeanor, and then effected the suave, confident businessman. But it wouldn't stick. He finally, after a long life of searching for it, became himself.

He activated the record function, and began to speak.

Terminator, Earth

In China, a country where the name Vambolent was respected without qualification, the phenomenon first started. Right after sunset on the second night of the New Moon, that planetary body began to turn color. As the terminator between night and day swept across those eastern longitudes, people stood outside and shone their lights on Luna. Laser pointers of every color, and white LED pump-torches were predominantly the light of choice for individuals. But those with searchlights, spot lights, and kleig lights turned their beams to the Moon also. There were few clouds, and all those uncounted billions of lumens aimed at the dark face of Luna turned it pink.

As Russia and India fell dark, people there too heard about the appeal by American space inventor Parks Vambolent. They too shone their lights, and the Moon became very noticeable. The phenomenon so excited people, made such good and quick press, that the effect scaled even as it raced towards Egypt. By sundown in Cairo the Earth's Moon was literally dancing with color and light. But there were high scattered clouds above that city, and the scene was lost on the residents who went outside to look.

In the Hababek household, a television was playing on the street-level floor. Hababek's youngest son was tending a perfume shop directly across from the Sphinx, and was just closing up when the news came on. He locked the doors, slammed the registers closed, and raced upstairs to his father's floor.

Now, the two of them watched live photographs of the moon.

"What is this?" Hababek was saying, interrupted in his evening ablutions. His arms were still wet, and he felt a chill.

Then, the television showed Parks Vambolent, and Hamid Hababek drew in a sharp breath through his jagged teeth. The translation came soon after, along with the text scrolled across the screen from right to left.

"Dear friends of our planet Earth. I need your help. We can no longer live on fossil fuels and nuclear power. They are killing us all. The planet is super-heating and generating superstorms. Poisons from our industry give us cancers and mutate our newborns. Our many civilizations are in imminent danger of environmental collapse. But I have a solution. Together with Afren Kwapili, we intend to create a solar-based hydrogen economy for the African Union. It will

be the first such large-scale effort in the world, but the news gets even better. With geometric manufacturing in space, and using my inventions, this clean, pure energy becomes ever less costly as you make more of it. My goal is to provide abundant, clean energy that will be too cheap to meter. It will be available to all mankind. But we are in danger of losing it.

Someone has a missile pointed at our facility, the largest orbital installation ever built by man. One missile can destroy it all. And they will. They have done it before.

Here is how you can help me: Light up Luna. If you have heard my dream, and if you want that kind of world, then I ask you to go outside tonight, just after sunset, and shine your light on the Moon. Make it as bright and as colorful as you can with as much light as you've got. If Luna stays dark, I will cease and desist, and give up on this dream forever. But if you light up Luna tonight, then maybe, just maybe, those people with the rocket will stand down. Maybe you can help them see that we are all in this together. Shine your light on Luna, and we shall see what happens. Inshallah. Al ham du lillah. Inshallah.*"*

At those last words, Hababek's demeanor softened. Something caught in his throat at the sincerity of those spoken Arabic words, sounding so natural, as often comes with a lifetime of practice. He looked straight into the eyes on his screen, and felt a connection as if the man were seated beside him. There was something in Vambolent's eyes that he had not seen when the man Jenks had visited. The moment stretched, and he saw deeper into Vambolent's eyes, down into the blacks, and he saw that the man had a good heart and could be trusted.

"Come, we must go to the roof," he said to his son, who took his elbow and helped him up. Laboriously, they climbed the concrete stairwell, one step after another up to a metal door with three locks. They entered onto the rooftop and looked up into the sky.

The sun's glow was showing redly through a gap in the clouds to the west, but the rest of the sky was cloudy. The two men moved to the edge, where a knee-high wall restrained them, and noticed many other people along the street below. On the plateau, thousands were streaming up towards the pyramids, looking dark against the silhouetting sky.

Then the clouds moved eastward and the moon was revealed as if a curtain slowly raised. A rose hue scintillated across the face of the New Moon. It was the start of Ramadan. Hamid Hababek sat down on the chair by the railing, motioning for his son to also sit.

"Son, I may have been wrong about this one particular business man."

He said no more, but watched the Moon dance over Giza, pearlescent, as hundreds of millions of small lights combined together to turn Luna the color of love. A collective cry of joy wafted over to them, echoing off the limestone enigmas. The son waited, frozen like the Sphinx against further orders, or possibly a reversal. But neither came. The old man had said exactly what was intended. He was engrossed by the scene before him, and seemed enraptured. The son left quietly and quickly, and went to go call the launch center.

*   *   *

Twenty miles into the Libyan Desert, west of the re-located mountainside of Abu Simbel, a sleek rocket stood belching vapor. In the bunkhouse, technicians stood aghast as they were ordered to stand down. Rage and frustration echoed inside the spare building as the supervisor directed them to change checklists. They had been launch-ready, waiting for the command to fire. To suddenly be ordered back from glory was a jarring dissonance with their expectations.

One of the technicians, planted by Jenks' team during the year of training and systems development, discretely sent an encrypted message through a hidden, ultra-secure link. The signal passed along the length of the Nile river to Alexandria, became wireless, and was finally relayed via a line-of-sight laser to a ship on the horizon north of Egypt's shores. Jenks had been watching the news, and was expecting the message.

He returned a message immediately, reversing the course of the first one. Within seconds, the operator in the bunkhouse received his reply, and initiated the override sequence.

Automated systems left standby mode and returned to the launch-ready state before they could be de-activated. The launch was set, dangerously, to 10 seconds. The cryogenic fuels would not be topped off perfectly, but there would still be just enough. Warning klaxons, disabled by the embedded spyware installed by the StarSys engineers months before, remained silent. Unless one of the other technicians calmed down enough to watch their monitors, they would miss the 4 second window of opportunity to abort the launch before the main engines turned on, and control was thrown to the bird.

When the engines lit, the bunkhouse denizens went stone silent. Then, screaming, they returned to their tasks. One man foolishly ran outside. They saw him raise an arm over his face at the incredible glow of the burning rocket fuel.

At 'all balls' the lock-downs exploded open and the rocket began to accelerate. Whining gyros stabilized the tall craft as it clawed the first, toughest inches from a stable, solid earth-bound launch pad. Once free, the heavy acceleration grew the velocity greater and greater and greater. On its way to escape velocity, the rocket covered the incredibly wide range of speeds, getting faster and faster, so it could fall around the planet without hitting, barely missing. After shedding the first stage, the second stage would boost up to GEO.

But the rocket would never quite make it to GEO. As designed by the StarSys team, the rocket's internal altimeter was wired to a small lump of plastique at the valve assembly throttle body where the two fuels mix. When it climbed to within 98 percent of geostationary orbit, the charge would blow, turning the remaining fuel into a bomb. The payload of diamond blades and black silicone gel were specially-designed to fracture, cut and coat silicon solar panels. The momentum that would have carried the intact rocket to 100 percent of GEO altitude would carry the expanding cloud of debris directly into the African Cluster.

Jenks went to the telescope on the roof of his yacht, turned his back to the still-scintillating moon, and trained the optics on the African Cluster. This was a show he did not want to miss.

GEO

Alarm klaxons sounded through the tin can where Lomish and Loray were watching the fantastic light show play across the face of the moon. Before they heard news of Vambolent's appeal, the had been half-expecting to be evacuated. The world-wide response was so overwhelmingly favorable, that rumors soon circulated that they were out of danger. That meant they could at last throw the switches to send power to the rectenna on Bateke Plateau.

The warnings thus came as a complete surprise. Voice activated switches connected them to their ground crew in Houston, Texas. Their CAPCOM gave the evacuation notice.

"What's coming our way?" Lomish asked, keeping a tight rein on his emotions.

"Looks to be a small rocket. Dneper-class, but modified," the CAPCOM told them through their earsets as the two men grabbed essential gear and cycled into the tight airlock. Their breathing was ragged in the close quarters, and a current of fear passed between them.

"How long do we have?" Loray asked their ground controller, trying to sound cool, keeping his voice low.

"We won't know the trajectory they're using until the second stage starts, which should be any second. But we got good intel from the folks in Cheyenne. You two have plenty of time to get out of the way. The explosion will be small. We just gotta get you out of the debris path."

"What about the facility?" Lomish asked, as they passed into their crew vehicle.

"Just get out of there, guys. The damage might not be as bad as we expect."

But they all knew it wasn't likely.

"Can NORAD tag it?" Lomish asked as Loray set up the vehicle to fly.

"Not unless they're stupid enough to overfly us," the CAPCOM said.

"Okay, we're set," Loray said. Lomish threw the bolt that separated them from the station – leaving it vacant.

"There wasn't time to properly power down all the systems," Lomish said into his throat mike. "Whoever comes next will have a mess on their hands."

Then, it hit him how absurd he was. Nobody was coming back. This place would be destroyed – turned into space junk.

Benza Loray put the pedal to the metal, and the powerful thrusters on their craft snapped them into their seats. After the first burn, he manualed the attitude jets and turned their ports to the cluster. It was large, and moved away slowly as they watched. Glittering regularity demarked the glossy crystal of the solar cells, still rotating to hide from the sun. The automation would keep working until the bitter end, oblivious to the impending danger.

As the station and the array passed away, they glided past the payload catchers. Those sturdy loops of superconductor were ganged to an LC circuit for harvesting back-emf as they received their iron canisters. Small altitude jets, using shuttered lasers and slag ablation targets fired on and off. The array of catchers made a curved line through space, and they traveled below it. Those independent satellites, together making up the receiving end of the transport system, were also doomed. Their flotsam would make this region of GEO a hazard for a long time to come.

"Hey guys, we got a call from the man himself. Here you go. Patching you in…"

"John, Benza, how are you? Are you out of danger?" It was Parks Vambolent's voice, sounding tired and lost.

"We're safe, Parks," Lomish answered. "But, how can we save the station?"

"It's more important that you two are safe."

"Wait, I got it!" Loray said, suddenly excited. "We turn on the microwave beam. We blast that little missile out of the sky. Melt the thing to a lump of slag before it even gets here!"

"And start the first war in space? I don't think so," Parks said.

"The antenna's not set up for that anyway," Lomish said. "We could warm it up a little, but it would hardly notice."

"Do we have any tools that can be used as weapons aboard this craft?" Lomish asked his partner and the CAPCOM. "We could fly over to it in this. Take a welding laser or something and disable the damned thing."

The CAPCOM broke it. "That's suicide, boys. Plus you don't have the delta-V to rendezvous."

"We'll throw something it its path. Blow it out of the sky!" Loray persisted.

"That's just what they want," CAPCOM told him. "If we blow it up *for* them, they don't even have to claim it will be an 'accident'. That's what the boys in sunglasses told us here at JSC. The debris will do their dirty work for them."

The comm channel was silent for several heartbeats. Only Parks' was still regular, ganged to an electronic brain.

"What if we use those?" John Lomish said, and pointed out the portal to the last of the superconductor rings, drifting past them like flotsam on a river.

"What does he mean?" Parks asked.

In the capsule, Benza Loray's gaze followed the outstretched finger. An amazed look crossed his face. "Is it made of iron?"

"What are you two talking about?" CAPCOM demanded.

"What's that rocket made of?" Lomish asked in an excited voice. "Is it magnetizable?"

"Oh, my God. Gotcha. I'll be right back."

"What?" Parks asked, then got it too, a little slower than the others. "You mean the payload catchers. You intend to use them on the missile!"

"You can't take them out of autopilot unless you power them down," CAPCOM reminded John. "You'd lose charge, and then they wouldn't work at all."

The spacers ignored him, exchanging rapid-fire technical details as they worked out how it might be done. Lomish switched off the feed absently, as they zeroed in on the solution. When they had finished their plan, and decided between them there was no other way, Loray turned the mikes back on. Hysterical shouting came from the speakers. He announced himself, and waited for the line to get quiet.

"We're going back," he told them, and overrode the flight controls to aim the ship back to the station. Over the strident objections from ground control, Lomish learned that the missile was indeed magnetizable, then shut off the comms again to concentrate. He gave a nod to Loray, who hammered the thruster, using up the reserves they would have needed to dock at LEO.

Mediterranean Sea

Jenks could see the missile's second stage coming back from the low polar orbit that would take it well beyond the US missile defense shield. Then, the engine ignited and his pupil contracted from the burst of magnified light. The day's cloud cover had blown over, and the skies were clear and bright with stars. He dialed the telescope's mount to point back to the SPS, hanging 32 degrees azimuth. It hummed and began gearing slowly to the new coordinates.

He sat back in his chair, sipping a lime and soda, and checked his watch. The anticipation mounted slowly. He knew that Vambolent would never recover from this blow. The man's company was already overextended, and no insurance company in the world would cover his SPS – Jenks and his associates in the organization had made sure of that.

The mount chimed, and Jenks checked the accuracy with a look at the LunarSource facility. He squinted at what he saw. A craft was approaching the station. It was their crew vehicle moving back towards the station! They weren't leaving. Well, they had just a little more than a half hour. But if they didn't leave well before that, the shrapnel would ruin their space ship. Dead astronauts at GEO would be an even graver consequence than what he planned. Jenks smiled at his own unintended pun.

Surely, they must know that. It didn't make sense. Actions that did not add up nagged at Jenks. He leaned back in his chair to think hard, to think through, and to think beyond.

What he didn't think about was the Special Forces team which had inserted just over the horizon from the Aswan High Dam, and were speeding by Humvees into the desert. As he mused, they laid waste to the security around the launch site, and quickly overcame the bunkhouse. An informant was coerced to reveal the hidden connection. A specialist on the team affixed a small electronics box to a console, removed a panel and jumpered two wires. He pressed a sequence of keys on his shoulder mike and then flipped up the safety on the switch box and pressed a black button.

Monitoring systems placed aerially and at certain junctions along the telephony systems of Egypt captured the custom-designed signal and traced its path. Another team in Cairo lifted into helicopters and flew straight to Alexandria. They landed in the vacant lots around the lone pillar of the Library, and swarmed outside. Two of them released airborne e-midges which spread out to give them 3-D electronic surveillance at this terminus of the signal.

Minutes later, they had identified the laser line. With optical devices triangulated by the swarm and aided by an overpass satellite, they quickly identified the yacht as the next terminus. The team leader opened an encrypted satellite link to his superiors and reported their findings. They were told to remain on station, and be battle ready.

GEO

"Hurry, you beasts!" Benza Loray was flying the coils. Each coil had a large chemical thruster, used as a back-up system to the laser/slag station keepers, and he was burning them full blast. The now ragged line of coils moved herky-jerky out of their alignment.

Lomish was programming. He would rather have re-wired each loop by itself, but they didn't have the time. Programming was part of his job, but he hated it. Sellenya had helped him become modestly proficient. Suddenly he sagged as he remembered his family. He ought to call them. No, better not to. Just get the job done.

Hacking into the automated systems was easy, but it took time. There was a lot you could screw up even with modular software, and especially when there was no room for error. Trying to stop what was basically a ballistic grenade brooked no misses.

"Okay, John, they're on their way. I'll get them lined up as they fly. But you've gotta give me the path of the rocket. And as soon as you've finished, you're gonna hafta call home." Benza Loray did not have any children that he knew of, but he was keenly aware of John's quiet but fierce devotion to his small family.

Lomish grunted, but did not look up from his work. He linked the station radar to the sensor arrays that orbited all around them for watching incoming lunar payloads. As he activated systems, their own cloud of microsats began turning and vectoring towards the incoming rocket. They had no idea when it was planned to detonate, but they surely wouldn't wait until it was right upon them. The bastards would blow it up before, they reasoned. So, the coils had to go down-orbit to intercept. It would be a faster closing velocity than the iron canisters shot from the L2 Relay Station. They might not have enough momentum to slow it.

"Got it, Benza. Here." He entered the final command, and released the controls to Loray's master console.

"Okay, patching in. Now make that call."

Lomish was about to re-activate the communications, which they had turned off after re-docking. "Wait a second. Can't we do the rest from the crew vehicle?"

"Sure, John. But where are we gonna go? We don't have enough fuel to go home."

"Shouldn't we just get out of the way?"

"It ain't that easy. To go lower, we have to basically fly right towards that missile. We could go higher, but then our chance of a rescue get smaller and smaller. So long as this crazy scheme works, we're best off staying right here until someone can ship us up some rocket fuel."

"Will it work?"

"Should!"

Lomish grimaced, the big frown creasing his face. "I guess I better call home now."

Benza Loray nodded gravely, then floated over to the portal and trained the telescope on the meeting about to occur in space.

* * *

Lomish was on the direct line, looking at Sellenya and Arty as the coasting missile encountered the first induction ring. Rising at greater than escape velocity, it passed directly through the center of the first ring. Powerful magnetic fields, released from stored electrical power, grabbed at the rocket. It still slipped through, but its passing dragged on the coil. Both bodies lost speed with respect to each other.

Another ring clutched at the missile. It slowed further. Then, a third ring passed around it. Three fatter, squat coil assemblies laid off the side of the missile's path tugged against the missile as it flew past, bending its trajectory. Another set of three pass-through rings quickly re-adjusted to the new alignment, their laser/slag thrusters firing rapidly and efficiently. The missile skewered and slowed even more. Further down the line, yanking on the missile and siphoning off its kinetic energy as a back-electromotive force, the line of payload catchers guided the deadly missile into a circular orbit lower than GEO. Already, it was falling behind the station.

"John," Loray whispered. "John! It worked."

"It did? Hey honey, it worked. I guess I can program after all!"

Sellenya hugged their baby and tears finally filled her eyes. She sniffled, trying to regain her composure. "John," she said, her voice choked with emotion. "The folks in Houston *really* want you to call them now."

John laughed, and it was a belly-jiggler. Through the vid, Sellenya's near-sob burst out as a laugh of relief. She hugged her baby tight to her cheek and the tears finally spilled over and trickled down into his hair.

Mediterranean Sea

Jenks watched all this with growing anger and rage. His lips pulled back in a sneering grimace. There was a manual override for missile and for the plastique, but his link to the launch facility had been dead now for several minutes. He had a secondary backup, but it was not secure. Using it meant revealing his hand, unless he was unrealistically lucky and nobody was listening in. The small dish antenna and controller case were stowed on the deck. He got up to open the case, and took two staggering steps backward.

The sleek conning tower of an attack submarine rose from the waters just off his starboard bow. Jenks stood transfixed, staring, as the menacing black monolith rose higher and higher until it was taller than he, standing on the top of his yacht. From the corner of his eye, he noticed the lights of watercraft converging on him from every point of the compass. He felt something give inside of him, and his bowels embarrassingly lost their tightness and quivered full open.

Epilogue

Parks Vambolent stood on the marble balcony overlooking the trading floor of the New York Stock Exchange. The most eagerly anticipated initial public offering of the century was Vambolent Interspace Industries, going public this day. In the last few weeks, just as the glow of his heart-felt global appeal had begun to fade, the world had reeled at the news from the Far Side Base. Their excavations of gravitational anomalies had uncovered an ore body not native to the moon. It was replete with even more platinum group metals than had been cast across the surface by asteroid 06-401A. In that shallowly-buried mother lode was also a rich source of uranium. With that metal, and the isotope separation capabilities of Parks' inventions, VII could readily enrich it to produce fuel for nuclear thermal rocket engines. And being on the lunar far side, they were technically outside cislunar space, and thus immune to the UNOOSA restrictions. This news now opened up the entire solar system for rapid transport. At long last, it would be possible to easily steer asteroids into useful orbits. Manned missions to Mars, in times short enough to avoid deleterious health effects from galactic cosmic rays, were possible for the first time. The people Parks had working on the moon, lead by space veteran Dallas Barstow, had discovered the legacy promised by Parks' late friend, Cloy Morgan.

They had found the Regulus.

Flanking Parks on the balcony were Elise and Mindy on one side and Eric Vambolent on the other. At precisely 9:30 am, he rang the opening bell, and the day's work began.

On the Big Board, shares of ticker VII quickly ran up over the pricing Vambolent had originally set. Offered shares sold quickly through his underwriter. On his PDA, Parks watched as his net worth approached thirteen figures.

* * *

Riding a suborbital jump jet from a spaceport in the Hudson River, the Vambolent family flew to Colorado Springs for a reception at the New White House. Until they were safely ensconced inside the craft, reporters had hounded their every step. All four of them breathed a collective sigh of release at the welcome privacy and security.

"Well, Eric," his father asked. "What did you think?"

"Way to go Dad. I always knew you could do it. You're a gazillionaire, right?"

Parks just smiled and rumpled his son's hair.

"Mindy?" he asked.

"Does this mean I get the house in Hawaii?"

Elise and Parks chuckled together. Both their children had fallen in love with the 50th state.

"As long as you visit us often in Indiana," Elise said.

The family talked pleasantly for the whole trip, which seemed to take just minutes.

* * *

He had done it. All that Parks Vambolent set out to do decades before had finally come to pass. He looked over at Elise. Her education system was being used in almost every country on the planet. All of Africa used it, and the system had grown beyond anything she originally imagined. She still worked her e-mail lists, being Aunty Lise to the younger children, and telling them about the wonders of outer space, even as she ran a viable space enterprise delivering precious metals to fuel the hydrogen revolution.

She smiled at Parks, and he smiled back. He reached for her hand and held it as they landed. The Vambolent family was reluctant to get out, and did so slowly, keeping close to one another.

At the other landing pad of the heliport stood a huge blue craft with its rotors already turning. The Great Seal was emblazoned on its side. A young man with sunglasses and a broad smile watched the family. He walked over to intercept them.

"Mr. Vambolent. Hello. I'm here to invite you and your family to the Whitehouse."

"Thank you," Parks said, shaking the man's hand.

"Actually, I'm not supposed to tell you, sir," the man said, holding his lips closed like it was an effort. Then, he blurted out: "But, I believe you are being awarded the National Medal of Technology today."

"Well, I'll be."

"It's about time," Mindy said under her breath.

"Right this way, please."

The tight family walked, bewildered, over to the waiting helicopter. They rose up into the dusky sky and banked away.

Parks' last thought as they approached New White House, was that he had one more promise to fulfill. It was a promise he made long ago, when the whole journey was just beginning. It was a last request, a death-bed promise, the most binding of all.

He wondered how the President would take the news of the revolution.

THE END

Made in the USA
Monee, IL
04 November 2026

67379965R00272